Energy Localisation and Transfer

ADVANCED SERIES IN NONLINEAR DYNAMICS*

Editor-in-Chief: R. S. MacKay *(Univ. Warwick)*

Published

Vol. 4 Hamiltonian Systems & Celestial Mechanics
eds. J. Llibre & E. A. Lacomba

Vol. 5 Combinatorial Dynamics & Entropy in Dimension One
L. Alseda, J. Llibre & M. Misiurewicz

Vol. 6 Renormalization in Area-Preserving Maps
R. S. MacKay

Vol. 7 Structure & Dynamics of Nonlinear Waves in Fluids
eds. A. Mielke & K. Kirchgässner

Vol. 8 New Trends for Hamiltonian Systems & Celestial Mechanics
eds. J. Llibre & E. Lacomba

Vol. 9 Transport, Chaos and Plasma Physics 2
S. Benkadda, F. Doveil & Y. Elskens

Vol. 10 Renormalization and Geometry in One-Dimensional and Complex Dynamics
Y.-P. Jiang

Vol. 11 Rayleigh–Bénard Convection
A. V. Getling

Vol. 12 Localization and Solitary Waves in Solid Mechanics
A. R. Champneys, G. W. Hunt & J. M. T. Thompson

Vol. 13 Time Reversibility, Computer Simulation, and Chaos
W. G. Hoover

Vol. 14 Topics in Nonlinear Time Series Analysis – With Implications for EEG Analysis
A. Galka

Vol. 15 Methods in Equivariant Bifurcations and Dynamical Systems
P. Chossat & R. Lauterbach

Vol. 16 Positive Transfer Operators and Decay of Correlations
V. Baladi

Vol. 17 Smooth Dynamical Systems
M. C. Irwin

Vol. 18 Symplectic Twist Maps
C. Gole

Vol. 19 Integrability and Nonintegrability of Dynamical Systems
A. Goriely

Vol. 20 The Mathematical Theory of Permanent Progressive Water-Waves
H. Okamoto & M. Shoji

Vol. 21 Spatio-Temporal Chaos & Vacuum Fluctuations of Quantized Fields
C. Beck

*For the complete list of titles in this series, please write to the Publisher.

ADVANCED SERIES IN NONLINEAR DYNAMICS
VOLUME 22

Energy Localisation and Transfer

Editors

Thierry Dauxois

Ecole Normale Supérieure de Lyon, France

Anna Litvak-Hinenzon & Robert MacKay

University of Warwick, UK

Anna Spanoudaki

National Technical University of Athens, Greece

World Scientific

NEW JERSEY • LONDON • SINGAPORE • SHANGHAI • HONG KONG • TAIPEI • BANGALORE

Published by

World Scientific Publishing Co. Pte. Ltd.
5 Toh Tuck Link, Singapore 596224
USA office: Suite 202, 1060 Main Street, River Edge, NJ 07661
UK office: 57 Shelton Street, Covent Garden, London WC2H 9HE

British Library Cataloguing-in-Publication Data
A catalogue record for this book is available from the British Library.

ENERGY LOCALISATION AND TRANSFER

ISBN 981-238-742-0

Printed in Singapore by World Scientific Printers (S) Pte Ltd

PREFACE

This volume contains lecture notes and discussion session summaries for a training school held in the Center of Physics at Les Houches, France, from 27 January to 1 February 2003. Under the heading "Energy Localisation and Transfer in Crystals, Biomolecules and Josephson Arrays", the lectures and discussions explored a variety of topics involving the localisation of energy in spatially discrete physical systems.

The aim of the school was to give an introduction to localised excitations in spatially discrete systems, from experimental, numerical and mathematical points of view. Also known as "discrete breathers", "nonlinear lattice excitations" and "intrinsic localised modes", these are spatially localised time-periodic motions in networks of dynamical units. Examples of such networks include molecular crystals, biomolecules, and arrays of Josephson superconducting junctions. The school addressed the formation of discrete breathers and their potential role in energy transfer in such systems.

As the lecturers have performed a special effort to make their material accessible and attractive to both experimentalists and theorists, it was natural to publish pedagogically written lecture notes. The contributions collected in this book will be useful for advanced graduate students, postdoctoral researchers and more generally for any researcher who would like to enter this field.

During the school, the lecture courses were complemented by discussion sessions on less established topics. Indeed, the recent development of new methodologies to approach the localization of energy in these systems has revealed its importance also in a trans-disciplinary perspective. New prospects for applications of nonlinear localised excitations to physics are therefore briefly presented. Particularly intriguing is the possibility of applying the concept to understand some aspects of molecular motion in biomolecules. Consequently, an introductory lecture on protein dynamics is included and a discussion of a few examples, with the aim of stimulating thoughts and hopefully further experimental studies or theoretical investigations paying particular attention to the experimental aspects.

The main lecturers of the Les Houches school were François Fillaux (CNRS, Thiais), Sergej Flach (MPIPKS, Dresden), Peter Hamm (University of Zurich), Robert MacKay (University of Warwick), Juan Mazo (University of Zaragoza), Yves-Henri Sanejouand (ENS, Lyon) and Alexey Ustinov (University of Erlangen). Discussion sessions were led by Serge Aubry (CEA, Saclay), Thierry Dauxois (ENS, Lyon), Anna Litvak-Hinenzon (University of Warwick), George Kopidakis (Heraklion), Michel Peyrard (ENS, Lyon), Nikos Theodorakopoulos (NHRF, Athens) and George Tsironis (Heraklion).

We would like to express our sincere gratitude to the lecturers and discussion leaders for all their efforts in preparing, presenting and writing up their lectures. The lecture notes and discussion summaries have also benefited from the questions during the conference and we would like to thank all the participants. Furthermore, we warmly acknowledge Michel Peyrard for his frequent helpful advice in the scientific organisation of this school, and the staff of the secretarial office of the school for their help in all aspects of the organisation.

Our thanks are also due to the sponsor of this conference, LOCNET, which is a European Commission Research and Training Network on "Localisation by Nonlinearity and Spatial Discreteness, and Energy Transfer, in Crystals, Biomolecules and Josephson Arrays" (EC contract HPRN-CT-1999-00163). Many exchanges, conferences, collaborations and increased understanding have been made possible thanks to the stimulating atmosphere of LOCNET's members and its funding.

Additional pedagogical lectures are available (together with many summaries of research presentations on this subject) in *Localization and Energy Transfer in Nonlinear Systems*, Proceedings of the Third Conference, San Lorenzo de El Escorial, Spain, 17–21 June 2002 by L. Vazquez, R. S. MacKay, M. P. Zorzano, Eds., World Scientific (2003).

T. Dauxois
ENS Lyon, France
Thierry.Dauxois@ens-lyon.fr

A. Litvak-Hinenzon
Warwick University, UK
litvak@maths.warwick.ac.uk

R. S. MacKay
Warwick University,UK
mackay@maths.warwick.ac.uk

A. Spanoudaki
National Technical University, Greece
annasp@central.ntua.gr

PARTICIPANTS*

- *Aubry Serge*, CEA-Saclay, France (16)
- *Barré Julien*, Ecole Normale Supérieure de Lyon, France (38)
- *Barthes Mariette* , Université de Montpellier, France
- *Benoit Jérome*, NHRF, Athens, Greece
- *Berger Arno*, Warwick University, UK (2)
- *Cuesta Lopez Santiago*, Universidad de Zaragoza, Spain (4)
- *Dauxois Thierry*, Ecole Normale Supérieure de Lyon, France (34)
- *Dorignac Jerome*, Warwick University, UK (27)
- *Dusuel Sébastien*, Ecole Normale Supérieure de Lyon, France (41)
- *Dyer Nigel*, Warwick University, UK (33)
- *Edler Julian*, University of Zürich, Switzerland (35)
- *Fillaux Francois*, CNRS-UMR, Thiais, France (26)
- *Flach Sergej*, Max-Planck-IPKS, Germany (23)
- *Gaididei Yuri*, Bogolyubov Institute for Theoretical Physics, Ukraine (10)
- *Gómez Jesús*, Universidad de Zaragoza, Spain (9)
- *Hamm Peter*, University of Zürich, Switzerland (28)
- *Kastner Michael*, Universitá di Firenze, Italy (21)
- *Katerji Caisar*, Universidad de Sevilla, Spain (7)
- *Katsuki Hiroyuki*, University of Zürich, Switzerland (24)
- *Kopidakis Giorgos*, University of Crete, Greece (13)
- *Larsen Peter Vingaard*, Technical University of Denmark (5)
- *Litvak-Hinenzon Anna*, Warwick University, UK (29)
- *MacKay Robert*, Warwick University, UK (14)
- *Maniadis Panagiotis*, CEA-Saclay, France (17)
- *Mazo Juan*, Universidad de Zaragoza, Spain (31)
- *Meister Matthias*, Universidad de Zaragoza, Spain (1)
- *Miroshnichenko Andrey*, Max-Planck-IPKS, Dresden, Germany (3)
- *Noble Pascal*, Université Paul Sabatier, Toulouse, France (37)
- *Oster Michael*, Technical University of Denmark (8)
- *Palmero Faustino*, Universidad de Sevilla, Spain (15)
- *Peyrard Michel*, Ecole Normale Supérieure de Lyon, France (40)
- *Pignatelli Francesca*, Universität Erlangen Nürnberg, Germany (6)
- *Pouthier Vincent*, Université de Franche-Comté, France (30)
- *Ruffo Stefano*, Universitá di Firenze, Italy (25)
- *Sanejouand Yves-Henri*, Ecole Normale Supérieure de Lyon, France (20)

*Numbers are referring to the picture.

- *Sepulchre Jacques-Alexandre*, CNRS Valbonne, France (36)
- *Sire Yannick*, INSA, Toulouse, France (18)
- *Soerensen Mads Peter*, Technical University of Denmark (32)
- *Spanoudaki Anna*, Ecole Normale Supérieure de Lyon, France (22)
- *Theodorakopoulos Nikos*, NHRF, Athens, Greece (19)
- *Tsironis Giorgos*, University of Crete, Greece (12)
- *Ustinov Alexey*, Universität Erlangen Nürnberg, Germany
- *Van Erp Titus*, Universiteit van Amsterdam, the Netherlands (39)
- *Zolotaryuk Yaroslav*, Technical University of Denmark (42)
- *Zueco David*, Universidad de Zaragoza, Spain (11)

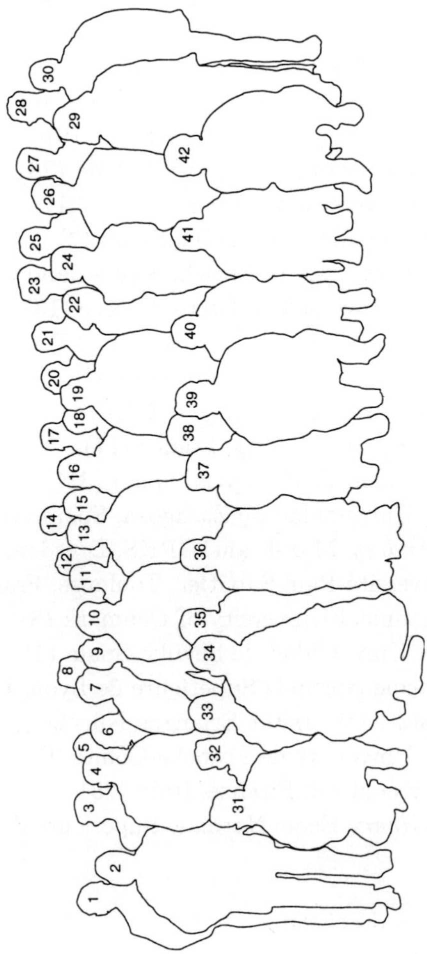

CONTENTS

CHAPTER 1

COMPUTATIONAL STUDIES OF DISCRETE BREATHERS

Sergej Flach

Max Planck Institute for the Physics of Complex Systems
Nöthnitzer Str. 38, D-01187 Dresden, Germany
E-mail: flach@mpipks-dresden.mpg.de

Dedicated to the Memory of Alexander Anatolievich Ovchinnikov

This chapter provides a description of the main computational tools for the study of discrete breathers. It starts with the observation of breathers through simple numerical runs, their study using targeted initial conditions, and discrete breather impact on transient processes and thermal equilibrium. Next we describe a set of numerical methods to obtain breathers up to machine precision, including the Newton method. We explain the basic approaches of computing the linear stability properties of these excitations, and proceed to compute wave scattering by discrete breathers, and to briefly discuss computational aspects of studying dissipative breathers. In a final part of this chapter we present computational approaches of studying quantum discrete breathers.

1. Introduction

The past decade witnessed remarkable developments in the study of nonlinear localized modes in different physical systems. One of the most exciting results has been the discovery of stable highly localized modes in spatial lattices,[1,2,3] coined discrete breathers (DB) or intrinsic localized modes.[4,5,6,7] The discreteness of space - i.e. the usage of a spatial lattice - is crucial in order to provide structural stability for spatially localized excitations. Spatial discreteness is a very common situation for various applications from e.g. solid state physics. Recent studies have shown that effects of spatial discreteness can be important in many other systems, including photonic crystals, coupled optical wave guides, coupled Josephson junctions, Bose-Einstein condensates in optically induced lattices and micromechanical cantilever systems (see the more detailed discussion below).

Discreteness is useful for avoiding resonances with plane wave spectra, which are bounded for spatial lattices, as opposed to the typical case of a space continuous field equation. DBs are spatially localized and time-periodic excitations in nonlinear lattices. Their structural stability and generic existence is due to the fact that all multiples of their fundamental frequency are out of resonance with plane waves. Thus localization is obtained in a system without additional inhomogeneities. Notably these excitations exist independent of the lattice dimension, number of degrees of freedom per lattice site and other details of the system under consideration (see Ref. 6 and references therein).

While during the first years studies of intrinsic localized modes have been mostly of mathematical nature, experimental results soon moved into the game. The discrete breather concept has been recently applied to various experimental situations. Light injected into a narrow waveguide which is weakly coupled to parallel waveguides (characteristic diameter and distances of order of micrometers, nonlinear optical medium based on GaAs materials) disperses to the neighboring channels for small field intensities, but localizes in the initially injected wave guide for large field intensities.[8] Notably the waveguides may be ordered both in a one-dimensional array as well as in a two-dimensional structure.[9] Furthermore it was shown in accord with theoretical predictions, that self-defocusing Kerr nonlinearities (which would not provide soliton formation in a spatially homogeneous medium) when combined with the spatial discreteness allow for the formation of DBs.[9] Bound phonon states (up to seven participating phonons) have been observed by overtone resonance Raman spectroscopy in PtCl mixed valence metal compounds.[10] Bound states are quantum versions of classical discrete breather solutions. Spatially localized voltage drops in Nb-based Josephson junction ladders have been observed and characterized[11] (typical size of a junction is a few micrometers). These states correspond to generalizations of discrete breathers in dissipative systems. Localized modes in anti-ferromagnetic quasi-one-dimensional crystals have been observed in Ref. 12. And finally recent observations of localized vibrational modes in micromechanical cantilever oscillators arrays have been reported in Ref. 13.

All these activities demonstrate that the concept of intrinsic localized modes, or discrete breathers, as predicted more than 10 years ago, has a strong potential for generalizations to and applications in various areas of science. At the same time we are facing a dramatic enlargement of physics research areas to artificial or man-made devices on the micrometer and nanometer scales (of both optical and solid state nature), together with a huge interest growing in the area of quantum information processing. We may safely expect interesting new developments in these areas, which will be connected in various ways to the understanding of the concept of nonlinear localized modes. One example is the recent connection of discrete breathers and the physics of Bose-Einstein condensates in optical traps.[14]

We stress here that the research on DBs was initially purely theoretical, while experiments moved into the game at a later stage. It turned out that it needs a bit of curiosity, a simple computer, and a bit of surprise after observing that localized excitations in perfectly ordered lattices do not decay into extended states. The reason why theory could evolve that fast and that far during a couple of years, is because the systems under study are described using coupled ordinary differential equations (ODE), and because the objects of interest are highly localized on the lattice, i.e. often a few lattice sites (or ODEs respectively) are enough to capture the main properties. The rest of the lattice (or of the many ODEs) can be taken into account using analytical considerations with reasonable approximations, which are always systematically tested afterwards in numerical simulations. This fruitful combination of analytical and numerical methods has lead to an enormous number of key results on DB properties. At the prominent edge of this spectrum we now find a whole set of rigorous methods to prove DB existence implicitly.[15,16,17,18,19,20,21] Remarkably even such rigorous mathematical existence proofs[15] have been immediately turned into highly efficient numerical tools for computing DB solutions with machine precision. A large part of the DB studies can be thus characterized truly as computational ones. This chapter is written in order to provide the interested reader with knowledge about the main computational tools to study DB properties. We will usually refer to the simplest model systems, and comment on expected or known problems which may occur when more complicated systems are chosen.

We implicitly assumed that the above discussion of computational methods is concerned with classical physics. Once DBs are identified for a given system or class of systems, a natural question is what sort of eigenfunctions of the corresponding quantum Hamiltonian operator may be coined quantum DBs. While the quantum problem seems to be just an eigenvalue problem, it is much harder to be studied numerically as compared to its classical counterpart. The reason is that in many cases even the Hilbert space of a single lattice site may be infinite dimensional. But even for finite local dimensions, the dimension of the lattice Hilbert space is typically growing exponentially with the system size. In addition straightforward solving of the quantum problem implies diagonalization of the Hamiltonian. So the success of computational studies of classical DBs ends abruptly when we enter the quantum world. Nevertheless the huge accumulated knowledge on classical DBs can be used to help formulate predictions for quantum DB properties. But to confirm these predictions we have to solve the quantum problem numerically, and are thus typically restricted either to small systems (two or three lattice sites, which makes the problem more an abstract model for molecules rather than for extended lattices) or to the low energy domain of larger lattices (however note that even in the case of a spin one-half lattice exact diagonalizations are restricted to a maximum of about twenty sites).

Let us set the stage now by choosing a generic class of Hamiltonian lattices:

$$H = \sum_l \left[\frac{1}{2} p_l^2 + V(x_l) + W(x_l - x_{l-1}) \right] . \tag{1}$$

The sum index integer l marks the lattice site number of a possibly infinite chain, and x_l and p_l are the canonically conjugated coordinate and momentum of a degree of freedom associated with site number l. The on-site potential V and the interaction potential W satisfy $V(0) = W(0) = V'(0) = W'(0) = 0$ and $V''(0), W''(0) \geq 0$. This choice ensures that the classical ground state $x_l = p_l = 0$ is a minimum of the energy H. The equations of motion read

$$\dot{x}_l = p_l \; , \; \dot{p}_l = -V'(x_l) - W'(x_l - x_{l-1}) + W'(x_{l+1} - x_l) \; . \tag{2}$$

If we linearize the equations of motion around the classical ground state, we obtain a set of linear coupled differential equations with solutions being small amplitude plane waves:

$$x_l(t) \sim \mathrm{e}^{\mathrm{i}(\omega_q t - ql)} \; , \; \omega_q^2 = V''(0) + 4W''(0) \sin^2 \left(\frac{q}{2} \right) \; . \tag{3}$$

The dispersion relation ω_q is shown in Fig. 1 for the case of an optical plane wave spectrum $V''(0) > 0$ and for an acoustic spectrum $V''(0) = 0$. While the first one is characterized by a nonzero frequency gap below the spectrum, the latter one is gapless due to the conservation of total mechanical momentum (at least for the linearized equations of motion). Both cases share the common and most important feature that the dispersion relation is periodic in the wave number q and possesses a finite upper bound.

Another important feature of this dispersion is the group velocity of plane

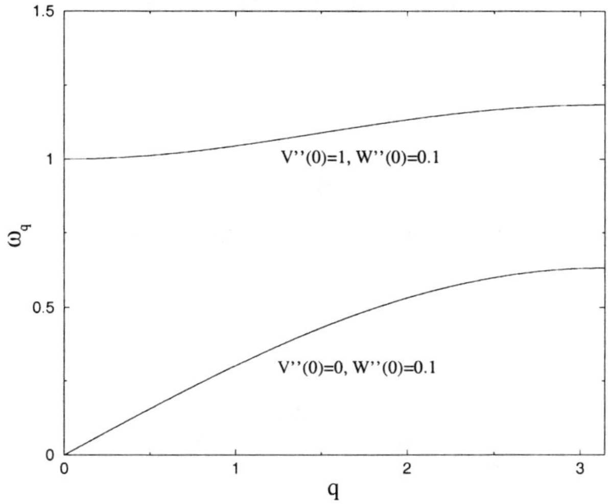

Fig. 1. The dispersion relation of small amplitude plane waves of model (1).

waves $v_g(q)$:

$$v_g(q) = \frac{\mathrm{d}\omega_q}{\mathrm{d}q} \,, \tag{4}$$

which vanishes at the nonzero band edges of ω_q.

When studying the properties of the original Hamiltonian problem (1) numerically for say N sites, we thus deal with a $2N$ dimensional phase space and as much coupled first order ODEs (2). The chosen system is rather simple. Nevertheless for most of the results discussed below complications like larger interaction range, increase of the lattice dimension, more degrees of freedom per site (or a better unit cell) are not of crucial importance and can be straightforwardly incorporated. We will provide with useful hints whenever such generalizations may lead to less trivial obstacles.

To give a flavour of what discrete breathers are in such simple models, we plot three different types of them schematically in Fig. 2. Case A corresponds to an acoustic chain with $V = 0$ and nonlinear functions W'. Typically simplest stable breathers involve two neighbors oscillating out of phase with large amplitudes. Case B is similar to A, but W is a periodic function. In this case roto-breathers exist, i.e. in the simplest case one degree of freedom is rotating, while the rest is oscillating. Finally case C

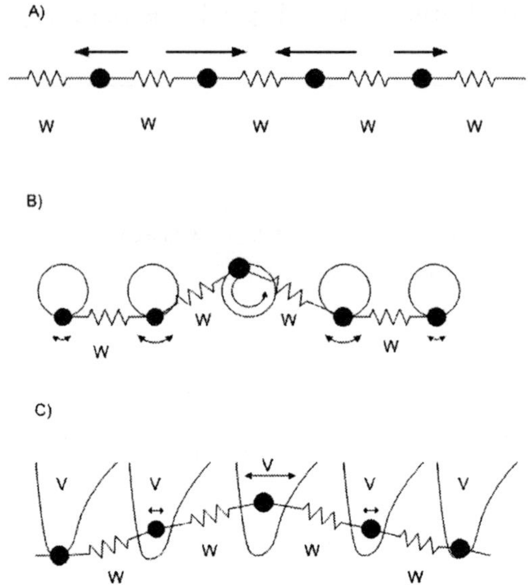

Fig. 2. Three different discrete breather types. See text for details.

corresponds to an optical chain with nonzero V. In this case each degree of freedom corresponds to an oscillator moving in V and coupled to nearest neighbors by W. A simple breather solution consists of one oscillator oscillating with a large amplitude. In all three cases the oscillations in the tails will have less amplitude with growing distance from the center, and vanish exactly if an infinite chain is considered. Note that similar excitations can be easily constructed for large lattice dimension.

2. A bit on numerics of solving ODEs

As mentioned in the introduction, DB studies in classical systems are mainly about solving coupled ODEs. So before coming to the actual topic of this chapter, let us discuss briefly some relevant informations concerning integrating ODEs. The basic problem is not the coupling between different ODEs, but first the integration of a single ODE. If we are heading for a specific solution like time-periodic oscillations, it may be appropriate to expand the yet unknown solution in a Fourier series and then to compute the solutions of the equations for the resulting Fourier coefficients. We will come to this aspect later. Here we are interested in a brute force integration of the ODEs without prior knowledge of what we may expect. In such a case the standard procedure is to replace the differentials by differences and to replace the continuous variable (say time t) by a set of grid points. While a good choice is to make the grid or mesh fine enough, there are still subtle choices one can make which are or are not appropriate depending on the concrete situation one is interested in.

For Hamiltonian systems or more general systems which preserve the phase space volume, a number of so called symplectic routines is available. For system (1) we may rewrite the Hamiltonian equations of motion (2) in a Newtonian way

$$\ddot{x}_l = -V'(x_l) - W'(x_l - x_{l-1}) + W'(x_{l+1} - x_l) \equiv f_l(x(t)) . \qquad (5)$$

In that case a standard symplectic routine is the so-called Verlet or leap-frog method:[22]

$$x_l(t+h) - 2x_l(t) + x_l(t-h) = \frac{1}{2}h^2 f_l(x(t)) . \qquad (6)$$

The time step h defines the grid in time, and the error per step is $O(h^4)$. The advantage of this method is that only one calculation of the force f_l is needed per step. A slight disadvantage is that we need not only the coordinates at some initial time t_0, but also the coordinates at the previous time step $t_0 - h$ when starting the integration. However this problem can be easily circumvented by using approximate expressions which connect the positions at various times and the velocities (or momenta), e.g. $p_l(t_0) = (x_l(t_0+h) - x_l(t_0-h))/2h$. Inserting this into (6) at time t_0 we obtain

$$x_l(t_0+h) - x_l(t_0) - hp_l(t_0) = h^2 f_l(x(t_0)) . \qquad (7)$$

While the error in this first step is of order $O(h^3)$, this is typically not crucial, as one should return to (6) after the first step.

A much more often used method is the Runge-Kutta method of 4th order.[23] The error per step is of order $O(h^5)$. This method integrates 1st order ODEs and is used also for dissipative systems without phase space volume conservation. However this method is not symplectic, so integration of Hamiltonian systems may lead in general to a systematic drift of conservation laws like energy on large time scales. Another disadvantage is that we need four force calculations per one time step, so routines may become computing-time consuming.

Before choosing a specific algorithm we should decide i) whether the total simulation time is large compared to the characteristic internal time scales or not, ii) what the maximum allowed error is, and iii) whether we do care about overall stability w.r.t. integrals of motion or not. Given the above choice of two algorithms the thumb rule would be to use the Verlet algorithm for long time simulations with maximum stability, and the Runge-Kutta algorithm for short time simulations or those where we do not care about overall stability.

Another set of related questions concerns finite temperature simulations. Here in addition to the choice of the algorithm we have to worry about the most efficient way to emulate a statistical ensemble. Typically there are two methods one may use - deterministic and stochastic ones.[22] Among deterministic methods there is the simple microcanonical simulation of a large enough system, and the so-called Nose-Hoover thermostat, which consists of coupling an additional artificial degree of freedom to the system of N degrees of freedom and performing the microcanonical simulation of the $(N+1)$ degrees of freedom system. Among the stochastic algorithms two main ones are Monte-Carlo methods (random sampling) and solving of Langevin equations obtained by extending the original equations which incorporate damping and random forcing. Typically one heads for the computation of averages, i.e. in the most general case for correlation functions which may depend both on space distance and on distance in time, e.g. the displacement-displacement correlation function

$$S_{lk}(t) = \langle x_l(t+\tau)x_k(\tau)\rangle_\tau \ . \tag{8}$$

Such functions are analyzed with the help of temporal and spatial transforms

$$A(\omega) = \int_0^\infty \cos(\omega t)A(t) \ , \ A_q = \sum_l \mathrm{e}^{\mathrm{i}q(l-k)}A_{lk} \ . \tag{9}$$

To decide which method is the most useful for a given problem, we have again to decide whether we head for short time correlations, i.e. for the statistics of excitations, or for long time correlations, i.e. for the properties of slow relaxations. Since stochastic methods unavoidably introduce cutoffs in the correlation times of the original dynamical system, these methods are best if one heads for the statistics of excitations, as they may replace the probably very slow relaxation of the dynamical system by a faster mixing due to the incorporated stochasticity. On the other hand, the statistics of slow relaxations of the dynamical system call for deterministic methods, as the optional additional stochasticity would have to become active anyway on much larger time scales than the internal relaxation times (such as not to spoil the statistics) and can be thus safely neglected all together.

Regarding the spatial correlations, we should carefully choose the system size such as to avoid finite size effects. A way to check this is to compute a correlation length

$$\xi^2 = -\frac{\left[\frac{\mathrm{d}^2}{\mathrm{d}q^2} S_q(t=0)\right]_{q=0}}{2S_{q=0}(t=0)} \tag{10}$$

and to compare it with the system size.

While the spatial transform in (9) is a simple sum, temporal transforms as in (9) are again integrals. For a correlation function which has a short time (high frequency) oscillatory contribution as well as a slow long time relaxation stretched over several decades, use the Filon integration formula[24]

$$\int_{t_0}^{t_{2n}} f(t)\cos(\omega t)\mathrm{d}t = h[\alpha(\omega h)\left(f_{2n}\sin(\omega t_{2n}) - f_0\sin(\omega t_0)\right) + \beta(\omega h)C_{2n} + \gamma(\omega h)C_{2n-1}] + O(nh^4 f^{(3)})$$

with

$$C_{2n} = \sum_{i=0}^{n} f_{2i}\cos(\omega t_{2i}) - \frac{1}{2}\left[f_{2n}\cos(\omega t_{2n}) + f_0\cos(\omega t_0)\right]$$

$$C_{2n-1} = \sum_{i=1}^{n} f_{2i-1}\cos(\omega t_{2i-1})\ ,\ \alpha(z) = \frac{1}{z} + \frac{\sin 2z}{2z^2} - \frac{2\sin^2 z}{z^3}$$

$$\beta(z) = 2\left(\frac{1+\cos^2 z}{z^2} - \frac{\sin 2z}{z^3}\right)\ ,\ \gamma(z) = 4\left(\frac{\sin z}{z^3} - \frac{\cos z}{z^2}\right)\ .$$

By dividing the whole accessible time interval into different sub-parts which are sampled with different grid points (with grid point distances which could vary by orders of magnitude) it is straightforward to compute a reproducible high-quality spectrum covering several decades in frequency.

Contrary, if we are concerned with the Fourier transform of an analytical time-periodic function

$$A(t) = A(t+T)\ ,\ \omega = \frac{2\pi}{T}$$

the simple trapezoidal rule[23] does the job with exponential accuracy, provided that the period T is exactly a multiple of the grid size h:

$$A(k\omega) = \int_0^T \cos(k\omega t)A(t)\mathrm{d}t = h\sum_{m=1}^{m=T/h}\cos(k\omega mh)A(mh) + O(\mathrm{e}^{-\cdots/h})\ .$$

3. Observing and analyzing breathers in numerical runs

3.1. *Targeted initial conditions*

For convenience we will sometimes use a Taylor expansion of the potentials in (1):

$$V(z) = \sum_{\alpha=2,3,\dots} \frac{v_\alpha}{\alpha} z^\alpha \,, \; W(z) = \sum_{\alpha=2,3,\dots} \frac{w_\alpha}{\alpha} z^\alpha \,. \tag{11}$$

Let us choose $v_2 = 1$, $v_3 = -1$, $v_4 = \frac{1}{4}$, $w_2 = 0.1$ with all other coefficients equal to zero. The on-site potential in this case has two wells separated by a barrier, and the interaction potential is a harmonic one. One of the simplest numerical experiments to observe localized excitations then is to choose initial conditions when all oscillators are at rest $p_l(0) = 0$, $x_{l\neq 0}(0) = 0$ except one at site $l = 0$ which is displaced by a certain amount $x_0(0)$ from its equilibrium position. Then we integrate the equations of motion e.g. using the Verlet method. We expect at least a part of the initially localized energy excitation to spread among the other sites. We choose a system size $N = 3000$. The maximum group velocity of plane waves (3) is of the order 0.1 here. Finite size effects due to recurrence of emitted waves which travel around the whole system and return to the original excitation point are thus not expected for times smaller than $t_{max} = 30000$. In other words, our simulation will emulaie the behavior of an infinite chain with the above initial conditions up to t_{max}. To monitor the evolution of the system we define the discrete energy density

$$e_l = \frac{1}{2} p_l^2 + V(x_l) + \frac{1}{2}(W(x_l - x_{l-1}) + W(x_{l+1} - x_l)) \,. \tag{12}$$

The sum over all local energy densities gives the total conserved energy. If DBs are excited, the initial local energy excitation should mainly remain at its initial excitation position. Thus defining

$$e_{(2m+1)} = \sum_{-m}^{m} e_l \tag{13}$$

by choosing a proper value of m in (13) we will control the time dependence of $e_{(2m+1)}$. If this function does not decay to zero or does so on a sufficiently slow time scale, the existence of a breather-like object can be confirmed. The term 'slowly enough' has to be specified with respect to the group velocities of small amplitude plane waves (3). We simply have to estimate the time waves will need to exit the half volume of size m which we monitor with (13). For the choice $m = 2$ we conclude that this time scale is of the order of $t_{min} \approx 20$. Thus the relevant times of monitoring the evolution of the system are still covering three decades

$$20 \ll t \leq 30000\,. \tag{14}$$

In Fig. 3 we show the time dependence of $e_{(5)}$ for an initial condition $x_0(t =$

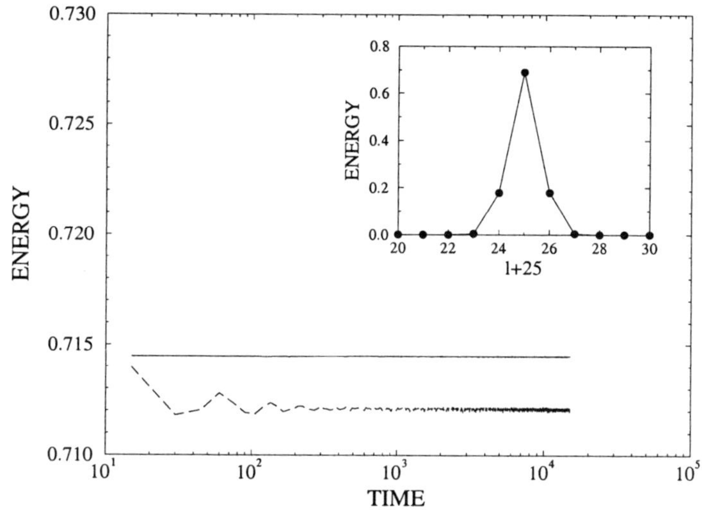

Fig. 3. $e_{(5)}$ versus time (dashed line). Total energy of the chain, solid line. Inset: energy distribution e_l versus particle number for the same solution measured for $1000 < t < 1150$.

$0) = 2.3456$.[25] Clearly a localized excitation is observed. After a short time period of the order of 100 time units nearly constant values of $e_{(5)}$ are observed. The breather-like object is stable over a long period of time with some weak indication of energy radiation. The energy distribution within the object is shown in the inset of Fig. 3. Essentially three lattice sites are involved in the motion, so we find a rather localized solution. While the central particle performs large amplitude oscillations, the nearest neighbors oscillate with small amplitudes. All oscillations take place around the groundstate $x_l = p_l = 0$. Note that due to the symmetry of the initial condition the left and right hand parts of the chain should evolve exactly in phase - a good test for the correctness of the used numerical scheme.

To get more insight into the internal dynamics of the found object, we perform a Fourier transform of $x_0(t)$ and $x_{\pm 1}(t)$ in the time window $1000 \leq t \leq 10000$ using the Filon algorithm.[25] The result is shown in

Fig. 4. We observe that there are essentially two frequencies determining

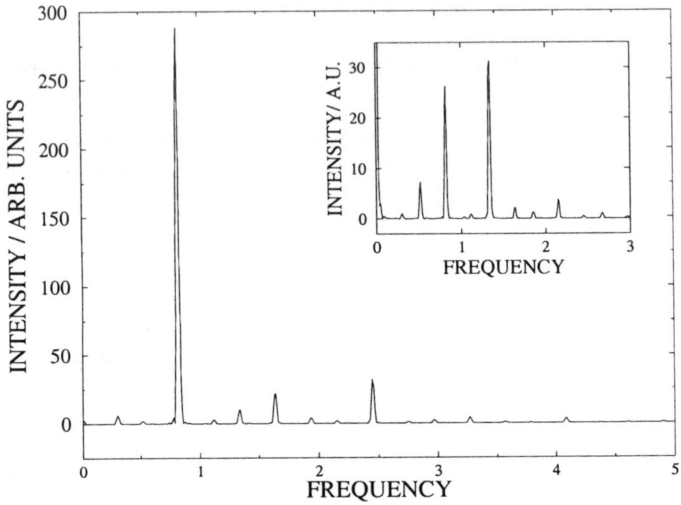

Fig. 4. Fourier transformed FT$[x_l(t \geq 1000)]\,(\omega)$ with initial condition as in Fig. 3 for $l = 0$. Inset: for $l = \pm 1$.

the motion of the central particle $\omega_1 = 0.822$, $\omega_2 = 1.34$. All peak positions in Fig. 4 can be obtained through linear combinations of these two frequencies. To that end we may conclude that we observe a long-lived strongly localized excitation with oscillatory dynamics described by quasi-periodic motion. To proceed in the understanding of the phenomenon, we plot in the inset in Fig. 4 the Fourier transformation of the motion of the nearest neighbor(s) to the central particle. As expected, we not only observe the two frequency spectrum, but the peak with the highest intensity is not at ω_1 as for the central particle, but at ω_2. Because of the symmetry of the initial condition the two nearest neighbors move in phase. Thus and because the other particles are practically not excited, we are left with an effective 2 degree of freedom problem (cf. inset in Fig. 3).

Instead of getting lost in the possibilities of initial condition choices for the whole system, we may now expect that as it stands the observed excitation must be closely related to a trajectory or solution of a reduced problem with a low-dimensional phase space. Indeed, fixing all but the three oscillators $l = -1, 0, 1$ at their groundstate positions reduces the dynamical problem to a three degree of freedom system, and restricting ourselves to the symmetric case $x_{-1} = x_1$ and $p_{-1} = p_1$ in fact to a two degree of freedom problem:

$$\ddot{x}_0 = -V'(x_0) - 2w_2(x_0 - x_{\pm 1}) \ , \tag{15}$$

$$\ddot{x}_{\pm 1} = -V'(x_{\pm 1}) - w_2(x_{\pm 1} - x_0) \ . \tag{16}$$

First we may choose the same initial condition in the reduced problem as done before in the full chain, and observe that indeed the two trajectories are very similar. Following this way of reduction we may then perform Poincaré maps of (15,16) and formally get full insight into the dynamical properties of this reduced problem. This has been done e.g. in Ref. 26. The same map has been then performed in the extended lattice itself, and the two results were compared. Not only was the existence of regular motion on a two-dimensional torus found in both cases, but the tori intersections for the reduced and full problems were practically identical.[25] Thus we arrive at two conclusions: i) the breather-like object corresponds to a trajectory in the phase space of the full system which is for the times observed practically embedded on a two-dimensional torus manifold, thus being quasi-periodic in time; ii) the breather-like object can be reproduced within a reduced problem, where all particles but the central one and its two neighbors are fixed at their groundstate positions, thereby reducing the number of relevant degrees of freedom.

Intuitively it is evident, that none of the observed frequencies describing the dynamics of the local entity should resonate with the linear spectrum (3), since one expects radiation then, which would violate the assumption that the object stays local without essential change. In truth the conditions are much stricter, as we will discuss below. Since the reduced problem defined above can not be expected to be integrable in general, we expect its phase space structure to contain regular islands filled with nearly regular motion (tori) embedded in a sea of chaotic trajectories. Note that this picture will strongly depend on the energy shell on which the map is applied. Chaotic trajectories have continuous (as opposed to discrete) Fourier spectra (with respect to time), and so we should always expect that parts of this spectrum overlap with the linear spectrum of the infinite lattice. Thus chaotic trajectories of the reduced problem do not appear as candidates for breather-like entities. The regular islands have to be checked with respect to their set of frequencies. If the island frequencies are located outside the linear spectrum of the infinite lattice, we can expect localization - i.e. that a trajectory with the same initial conditions if launched in the lattice will essentially form a localized object. Islands which do not fulfill this nonresonance criterion should be rejected as candidates for localized objects. Thus we arrive at a selection rule for initial conditions in the lattice by studying the low-dimensional dynamics of a reduced problem. This conjecture has been successfully tested in Ref. 26. In Fig. 5 we show a representative

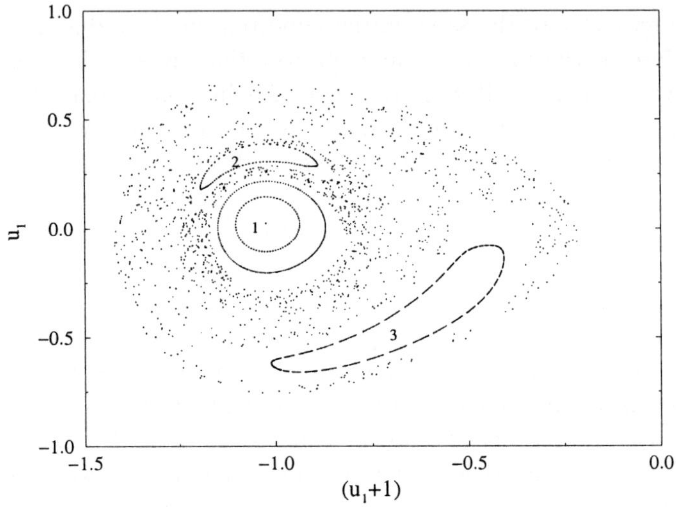

Fig. 5. Poincaré intersection between the trajectory and the subspace $[\dot{x}_1, x_1, x_0 = 0, \dot{x}_0 > 0]$ for the symmetric reduced three-particle problem and energy $E = 0.58$. Note that u instead of x is used in the axis labels.

Poincaré map of the reduced problem. In Fig. 6 the time dependence of the above defined local energy $e_{(5)}(t)$ is shown for different initial conditions which correspond to different trajectories of the reduced problem. The initial conditions of regular islands 1,2 of the reduced problem yield localized patterns in the lattice, whereas regular island 3 and the chaotic trajectory, if launched into the lattice, lead to a fast decay of the local energy due to strong radiation of plane waves. It is interesting to note that the energy decay of the latter objects stops around $e_{(5)} = 0.35$. In Ref. 26 it was noted that the fraction of chaotic trajectories in the reduced problem practically vanishes for energies below that value.

Another observation, which comes from this systematic analysis is that the fixed points in the Poincaré map of the reduced problem (in the middle of the regular islands in Fig. 5) correspond to periodic orbits. A careful analysis of the decay properties in Fig. 6 has shown that all objects were slightly radiating - but some stronger and some less. The objects corre-

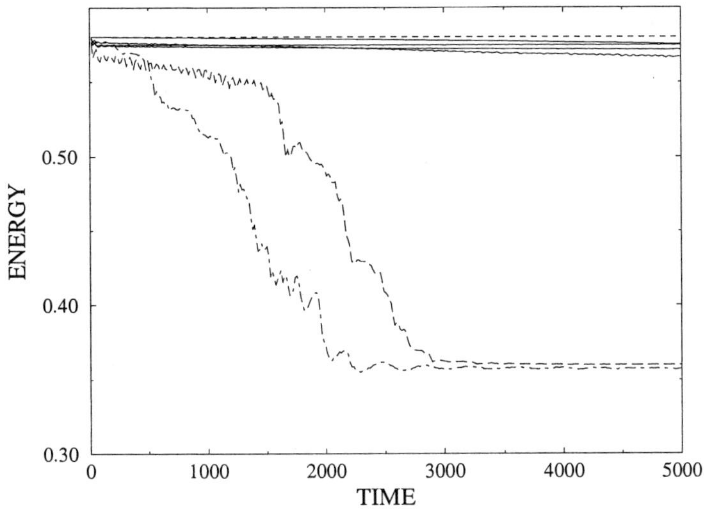

Fig. 6. $e_{(5)}(t)$ dependence. Upper short dashed line - total energy of all simulations; solid lines (4) - initial conditions of fixed points in islands 1,2 from Fig. 5 and larger torus in island 1 and torus in island 2 from Fig. 5; long dashed line - initial condition of torus in island 3 in Fig. 5; dashed-dotted line - initial condition of chaotic trajectory in Fig. 5.

sponding to the periodic orbits of the regular islands 1,2 of the reduced problem showed the weakest decay.[26] Thus we arrive at the suggestion that time-periodic local objects could be free of any radiation - i.e. be exact solutions of the equations of motion on the lattice! It makes then sense to go beyond the present level of analysis and to look for a way of understanding why discrete breathers can be exact solutions of the dynamical equations - provided they are periodic in time. Further the question arises, why their quasi-periodic extensions appear to decay - i.e. why do quasi-periodic discrete breathers seem not to persist for infinite times. We can also ask: suppose quasi-periodic DBs do not exist - what are then their patterns of decay; what about their life-times; what about moving DBs (certainly they can not be represented as time-periodic solutions)? And we may already state, that if time-periodic DBs are exact localized solutions, then they may be also stable with respect to small perturbations, as observed here.

The linear spectrum of the model used for the numerical results here is optical-like, with a ratio of the band width to the gap of about 1/10. However this does not imply that the discrete breathers exist merely due to some weakness of the interaction. An estimation of the energy part stored in the interaction of the DB object presented here yields a value of 0.4. Compare that to the full energy $E \approx 0.7$. Roughly half of the energy is stored in the interaction. By no means we can describe these excitations by completely neglecting the interaction among the different lattice sites.[26]

Since breather-like excitations can be described by local few-degree-of-freedom systems (reduced problem), there is not much impact one would expect from increasing the lattice dimension. We will have an increase in the number of nearest neighbors, which implies simply some rescaling of the parameters of the reduced problem. To see whether that happens, the above described method was applied to a two-dimensional analog of the above considered chain. The interested reader will find details in Ref. 27. Here we shorten the story by stating that practically the whole local ansatz can be carried through in the two-dimensional lattice. An analog of Fig. 3 for the two-dimensional case is shown in Fig. 7 where the energy distribution

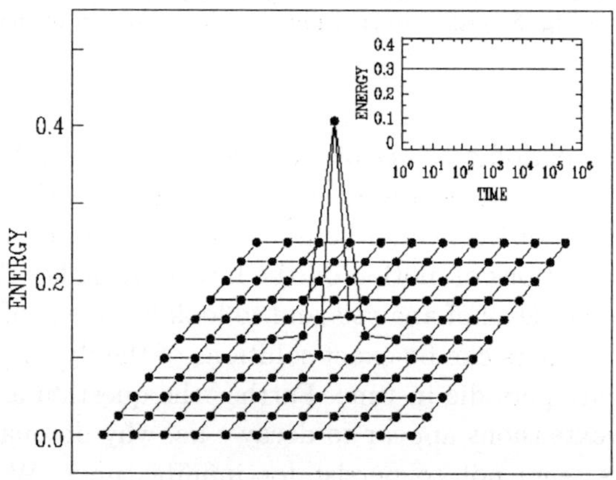

Fig. 7. Energy distribution for the breather solution with initial energy $E = 0.3$ after waiting time $t = 3000$. The filled circles represent the energy values for each particle; the solid lines are guides to the eye. Inset: Time dependence of the breather energy $e_{(5)}$.

in a discrete breather solution is shown, and the inset displays the time dependence of a local energy similar to $e_{(5)}(t)$. The reader will ask how we deal with radiation in this case. Indeed, the system in Fig. 7 has dimension 20×20 (only a subpart of size 10×10 is actually shown), which implies a characteristic time $t_{max} \approx 100$. The necessary trick is to add to the Hamiltonian part of the lattice a dissipative boundary, here of 10 more sites on each edge, increasing the total size of the system to 40×40. In these dissipative boundaries simple friction is applied in order to dissipate as much energy radiation as possible. Since both zero and infinite friction will lead to total reflection of waves instead of absorption, the next step is to impose a friction gradient from small to large values as one penetrates the dissipative layer coming from the Hamiltonian core. By simple variation of the friction gradient and the maximum friction value it is possible to optimize the absorption properties of this layer.[27]

3.2. *Breathers in transient processes*

If breather-like states are easily excited by a local perturbation, then we expect that these objects may be also relevant in systems with a nonzero energy density which is nonuniformly distributed among the lattice. One possibility is to excite a uniform energy density distribution which is however unstable with respect to small perturbations - something known as modulational instability, Benjamin-Feir instability etc. Analytical predictions for such instabilities can be obtained by finding an exact solution of a plane wave of nonzero amplitude and linearizing the equations of motion around the solution. If the result indicates instability, it can be easily implemented numerically by taking initial conditions which correspond to such a plane wave and adding a weak noise to them. Typically the outcome is the evolution of the energy density into spatially nonuniform patterns. Even if the outcome of a very long time simulation would not show up with breather-like states, the transient into such equilibria may take a lot of time, and on this path breathers can be observed. The formation of breather-like states through modulational instability was reported in several publications.[28,29,30,31,32]

While a number of publications has been devoted to these problems, for reasons of coherence (staying within one model class) below we will show recent numerical results done by Ivanchenko and Kanakov.[33] The model parameters are $v_2 = 1$, $v_4 = 0.25$ and $w_2 = 0.1$. The initial conditions can be encoded as

$$x_l(0) = (a + \xi)\cos(ql) \ , \ \dot{x}_l(0) = \omega(a + \xi)\sin(ql) \tag{17}$$

for the one-dimensional case with $\omega^2 = \omega_q^2 + 0.75a^2$, the wave number $q = 3\pi/4$, the amplitude $a = 0.5$ and the noise ξ being uniformly distributed in the interval $0 \leq \xi \leq 0.001$. The system size is $N = 400$, and periodic boundary conditions are used. In Fig. 8 we plot the energy density

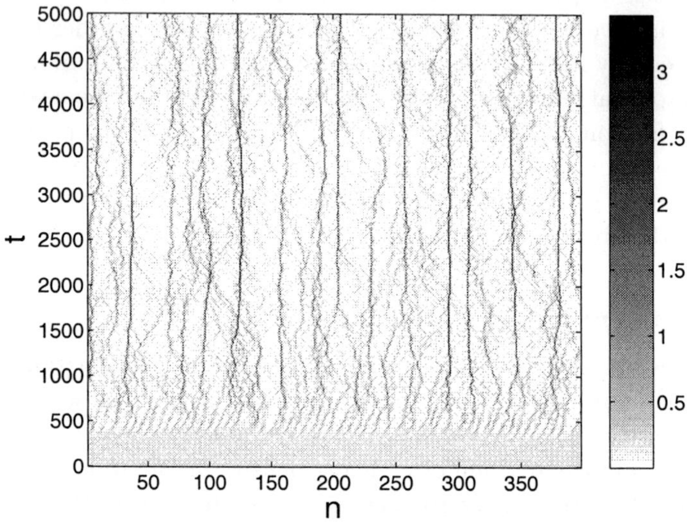

Fig. 8. Energy density evolution in a chain with parameters given in the text. Horizontal axis - chain site, vertical axis - time. Energy density is plotted in a gray scale coding from white (zero) to maximum observed values (black).

evolution up to a time $t = 5000$. Note that on short time scales the modulational instability is observed, both with a characteristic regular distance between the evolving maxima of the energy density and with a characteristic shift of the maxima positions in time due to the nonzero group velocity of the plane wave. Discrete breather-like objects are formed in the next part of the evolution, when some of these energy lumps start to collide and exchange energy,[34] leaving the system over long times with immobile highly localized excitations, which coexist with a diluted gas of plane waves or small amplitude solitons. These plane waves and solitons are observed to sometimes scatter from a breather, sometimes penetrate it, and surely their presence will lead to a further thermalization of the lattice on much larger time scales than the numerically studied. Indeed extending the observation time by two orders of magnitude we observe further focusing of energy in

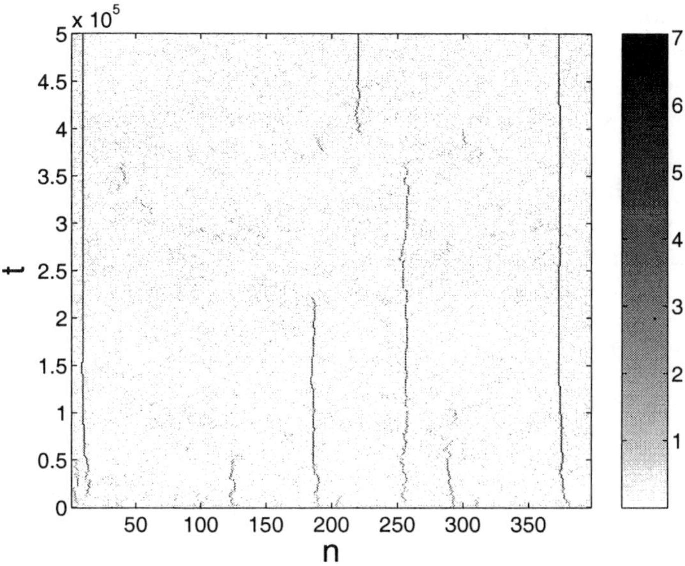

Fig. 9. Energy density evolution in a chain with parameters given in the text. Horizontal axis - chain site, vertical axis - time. Energy density is plotted in a gray scale coding from white (zero) to maximum observed values (black).

high energy breathers (Fig. 9). Note that the results of Fig. 8 are not observable here because they cover one percent of time here, and because the gray scale coding is significantly changed.

In some studies thermalization leads ultimately to a disappearance of large amplitude breathers (or better to a negligible probability to observe formation again). In other cases (see below) breather formation is even observed in what is believed to be thermal equilibrium. The outcome sensitively depends both on model parameters but most importantly on the temperature, which is implicitly defined by the average energy density of the initial conditions. Too low temperature will on one hand still show modulational instability and breather formation, and very long transient times into a final equilibrium state without breathers, but only plane waves. Intermediate temperatures will again provide with modulational instability, but transient times are shorter, and breathers may now be expected even in thermal equilibrium (simply because probability of large local fluctuations increases). Note that in general the temperature, i.e. the average energy density, is given by both the amplitude of the plane wave and the way the initial conditions are noised. Here we assume that the noise contribution is always weak, so the energy density is mainly given by the plane wave amplitude.

The same scenario can be also observed in two-dimensional lattices.[33] With the same parameters as above but replacing the argument (ql) by $(q(l+m))$, where l and m are the lattice indices of a square lattice of size 80×80 with periodic boundary conditions, we show the energy density distributions at four different times in Fig. 10. Note the increasing grey

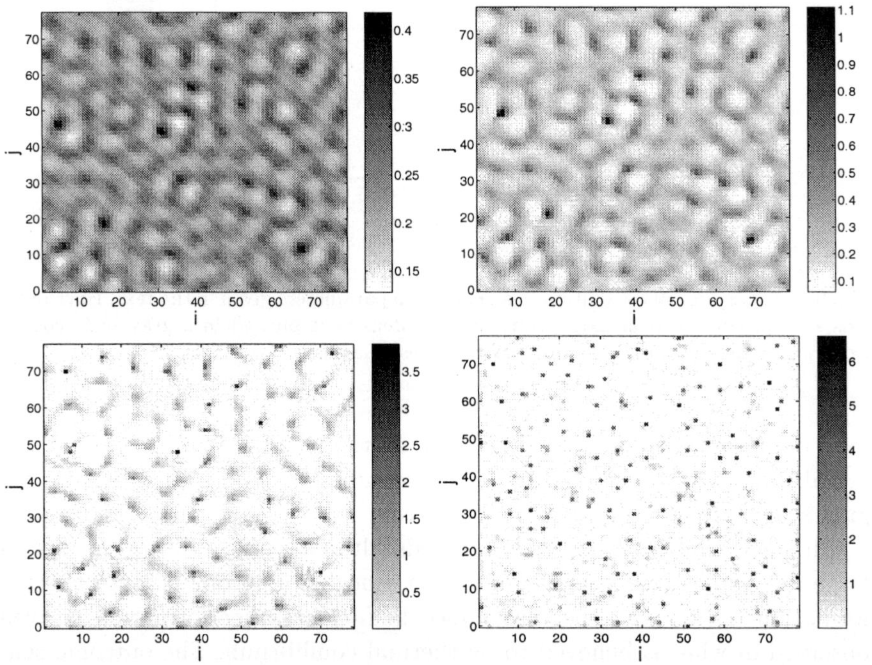

Fig. 10. Energy density distribution in a square lattice with parameters given in the text. Energy density is plotted in a gray scale coding from white (zero) to maximum observed values (black). Times of observation are $t = 400, 450, 500, 5000$.

scale coding limit due to more energy getting attracted into high energy breathers.

Another way to observe breathers in transient processes is to randomly excite a given sub-part of a lattice, with the rest of the lattice being not excited. Then, as in the case of targeted initial conditions, one may expect that all plane waves will be radiated into the infinite nonexcited part, and only breathers will stay.[35] In that sense one could even try to measure the energy fraction stored in breathers for a given lattice at a given temperature. We show experiments for a system size 50×50 plus friction boundaries, with model parameters as above.[33] In Fig. 11 the energy density distribution is shown at four times $t = 0, 4900, 11900, 19900$ and $c = 3$. We observe that even at these low temperatures about 5% of the total energy was stored on long-lived breathers, simply due to fluctuations in the initial conditions. The

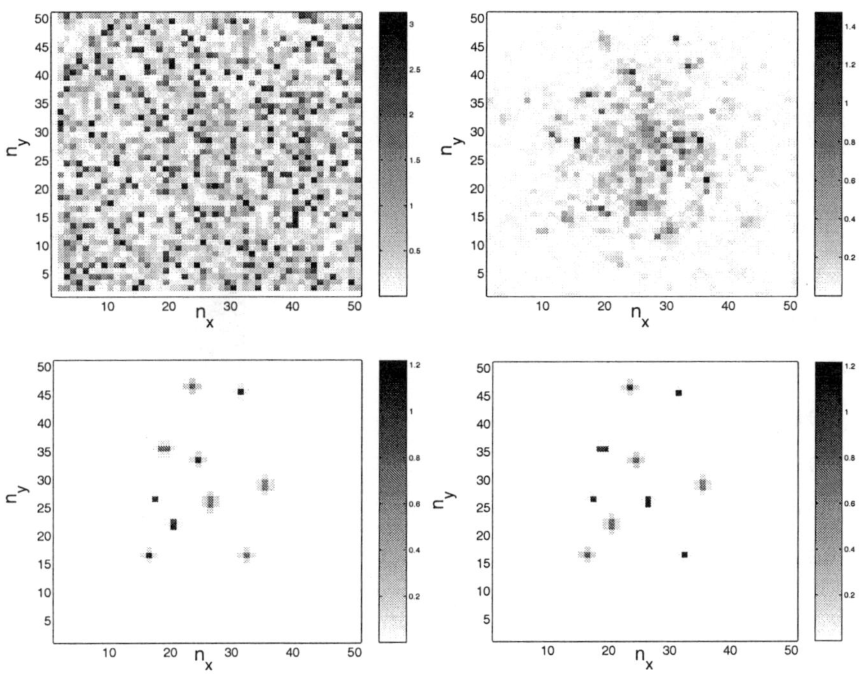

Fig. 11. Energy density evolution in a two-dimensional lattice. See text for details.

above mentioned method, however, cannot be applied to one-dimensional lattices. The reason is that while a single breather-like excitation by one local perturbation is easily detectable (see 3.1) in one-dimensional systems, we have to worry about the interaction between breathers and plane wave radiation when exciting the whole lattice or a big part of it. It turns out (see section 6.2) that breathers in one-dimensional systems usually very effectively backscatter plane waves. Consequently exciting e.g. two breathers in a one-dimensional system and some plane waves between them, will lead to a trapping of the radiation between the two breathers and also to some enhanced retarded interaction between the breathers mediated by the radiation. In contrast, in systems with dimension $d \geq 2$ breathers as point-like (zero dimensional) objects may scatter plane waves but not trap them. Consequently plane waves will still easily exit the excited lattice volume, and breathers left will practically not interact with each other (the only interaction channel left are spatially decaying breather tails, which may become exponentially small with growing distance from a breather core). Indeed,

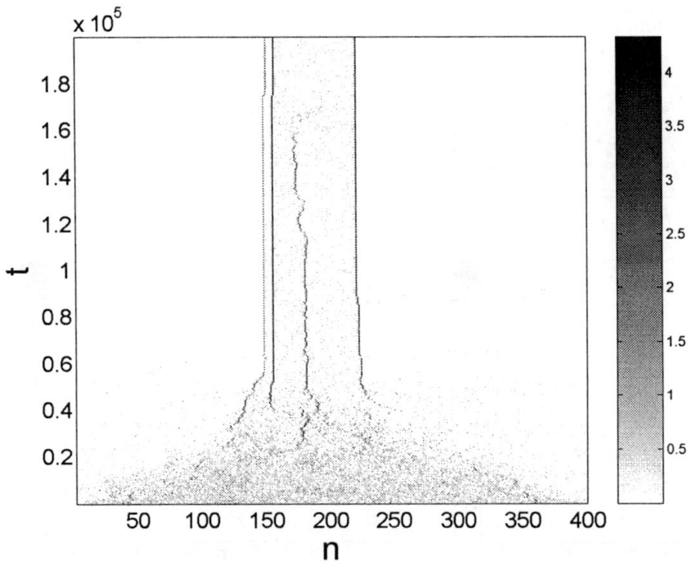

Fig. 12. Energy density evolution in a one-dimensional lattice. See text for details.

repeating the above experiment in a one-dimensional analog (same parameters except $c = 2$) we find in Fig. 12 that the energy distribution is trapped between two large amplitude breathers (see also Refs. 36 and 37). With increasing time some radiation escapes, and the two guarding breathers are slowly shifting towards each other.

3.3. *Breathers in thermal equilibrium*

Finally breathers have been also observed in thermal equilibrium.[32,38,39,40,41] In Fig. 13 we show the evolution of a one-dimensional chain with same parameters as in the preceding section. Periodic boundary conditions are applied, and the initial conditions for x_l and $\dot{x}_l$ being randomly uniformly distributed between $-c/2$ and $c/2$. We clearly observe the formation of breather-like highly localized objects, and more of them for larger energy densities. The same procedure

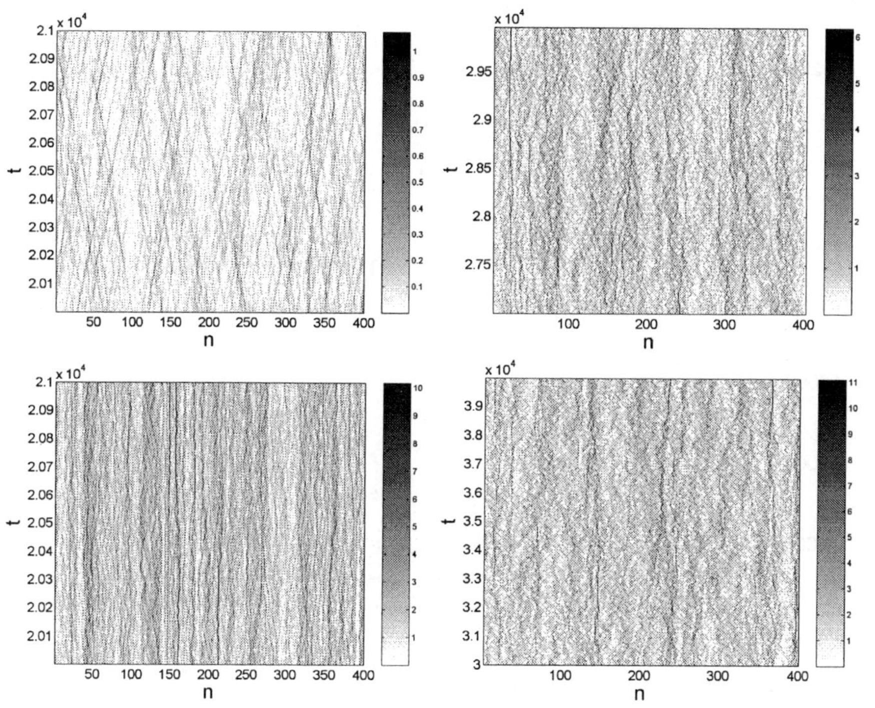

Fig. 13. Energy density evolution in a one-dimensional lattice for a time window after giving the system time to equilibrate. Left upper picture - $c = 1$, right upper picture - $c = 3$, second row - $c = 4$.

can be applied to a similar two-dimensional square lattice. In Fig. 14 we show the evolution of the energy density distribution using a simple cut procedure, where black dots are plotted if the energy density at a given lattice point exceeds five times the average energy density. Nearly all the observed spots and especially the long vertical lines correspond to breather excitations. All these results confirm that breather-like objects are easily

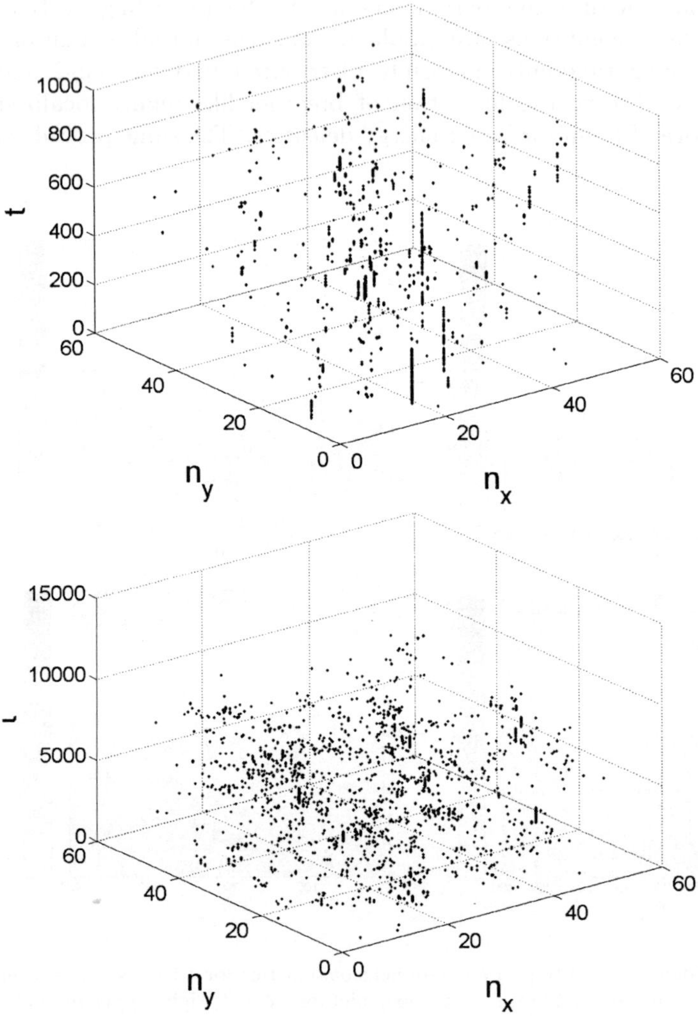

Fig. 14. Energy density evolution in a two-dimensional lattice for various time windows. For both cases $c = 5$ was chosen.

excited in lattices, that they can be obtained both with targeted initial conditions, during transient processes and in thermal equilibrium. We are only beginning to develop a reliable quantitative way to compute their statistical contribution and weights. Another important aspect - interaction between breathers - is also waiting further clarification. Already such straightforward studies as the ones discussed show that this problem depends both on the dimensionality of the system and on the relative contributions of phonon mediated interaction and tail-tail interactions.

4. Obtaining breathers up to machine precision: Part I

From section 3 we learned that breather-like objects exist due to weak resonance with the plane wave spectrum ω_q. Also these studies suggested that time-periodic breathers could be exact solutions, i.e. do not radiate at all. If so, let us try to obtain a time-periodic solution with period $T_b = \frac{2\pi}{\Omega_b}$ which is localized in space

$$x_l(t) = x_l(t + T_b) \ , \ x_{|l| \to \infty} \to 0 \ . \tag{18}$$

By definition we can expand it into a Fourier series

$$x_l(t) = \sum_k A_{kl} \mathrm{e}^{ik\Omega_b t} \ . \tag{19}$$

The Fourier coefficients by assumption are also localized in space

$$A_{k,|l| \to \infty} \to 0 \ . \tag{20}$$

This ansatz has to be inserted into the equations of motion of (1,2) which we rewrite in the following form

$$\ddot{x}_l = -v_2 x_l - w_2(2x_l - x_{l-1} - x_{l+1}) + F_l^{nl}(x_{l'}) \ . \tag{21}$$

Here we have introduced the force term F_l^{nl} which incorporates all nonlinear terms of the equations of motion. For (1,11) it takes the form

$$F_l^{(nl)} = - \sum_{\alpha=3,4,\ldots} \left[v_\alpha x_l^{\alpha-1} + w_\alpha((x_l - x_{l-1})^{\alpha-1} - (x_{l+1} - x_l)^{\alpha-1})\right] \ . \tag{22}$$

With ansatz (19), F^{nl} can be also expanded into a Fourier series:

$$F_l^{(nl)}(t) = \sum_{k=-\infty}^{+\infty} F_{kl}^{(nl)} \mathrm{e}^{ik\Omega_b t} \ . \tag{23}$$

Thus we arrive at a set of coupled nonlinear algebraic equations for the Fourier coefficients A_{kl} of the breather solution we search for:

$$k^2 \Omega_b^2 A_{kl} = v_2 A_{kl} + w_2(2A_{kl} - A_{k,l-1} - A_{k,l+1}) + F_{kl}^{(nl)} \ . \tag{24}$$

If a breather solution exists, then in its spatial tails all amplitudes are small. Thus we can assume that the nonlinear terms in (24) are negligible in the tails of a breather. We are then left with the linearized equations

$$k^2\Omega_b^2 A_{kl} = v_2 A_{kl} + w_2(2A_{kl} - A_{k,l-1} - A_{k,l+1}) . \tag{25}$$

These equations are not much different from the linearization of the equations of motion as discussed in 1 which lead to the dispersion relation ω_q for small amplitude plane waves. All it would need is to replace $k^2\Omega_b^2$ in (25) by ω_q^2. Consequently, if $k^2\Omega_b^2 = \omega_q^2$, small amplitudes of (25) will not decay in space, in contrast to our initial assumption. However, if $k^2\Omega_b^2 \neq \omega_q^2$ for any q, no plane waves exist, and instead we can obtain localization. In the considered case it is exponential

$$A_{kl} \sim \mathrm{e}^{-\xi_k |l|} \; , \; k^2\Omega_b^2 = v_2 + 2w_2(1 - \cosh \xi_k) . \tag{26}$$

Thus we arrive at a generically necessary nonresonance condition for the existence of breathers:[15,42]

$$k^2\Omega_b^2 \neq \omega_q^2 \tag{27}$$

for all integer k and any q. Clearly such a condition can be in principle fulfilled for any lattice, since ω_q^2 is bounded from above (in contrast to space continuous systems). The upper bound or cutoff is a result of the discreteness of the system. Right on the spot we may also conclude, that quasi-periodic in time and spatially localized excitations will not be exact solutions generically, since they will always radiate energy due to resonances. Indeed there is always an infinite number of pairs of integers k_1, k_2 which for any choice of incommensurate frequencies Ω_1, Ω_2 will lead to resonance $k_1\Omega_1 + k_2\Omega_2 = \omega_q$. So we have already an explanation for the weak but nonzero radiation observed in 3.1 for quasi-periodic excitations.

Returning to the time-periodic solutions, all we need is to tune the breather frequency and all its multiples out of resonance with ω_q. The nonlinear terms in the equations of motion will be responsible for that.

4.1. *Method No.1 - designing a map*

We will now design a map to find breather solutions up to machine accuracy. This method No.1 is one of the first which have been used to perform high precision computations of DBs. It is instructive that one can accomplish the task with using a bit of intuition and luck.[42,43]

Let us rewrite (24) as a map in two different ways. Map A:

$$A_{kl}^{(i+1)} = \frac{1}{k^2\Omega_b^2}\left[(v_2+2w_2)A_{kl}^{(i)} - w_2(A_{k,l-1}^{(i)} + A_{k,l+1}^{(i)}) + F_{kl}^{(nl)}(A_{k'l'}^{(i)})\right] , \tag{28}$$

with

$$\lambda_{kl} = \frac{v_2}{k^2\Omega_b^2}$$

and Map B:

$$A_{kl}^{(i+1)} = \frac{1}{v_2}\left[(k^2\Omega_b^2 - 2w_2)A_{kl}^{(i)} + w_2(A_{k,l-1}^{(i)} + A_{k,l+1}^{(i)}) - F_{kl}^{(nl)}(A_{k'l'}^{(i)})\right] , \tag{29}$$

with

$$\lambda_{kl} = \frac{k^2\Omega_b^2}{v_2} .$$

We can define a lattice map by using any of the two maps for any k and l, and a solution of (24) will be always a fixed point of the chosen lattice map. Two questions arise: is the breather solution a stable fixed point for the chosen lattice map, and what is a good initial guess? Instead of being worried about stability as one normally should, we may also approach the problem inversely. We know that we want to find a breather with frequency Ω_b located e.g. at site $l = 0$. Let us then put initially all Fourier amplitudes to zero except $A_{\pm 1,0}$ which is small but nonzero. For $k = \pm 1, l = 0$ we will choose the map with $\lambda_{\pm 1,0} > 1$ and the map with $\lambda_{kl} < 1$ for all other coefficients. Thus we will impose a local instability (growth) at $k = \pm 1, l = 0$ when we start the iteration. At the same time all other coefficients will tend to stay at zero, since their maps are chosen to be locally stable around the value zero. Thus we expect a breather to grow during the iteration. All we now have to do is to hope that the breather solution is a stable fixed point. For low order polynomial potential functions we can compute

$$F_{kl}^{(nl)} = \sum_{\alpha=3,4,\ldots} v_\alpha \sum_{k_1,k_2,\ldots,k_{\alpha-1}=-\infty}^{+\infty} A_{k_1 l}A_{k_2 l}\ldots A_{k_{\alpha-1} l}\delta_{k,(k_1+k_2+\ldots+k_{\alpha-1})} \tag{30}$$

very efficiently during each iteration. Otherwise we can take all $A_{kl}^{(i)}$ at a given step, compute $x_l(t)$ and by numerical integration obtain

$$F_{kl}^{(nl)} = \frac{1}{T_1}\int_{-T/2}^{T_2} F_l^{(nl)}(t)\mathrm{e}^{-ik\omega_1 t}\mathrm{d}t . \tag{31}$$

Of course we have to impose a cutoff in k-space, which can be justified afterwards by checking that the Fourier amplitudes close to the cutoff are reasonably small. The iteration can be stopped when e.g.

$$\sum_{k,l} |A_{kl}^{(i)} - A_{kl}^{(i-1)}| < 10^{-10} . \tag{32}$$

The following results have been obtained along these lines for a breather with frequency $\Omega_b = 1.3$. In Fig. 15 the solution for the Fourier coefficients

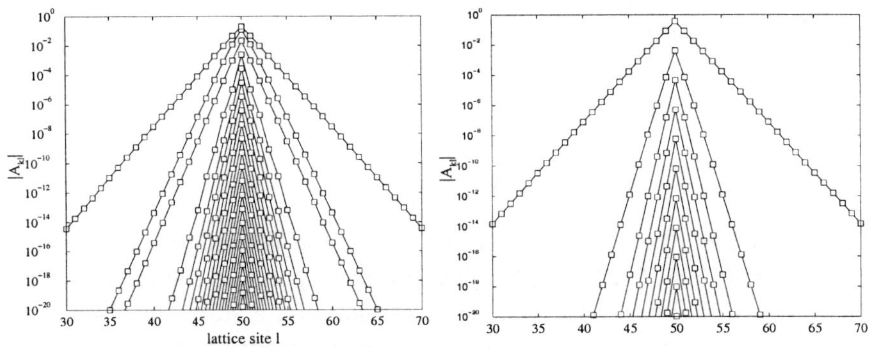

Fig. 15. Breather solution by method No.1. Left picture: $v_2 = 2$, $v_3 = -3$, $v_4 = 1$, $w_2 = 0.1$; right picture: $v_2 = 1$, $v_4 = 1$, $w_2 = 0.1$.

is plotted for two different systems. Absolute values of A_{kl} are plotted on a logarithmic scale versus lattice site number l. The non-filled squares are the actual numerical data. Coefficients with same values of k are connected with lines. We find the expected exponential decay in space, with exponents (slopes) clearly being dependent on k. A surprising numerical fact is that the computed amplitudes seem to be correct down to values 10^{-20}, although the Fortran compiler uses double precision floating point numbers (16 decimal digits). Moreover, the limit of the computation here would be actually at 10^{-307}. The reason is that we search for solutions which are localized around zero, and the issue is not numerical precision, but the encoding of small numbers. If however we would shift the classical ground state position to say $x_l = 1$, then the same computation would be restricted by the numerical precision.

To check whether the numerically computed exponential decay in space is in accord with the predicted one (26) from the linearized equations (25) we simply measure the slopes in Fig. 15 and compare them with the solutions of (26) for the left picture in Fig. 15

k	num.result	linearization
0	−1.3202	−1.3415
1	−0.6904	−0.6898
2	−1.3796	−1.6588
3	−2.0748	−2.1143
4	−2.3957	−2.3951
5	−2.6018	−2.6026
6	−2.7663	−2.7682
⋮	⋮	⋮

While most of the numbers do coincide, clear deviations are observed for $k = 2, 3$. Note that the numerical slope is weaker than the predicted one. The obvious reason is that for these Fourier numbers weakly decaying nonlinear corrections have to be taken into account,[43] which decay slower than the predicted linearized result. Here these corrections are simply $\sim A_{1l}^2$ for $k = 2$ and $\sim A_{1l}^3$ for $k = 3$. The analytically predicted slopes are then simply $2 \cdot 0.6898 = 1.3796$ for $k = 2$ and $3 \cdot 0.6898 = 2.0694$ for $k = 3$. A full treatment of nonlinear corrections is given in Ref. 43. Note that the nonresonance condition (27) is not affected by these corrections. Also important is, that the Fourier amplitude with the weakest spatial decay is always correctly described by the linearized equations in the breather tails.

For the right picture in Fig. 15 we find respectively

k	num.result	linearization
1	−0.6722	−0.6709
3	−1.9910	−2.1464
5	−2.6103	−2.6133
7	−2.9114	−2.9117
9	−3.1324	−3.1325
⋮	⋮	⋮

Only the $k = 3$ values differ, and the correct slope is again given by terms $\sim A_{1l}^3$: $3 \cdot 0.6709 = 2.0127$.

4.2. Method No.2 - saddles on the rim with space-time separation

A subclass of systems (1) is characterized by space-time separation (see Refs. 44, 16 and 45). Consider

$$H = \sum_l \left[\frac{1}{2}p_l^2 + \frac{v_2}{2}x_l^2\right] + POT \ , \tag{33}$$

with

$$POT = \sum_l \left[\frac{v_{2m}}{2m}x_l^{2m} + \frac{w_{2m}}{2m}(x_l - x_{l-1})^{2m}\right] \ , \ m = 2, 3, 4, ... \tag{34}$$

being a homogeneous function of the coordinates. The equations of motion take the form

$$\ddot{x}_l + v_2 x_l = -v_{2m}x_l^{2m-1} - w_{2m}(x_l - x_{l-1})^{2m-1} + w_{2m}(x_{l+1} - x_l)^{2m-1} \ . \tag{35}$$

These systems allow for time space separation for a sub-manifold of all possible trajectories:

$$x_l(t) = A_l G(t) \ . \tag{36}$$

Inserting (36) into (35) we obtain

$$\frac{\ddot{G} + v_2 G}{G^{2m-1}} = -\kappa \ , \tag{37}$$

$$-\kappa = \frac{1}{A_l}\left[-v_{2m}A_l^{2m-1} - w_{2m}(A_l - A_{l-1})^{2m-1} + w_{2m}(A_{l+1} - A_l)^{2m-1}\right] \ . \tag{38}$$

Here $\kappa > 0$ is a separation parameter, which can be chosen freely. The master function G obeys a trivial differential equation for an anharmonic oscillator

$$\ddot{G} = -v_2 G - \kappa G^{2m-1} \ . \tag{39}$$

Its solution sets the temporary evolution of the breather.

The spatial profile is given by

$$\kappa A_l = \frac{\partial POT}{\partial x_l}|_{\{x_{l'} \equiv A_{l'}\}} \ , \tag{40}$$

or better by the extrema of a function S:

$$\frac{\partial S}{\partial A_l} = 0 \ , \ S = \frac{1}{2}\kappa \sum_l A_l^2 - POT(\{x_l' \equiv A_l'\}) \ . \tag{41}$$

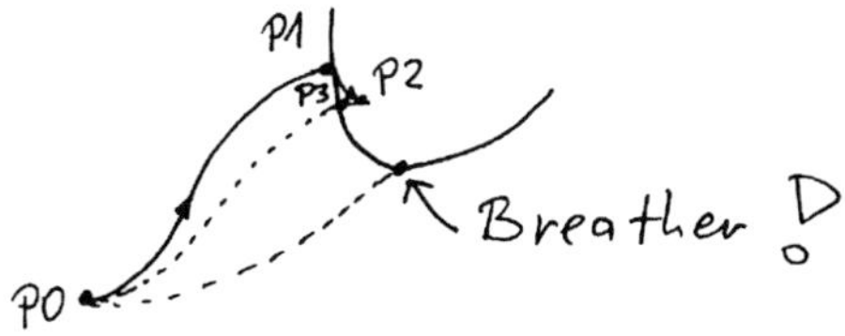

Fig. 16. Schematic representation of function S (41) and the pathway to a breather being a saddle.

Let us discuss some properties of S. This function has a minimum at $A_l = 0$ for all l with height $S = 0$ (point $P0$ in Fig. 16). When choosing a certain direction in the A_l space starting from $P0$, S will first increase, then pass through a maximum and further decrease to $-\infty$. So there is a rim surrounding the minimum $A_l = 0$. Since breathers are spatially localized solutions, variation of the amplitudes A_l in the tails of a breather around zero will increase S. At the same time the breather corresponds to an extremum of S, but there is only one trivial minimum of S located at $P0$. Thus breathers are saddles of S.

It is remarkably easy to compute such a saddle. First choose direction in the N-dimensional space of all A_l, e.g. (...0001000...) , (...0001001000...) etc. Then start from space origin $P0$, $A_l = 0$, depart with small steps in the chosen direction, compute S. It will first increase and then pass through a maximum $P1$. Now we are on the rim. Compute the gradient of S here and make a small step in opposite direction, to arrive at $P2$. Maximize S on the line $P0 - P2$ to be on the rim again. Repeat until you reach a saddle with required accuracy.

This method has been used to compute various types of breathers and multi-breathers. Note that it is very simple to extend the computation to two- or three-dimensional lattices.[45]

4.3. *Method No.3 - homoclinic orbits with time-space separation*

Using again the time-space separability as discussed in 4.2, breathers can be considered as homoclinic orbits of a two-dimensional map.[16] Indeed, we may rewrite (38) in the following way:

$$A_{l+1} = A_l + \left[v_{2m} A_l^{2m-1} + w_{2m}(A_l - A_{l-1})^{2m-1} - \kappa A_l\right]^{\frac{1}{2m-1}} \qquad (42)$$

where we can compute a given amplitude profile starting with a given pair of nearest neighbor amplitudes (both to the right and to the left of course). Using a two-dimensional vector

$$\vec{R}_l = (x_l, y_l) = (A_{l-1}, A_l) \tag{43}$$

the procedure can be cast into the form of a two-dimensional map with

$$x_{l+1} = y_l \tag{44}$$

$$y_{l+1} = y_l + \left[v_{2m} y_l^{2m-1} + w_{2m}(y_l - x_l)^{2m-1} - \kappa y_l\right]^{\frac{1}{2m-1}} \tag{45}$$

This map (Fig. 17) has a fixed point $\vec{R}_F = (0,0)$. The fixed point be-

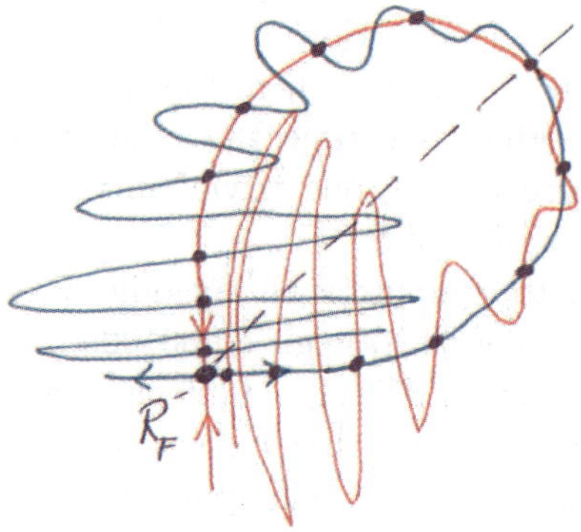

Fig. 17. Schematic representation of the map (44,45). Red line - stable invariant manifold, green line - unstable invariant manifold, black spots - intersection points of both manifolds for a given breather solution. Dashed blue line - diagonal $x = y$.

longs both to a stable (red) and unstable (green) one-dimensional invariant manifolds. Taking a point on the stable manifold and iterating forward, we will approach the fixed point. The same happens with a point on the unstable manifold when iterated backwards. These manifolds intersect in many points. By definition any of these intersection points, when iterated either forward or backward, will converge to $\vec{R}_F$ and thus corresponds to a breather solution. Such map trajectories are also called homoclinic orbits. Note that many intersection points belong to the same homoclinic orbit or to the same breather, as indicated by the ones marked with black spots in

Fig. 17. However since the above map is locally (around $\vec{R}_F$) volume preserving, the structure of the invariant manifold lines will generically show up with horseshoe patterns (wiggles in Fig. 17). These patterns generate additional intersection points. Consequently there will be an infinite number of different homoclinic orbits and thus breathers. They will differ by the amplitude distribution inside the breather core, which can become arbitrary complicated, and an exponential tail outside. Thus we already at this stage arrive at the conclusion that in addition to single site breathers discussed so far also so-called multi-breather solutions can exist, i.e. localized excitations with a complicated pattern of energy distribution inside the breather core (see also Ref. 15).

Due to the space-reflection symmetry of the map there will be always one intersection point on the line $x = y$. The position of this point will depend only parametrically on κ. Thus it is possible to design simple search routines by e.g. fixing $x_0 = y_0$ and varying κ (see Ref. 16). The numerical scheme has been even used for a formal existence proof of breathers as homoclinic orbits.[16]

5. Obtaining breathers up to machine precision: Part II

So far we have searched for discrete breather periodic orbits as solutions of algebraic equations. The variables were either Fourier coefficients or simply the amplitudes at a given site. Also the methods of solving these equations have been quite special, using some particular properties of the system. What if we don't know or do not want to know any particular system properties we could use? We could of course use more general methods of solving algebraic equations, e.g. various gradient methods or Newton routines.[23] For them to converge we need always a good initial guess. This usually implies that we should start computations close to a case where we know the solution, and then depart from this limit with small parameter steps.

Gradient methods are more sophisticated in programming, while Newton routines may suffer from the long times that may be needed to invert matrices, and also from the danger of coming close to a noninvertible case due to bifurcations. Recall here that the Newton map for finding the zero of a known function $f(x)$ (meaning that we can compute its value) is given by $f(x = s) = 0$, $f(x) = f(x_0)+f'(x_0)(x-x_0)+...$, $x_{n+1} = x_n-f(x_n)/f'(x_n)$. In our cases f will be a vector function and its derivative a matrix.

Instead of solving algebraic equations for amplitudes, we may also try to compute the periodic breather orbit directly in the phase space of our system. Recall that a periodic orbit (PO) is a loop in phase space. Generic POs of generic nonintegrable Hamiltonian systems are isolated ones, i.e. in a small neighborhood in phase space we will generically not find other slightly deformed POs with identical values of conserved quantities like energy. This is in contrast to POs on resonant tori of integrable Hamiltonian systems. However isolated POs have generically slightly deformed POs in

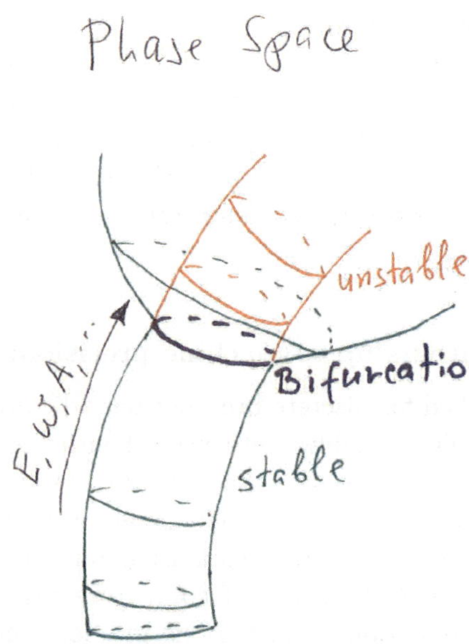

Fig. 18. Schematic representation of a family of isolated POs. Green sector - stable POs, red sector - unstable POs, blue line - bifurcation location of additional PO family detaching.

their neighborhood with slightly different values of conserved quantities (see Fig. 18). So we can think of isolated POs residing on cylinders in phase space, where each point on a cylinder belongs to a closed loop which is a PO. Sliding along the cylinder we change all the parameters of the PO. In particular, a PO can turn from stable to unstable, due to a bifurcation, possibly resulting in new families of POs, as indicated in Fig. 18.

5.1. *Method No.4 - Newton in phase space*

Now we may proceed in describing the most popular method of finding discrete breathers - a Newton map in phase space.[46] Let us integrate a given initial condition $\vec{\tilde{R}}$ with

$$x_l(t=0) \equiv X_l \ , \ p_l(t=0) \equiv P_l \tag{46}$$

over a certain time T:

$$x_l(T) \equiv I_l^x(\{X_{l'}, P_{l'}\}, T) \ , \tag{47}$$

$$p_l(T) \equiv I_l^p(\{X_{l'}, P_{l'}\}, T) \ . \tag{48}$$

Consider the functions

$$F_l^x = I_l^x - X_l \ , \ F_l^p = I_l^p - P_l \ . \tag{49}$$

If $\vec{\tilde{R}}$ belongs to a PO with period T then

$$F_l^x = F_l^p = 0 \ . \tag{50}$$

Now we can implement a Newton map such that all functions in (50) will vanish. Our variables are simply the phase space variables which define the initial conditions. Since the Newton map needs inversion of a derivative matrix, we have to remove all possible degeneracies which lead to zero eigenvalues of the newton matrix. Indeed, if $\vec{\tilde{R}}$ belongs to the PO, then a 1d manifold of points belong to the PO. This is a degeneracy due to the phase of the PO. It can be removed by one additional condition, e.g.

$$P_M = 0 \ . \tag{51}$$

So for N degrees of freedom we will search for zeros in $2N-1$ coupled equations of $2N-1$ variables.

A less obvious obstacle we have to take care of is to make sure that a zero of these $2N-1$ equations with the additional initial condition $P_M = 0$ uniquely fixes $p_M(T) = 0$, e.g. through energy conservation. If that will be not the case, we can not ensure that our procedure computes a PO.

Let us define

$$\vec{R} = (X_1, X_2, ..., X_M, ..., X_N, P_1, ..., P_{M-2}, P_{M-1}, P_{M+1}, P_{M+2}, ..., P_N) \ , \tag{52}$$

$$\vec{F} = (F_1^x, F_2^x, ..., F_M^x, ..., F_N^x, F_1^p, ..., F_{M-2}^p, F_{M-1}^p, F_{M+1}^p, F_{M+2}^p, ..., F_N^p) \ , \tag{53}$$

$$\vec{F} = \vec{R}(T) - \vec{R} \ . \tag{54}$$

Given an initial guess $\vec{R}^{(0)}$ expand

$$F_n(\vec{R}) = F_n(\vec{R}^{(0)}) + \sum_m \frac{\partial F_n}{\partial R_m}|_{\vec{R}^{(0)}} (R_m - R_m^{(0)}) \ , \tag{55}$$

$$\vec{F}(\vec{R}) = \vec{F}(\vec{R}^{(0)}) + \mathcal{M}(\vec{R} - \vec{R}^{(0)}) \ , \tag{56}$$

$$\mathcal{M}_{nm} = \frac{\partial F_n}{\partial R_m}|_{\vec{R}^{(0)}} = \frac{\partial R_n(T)}{\partial R_m}|_{\vec{R}^{(0)}} - \delta_{nm} \ . \tag{57}$$

Now we may perform one Newton step, i.e. find an $\vec{R}$ such that $\vec{F} = 0$:

$$\vec{R} = \vec{R}^{(0)} - \mathcal{M}^{-1}\vec{F}(\vec{R}^{(0)}) \ . \tag{58}$$

This procedure can be repeated until some precision is obtained: $|\vec{F}| < \epsilon$ or $max|F_n| < \epsilon$.

What remains is to explain how to compute the Newton matrix $\mathcal{M}$. For the special case of a two-dimensional space of variables the notations in Fig. 19 will help to understand the following points. Given an initial guess

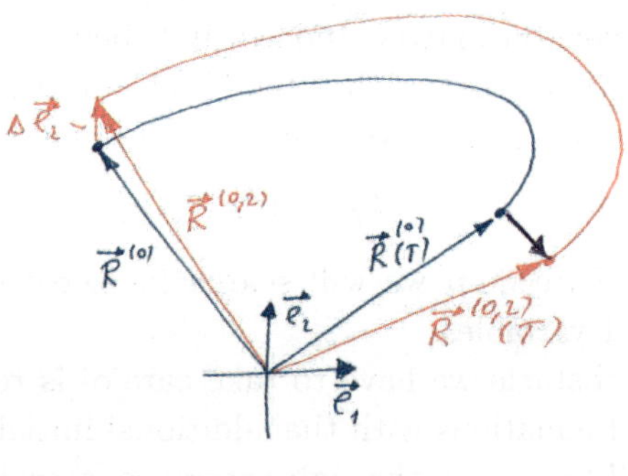

Fig. 19. Schematic representation of the computation of the Newton matrix in a two-dimensional space of variables. See text for details.

$\vec{R}^{(0)}$ and integrating over time T, we arrive at $\vec{R}^{(0)}(T)$. Generally the two points will differ in phase space. Now we perturb $\vec{R}^{(0)}$ in the direction m by Δ:

$$\vec{R}^{(0,m)} = \vec{R}^{(0)} + \Delta\vec{e}_m \ . \tag{59}$$

Here $\vec{e}_m$ denotes a unit vector in direction m. Integrating $\vec{R}^{(0,m)}$ over the period T we arrive at $\vec{R}^{(0,m)}(T)$. Then the Newton matrix elements are given by

$$\mathcal{M}_{nm} = \frac{1}{\Delta}\left(F_n(\vec{R}^{(0,m)}) - F_n(\vec{R}^{(0)})\right) . \tag{60}$$

For computational purposes it might be more convenient to use the alternative expression directly through the vectors:

$$\mathcal{M}_{nm} = \frac{1}{\Delta}\left(R_n^{(0,m)}(T) - R_n^{(0)}(T)\right) - \delta_{nm} . \tag{61}$$

The advantages of Newton maps are that they are relatively easy to program once we already have a good integrator. The map converges exponentially fast. Furthermore we may use one Newton matrix for several iterations, which may be useful when matrices get large. Disadvantages of Newton maps may be due to relatively large computational time $\sim N^2$ because of matrix inversion. Matrix inversions are sensitive to bifurcations, because at bifurcations additional degeneracies take place, which may lead to zero eigenvalues of $\mathcal{M}$. Sometimes we may need more subtle inversion routines using singular value decomposition etc. Note that at some point the efforts of removing all the obstacles from a Newton map approach might be equivalent to the ones of using alternative methods.

As always we need a good initial guess. Probably we have to deform our system parameters such that a known solution can be used, and afterwards system parameters are changed by small steps, tracing the solution. We should also keep in mind that other specific methods may deal with a certain limiting case easily, so a known solution must not be one we obtained analytically, but also numerically with various other methods at hand.

5.2. *Method No.5 - steepest descent in phase space*

Similar to the Newton map we may also use a steepest descent method in phase space.[47] Define the nonnegative function

$$g(\vec{\tilde{R}}) = \sum_l \left[F_l^x F_l^x + F_l^p F_l^p\right] \tag{62}$$

and its gradient with components

$$(\nabla g)_n = \frac{\partial g}{\partial \tilde{R}_n} . \tag{63}$$

Now we simply start at some point in phase space, compute the gradient, and descent in the direction opposite to the gradient. Then we again compute the gradient etc. A breather solution is found if g comes close enough to zero.

The advantages of steepest descent are that the computational time grows with $\sim N$. Furthermore the method is insensitive to bifurcations. Disadvantages of steepest descent are that it is more clumsy to program, that the convergence is slower than that of Newton maps and that it may be hard to distinguish zero minima from nearly zero minima.

5.3. *Symmetries*

Very often the equations of motion are invariant under some symmetry operations, e.g. the continuous time-shift symmetry $t \to t+\tau$, the time reversal symmetry $t \to -t$, $p_l \to -p_l$, some parity symmetry $x_l \to -x_l$, $p_l \to -p_l$, the discrete translational symmetry on the lattice and probably other discrete permutational lattice symmetries which leave the lattice invariant, like spatial reflections etc.

Each discrete symmetry implies that given a trajectory in phase space, a new trajectory is generated by applying the symmetry operation to the manifold of all points of the original trajectory. If the new manifold equals the original one, then the trajectory is invariant under the symmetry, and otherwise it is not invariant.

In linear equation systems symmetry breaking is possible only in the presence of degeneracies. In nonlinear equation systems symmetry breaking is a common feature. For example, a plane wave in a harmonic chain is not invariant under time reversal symmetry, because of degeneracy (of left and right going waves $\omega_q = \omega_{-q}$).

A breather is by definition not invariant under discrete translational symmetry. If however it is invariant under other symmetries, this can be used to substantially lower the numerical effort of computing the solution.[6]

For time-reversal breathers it is possible to find an origin in time when $x_l(t) = x_l(-t)$, $p_l(t) = -p_l(-t)$, which saves 50% of computational time. For time-reversal parity-invariant breathers $x_l(t + T/2) = -x_l(t)$, $p_l(t + T/2) = -p_l(t)$ we may save 75% of computational time.

Higher dimensional lattices may allow for further symmetries. Computing lattice permutational invariant breathers may substantially lower the computational effort by finding the irreducible breather section.

At the same time even in the presence of additional symmetries breather solutions may be found which lack these symmetries. The simplest example is again discrete translational symmetry, but also lattice reflection symmetries may be broken. Even breathers which are not invariant under time reversal and thus possess a nonzero energy flux do exist, except for one-dimensional systems.[48]

6. Perturbing breathers

Suppose we found a breather solution $x_l(t)$. Let us address the question of stability and interaction with plane waves. First we add a perturbation $\epsilon_l(t)$ to the breather solution. What can we say about the evolution of this perturbation? Evidently, if the amplitude of the perturbation is large, we may expect generic dynamical features of a nonintegrable system, which are usually rather complicated and hard to be addressed analytically. If however the perturbation size is small, we may linearize the resulting equations for $\epsilon_l(t)$:[5,49]

$$\ddot{\epsilon}_l = -\sum_m \frac{\partial^2 H}{\partial x_l \partial x_m}|_{\{x_{l'}(t)\}} \epsilon_m \ . \tag{64}$$

This problem corresponds to a time-dependent Hamiltonian $\tilde{H}(t)$

$$\tilde{H}(t) = \sum_l \left[\frac{1}{2}\pi_l^2 + \frac{1}{2}\sum_m \frac{\partial^2 H}{\partial x_l \partial x_m}|_{\{x_{l'}(t)\}} \epsilon_l \epsilon_m \right] \ , \tag{65}$$

$$\dot{\epsilon}_l = \frac{\partial \tilde{H}}{\partial \pi_l} \ , \ \dot{\pi}_l = -\frac{\partial \tilde{H}}{\partial \epsilon_l} \ . \tag{66}$$

The evolution of this time-dependent Hamiltonian is characterized by a conservation law $\dot{I} = 0$, where the symplectic product I is formed between two trajectories (with and without prime respectively):

$$I = \sum_l \left[\pi_l'(t)\epsilon_l(t) - \pi_l(t)\epsilon_l'(t) \right] \ . \tag{67}$$

The reader can verify that I is constant in time by straightforward differentiation with respect to time and by using the equations of motion (66). Let us briefly discuss the consequences of this conservation law.

For simplicity we drop the lattice index for the next lines. Define the matrix $\mathcal{J}$

$$\mathcal{J} = \begin{pmatrix} 0 & 1 \\ -1 & 0 \end{pmatrix} \tag{68}$$

and the evolution matrix $\mathcal{U}(t)$

$$\begin{pmatrix} \pi(t) \\ \epsilon(t) \end{pmatrix} = \mathcal{U}(t) \begin{pmatrix} \pi(0) \\ \epsilon(0) \end{pmatrix} \tag{69}$$

which maps the phase space of the perturbations onto itself by integrating each point over a given time t. It follows that we can express I in the following form

$$I = (\pi(t), \epsilon(t))\, \mathcal{J} \begin{pmatrix} \pi'(t) \\ \epsilon'(t) \end{pmatrix} \tag{70}$$

and using (69) as

$$I = (\pi(0), \epsilon(0))\, \mathcal{U}^T(t) \mathcal{J} \mathcal{U}(t) \begin{pmatrix} \pi'(0) \\ \epsilon'(0) \end{pmatrix} . \tag{71}$$

Since I is conserved, and $\mathcal{U}(t=0)$ is the identity matrix, we conclude

$$\mathcal{U}^T(t) \mathcal{J} \mathcal{U}(t) = \mathcal{J} . \tag{72}$$

We have obtained that $\mathcal{U}(t)$ is symplectic. Then it follows (and can be easily derived with the help of the obtained relations) that if y is an eigenvector of $\mathcal{U}$ with eigenvalue λ

$$\mathcal{U} y = \lambda y \; , \; \mathcal{U}^T y = \lambda y \; , \tag{73}$$

then y' is a related eigenvector with eigenvalue $1/\lambda$:

$$\mathcal{U} y' = \frac{1}{\lambda} y' \; , \; y' = \mathcal{J}^{-1} y = -\mathcal{J} y \; . \tag{74}$$

If $\mathcal{U}$ is real and (λ, y) are an eigenvalue and eigenvector, so are

$$(\lambda^*, y^*) \; , \; (\frac{1}{\lambda}, \mathcal{J} y) \; , \; (\frac{1}{\lambda^*}, \mathcal{J} y^*) \; . \tag{75}$$

Note that even though $\mathcal{U}$ is real, both eigenvectors and eigenvalues will be complex in general.

6.1. *Linear stability analysis*

Consider now the mapping over one period for a breather, which defines the real valued Floquet Matrix $\mathcal{F}$

$$\mathcal{U}(T_b) \equiv \mathcal{F} \; . \tag{76}$$

The eigenvalues and eigenvectors of $\mathcal{F}$ completely define the dynamics of small amplitude perturbations of a breather, or the dynamics of the linearized phase space flow around a breather solution. We can now study whether a breather is unstable or stable, how strongly plane waves are scattered by the breather, etc.

Before starting to address these questions, let us discuss the meaning of a nondegenerate complex eigenvalue λ and eigenvector y of $\mathcal{F}$ for the dynamics of the real valued phase space variables ϵ_l, π_l. For that purpose we write

$$\lambda = \lambda_r + i\lambda_i \ , \ y = y_r + iy_i \tag{77}$$

where $\lambda_r, \lambda_i, y_r, y_i$ are the real and imaginary parts of the eigenvalue and eigenvector. Then using $\mathcal{F}y = \lambda y$ we obtain

$$\mathcal{F}y_r = \lambda_r y_r - \lambda_i y_i \ , \tag{78}$$

$$\mathcal{F}y_i = \lambda_i y_r + \lambda_r y_i \ . \tag{79}$$

Thus taking any linear combination of y_r and y_i as an initial condition for ϵ_l, π_l, the Floquet map will perform some unitary transformation in the subspace spanned by y_r and y_i, and in addition change the length of the new vector by $|\lambda|$. We also know that if both y_r and y_i are nonzero, so are λ_r and λ_i. Then there exists another eigenvalue with λ_r and $-\lambda_i$ and y_r and $-y_i$. But from the point of view of the dynamics of the real-valued phase space variables this complex conjugated eigenstate does not add much new results. So we conclude that if a pair of complex eigenvectors y and y^* has been computed, their real and imaginary parts span a two-dimensional subspace in the phase space (of the perturbations) which is invariant under applying the Floquet mapping. The mapping performs simply a rotation only if $|\lambda| = 1$, otherwise it adds a contraction $|\lambda| < 1$ or an expansion $|\lambda| > 1$.

If there is an eigenvalue with $|\lambda| < 1$, due to (73,74) there is an eigenvalue with $|\lambda| > 1$ and vice versa. Consequently whenever we find eigenvalues with $|\lambda| \neq 1$, there are directions in the phase space of perturbations where we will observe growth, which implies linear instability. So we conclude that the only possibility for breathers to be marginally stable is to have all Floquet eigenvalues being located on the unit circle $|\lambda| = 1$.

All eigenstates which reside on the unit circle fulfill Bloch's Theorem, i.e. eigenstates with $\lambda = \mathrm{e}^{i\omega_\nu T_b}$ when taken as initial conditions correspond to

$$\epsilon_l(t) = \mathrm{e}^{i\omega_\nu t}\Delta_l^{(\nu)}(t) \ , \ \Delta_l^{(\nu)}(t) = \Delta_l^{(\nu)}(t + T_b) \ . \tag{80}$$

One Floquet eigenvalue is always located at $\lambda = +1$. Its eigenvector is tangent to the periodic orbit of the original breather. As eigenvalues come in pairs, there is another eigenvalue at $\lambda = +1$. It corresponds to perturbations tangent to the breather family of POs. Upon changing a control parameter the other Floquet eigenvalues may move on the unit circle, collide and leave the circle. Then a breather turns from being linearly stable to linearly unstable. A schematic outcome of the Floquet eigenvalues for a marginally stable and unstable breather solutions is shown in Fig. 20.

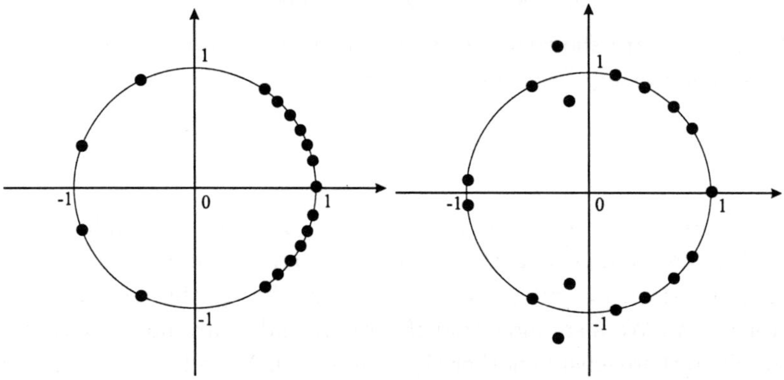

Fig. 20. Schematic view of an outcome of the Floquet analysis of a breather. Floquet eigenvalues (filled circles) and the unit circle are plotted in the complex plane. Left picture: marginally stable breather (all eigenvalues are located on the unit circle). Right picture: unstable breather (two eigenvalues are located outside the unit circle). Note that the group of closely nearby lying eigenvalues on the unit circle correspond to the plane wave continuum (extended Floquet eigenstates), while the separated eigenvalues on the circle correspond to localized Floquet eigenstates.

Floquet eigenvectors (i.e. the perturbations at time $t = 0$: $\vec{F} = (\epsilon_1, \epsilon_2, ..., \epsilon_N, \pi_1, \pi_2, ..., \pi_N))$ can be localized or delocalized in the lattice space. Because the breather is localized, for large enough lattice size N there will be a large number $\sim 2N$ of delocalized Floquet eigenvectors, and only a finite number of localized ones. Delocalized Floquet eigenstates correspond to plane waves far from the breather core.

The numerical computation of a Floquet matrix is similar to the above described way to compute the Newton matrix.[50] Using the results of 5.1 we choose a starting point on the breather orbit $\vec{R}^{(b)}$ with $\vec{R} = (X_1, X_2, ..., X_{N-1}, X_N, P_1, P_2, ..., P_{N-1}, P_N)$ and compute in analogy to (61)

$$\mathcal{F}_{nm} = \frac{1}{\Delta}\left(R_n^{(b,m)}(T_b) - R_n^{(b)}(T_b)\right) , \tag{81}$$

keeping in mind that all $2N$ phase space directions are used here. Note that most of the elements of the Floquet matrix are also contained in the Newton matrix of the last step of a Newton map, i.e. when being reasonably close to an exact DB solution.

Before diagonalizing $\mathcal{F}$ we could check all possible symmetries in order to reduce the Floquet matrix to its noninteracting irreducible parts. A good test of the quality of the numerically obtained spectrum is to confirm the double degeneracy of $\lambda = 1$ and the relations (75). The results are used in order to characterize stability of a given breather, to trace bifurcations of breathers, to make contact with possible moving breathers etc.

6.2. *Plane wave scattering*

The knowledge of the Floquet eigenvalues provides with stability information, and the Floquet eigenvectors tell us which directions in phase space are causing possible instabilities, and the nature of the eigenvector (localized or delocalized) provides with further information. However there is another information hidden in the extended eigenstates, namely their phases. These phases provide with information about the scattering of plane waves by discrete breathers. Such a scattering has been indeed observed in simple numerical runs, when an extended plane wave was sent into a breather, to show up with an energy density distribution as the one in Fig. 21.[51] We observe that most of the plane wave coming from the left is reflected back, and only a small fraction of about one percent is transmitted through the breather. This implies that breathers may act as very strong scattering centers. Computational studies of wave scattering have been so far done for one-dimensional lattices. [52,53,54,55,56] This is caused on one hand by the fact that scattering in higher lattice dimensions is more hard to be handled. On the other hand breathers in higher lattice dimensions are interacting much weaker with radiation.

For one-dimensional lattices we need to find the transmission coefficient as a function of the wave number of a plane wave which is sent into the breather from say the left end of the system. Since such a plane wave corresponds to an extended Floquet eigenstate, we may write it in its Bloch representation as

$$\epsilon_l(t) = \sum_{k=-\infty}^{\infty} e_{lk} \mathrm{e}^{\mathrm{i}(\omega_q + k\Omega_b)t} \,. \tag{82}$$

We find that inside the breather new frequencies $\omega_q + k\Omega_b$ are generated. These new frequencies are also frequently coined as channels (see Fig. 22

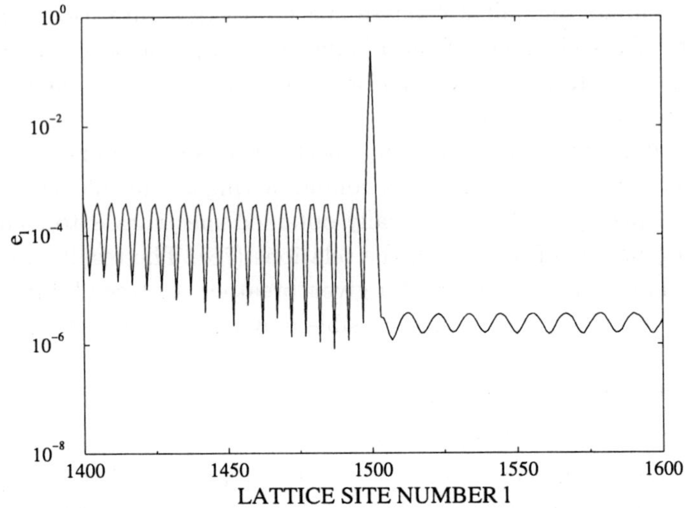

Fig. 21. Scattering of a plane wave with $q = 0.2\pi$ by a breather located at site 1500. The energy density distribution is shown. The incident wave comes from the left. The standing wave pattern on the left side of the DB is due to interferences between the incident and reflected waves.

for a schematic view). Can any of these new channels again resonate with

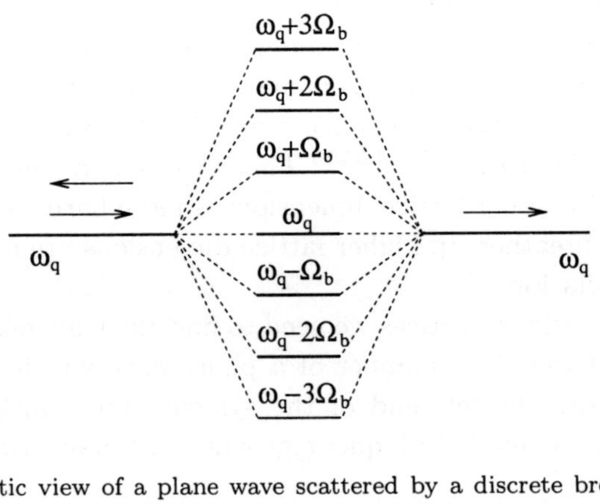

Fig. 22. Schematic view of a plane wave scattered by a discrete breather. The plane wave with frequency ω_q is injected from the left. Inside the breather new frequency channels are excited.

the spectrum $\pm\omega_q$ (note the $\pm$ sign indicating that we have to consider the frequency spectrum itself and not its squared analog or absolute values)?

Since the breather frequency Ω_b has to be in general larger than the width of the band ω_q, at maximum one of the additional channels can resonate with another plane wave frequency $-\omega_{q'} = \omega_q + k\Omega_b$. Such a case is called two-channel scattering, and channels which match plane wave frequencies are called open channels, while all others are called closed channels. It is straightforward to see that for m different plane wave bands at most $2m$ channels can be open. Returning to the case $m = 1$, two-channel scattering can be obtained under certain circumstances, but it is much easier to realize one-channel scattering, when all of the additionally generated channels inside the breather are closed. We also note here that one-channel scattering is always elastic, i.e. the energy flux of the outgoing waves (transmitted and reflected) equals the energy flux of the incoming wave.[52] Two-channel scattering is inelastic, with more energy carried away from the breather than sent inside. Thus in a real simulation two-channel scattering will lead to a linear in time decrease of the breather energy.[52] In the following we will focus on the case of elastic one-channel scattering only.

To compute the transmission coefficient for a plane wave, we need to know how large our chosen system should be. The system size N should be large compared to the localization length $1/\xi_k$ in (26) for any k. In addition we have to compute the localization length $1/\varsigma_k$ of all closed channels in a similar way

$$(\omega_q + k\Omega_b)^2 = v_2 + 2w_2(1 - \cosh \varsigma_k) \tag{83}$$

for all nonzero k and request that the system size is larger. Then we can approximate the extended Floquet state (82) by a simple plane wave for larger distances from the breather, with exponential accuracy. Assuming that this is done, we choose the labeling of the sizes of our finite system

$$-N, (-N+1), ..., -1, 0, 1, ...(N-1), N \tag{84}$$

where the breather is located in the center around site $l = 0$.

Solving the Floquet problem would provide only with a discrete set of extended eigenstates due to the finiteness of the system. Also we do not need all Floquet states, but are interested only in the transmission properties of a given extended state. Thus we simply emulate an infinite system by imposing the following boundary conditions:

$$\epsilon_{N+1} = e^{-i\omega_q t} \ , \ \epsilon_{-N-1} = (A + iB)e^{-i\omega_q t} \ . \tag{85}$$

While we assume that the transmitted wave on the right end has amplitude $|\epsilon_{N+1}| = 1$, the amplitude and relative phase on the left end are still undetermined and implicitly encoded in the real numbers A and B. Let us fix these numbers in some arbitrary way. The next step is to perform a Newton map (not a Floquet calculation!) in order to find the zeros of $\mathbf{G}$ which is defined as

$$\mathbf{G}(\vec{\epsilon}(0), \dot{\vec{\epsilon}}(0)) = \begin{pmatrix} \vec{\epsilon}(0) \\ \dot{\vec{\epsilon}}(0) \end{pmatrix} - e^{i\omega_q T_b} \begin{pmatrix} \vec{\epsilon}(T_b) \\ \dot{\vec{\epsilon}}(T_b) \end{pmatrix} . \tag{86}$$

Contrary to the Floquet approach, we thus obtain an extended Floquet eigenstate with an eigenvalue being located exactly on the unit circle. Moreover, in the ideal case we need only one step of the Newton map to converge to the solution, because the equations of motion are linear. Sometimes a second step is needed due to numerical errors done during the first step.

The obtained eigenstate is however in general not corresponding to the desired scattering setup, since we do not know whether on the right end of our system the obtained state corresponds to plane wave traveling to the right only. The reason for that is that extended Floquet eigenstates are two-fold degenerated for infinite systems. In order to proceed we add another Newton map with just two variables A and B such that the eigenstate solution from the first map satisfies

$$\epsilon_N = e^{-iq-i\omega_q t} , \tag{87}$$

which now implies that we have selected a Floquet eigenstate which corresponds to a plane wave traveling to the right at the right end of our system, and thus satisfying our scattering setup. With the notation

$$\epsilon_l(t) = \zeta_l(t)e^{-i\omega_q t} \tag{88}$$

and remembering that at the ends of our system ζ_l is a time-independent complex number, the transmission coefficient can be expressed through the obtained numbers A and B:

$$t_q = \frac{4\sin^2 q}{|(A+iB)e^{-iq} - \zeta_{-N}|^2} . \tag{89}$$

The described method[55] is remarkably easy to handle, provides with machine precision computations, doesn't care about any symmetry and structure of the underlying breather solution and can be applied as well to any related problem of scattering by a time-periodic scattering potential. In Fig. 23 we plot[55] the computed transmission coefficient versus q and Ω_b for an acoustic system with $V = 0$ and $w_2 = w_4 = 1$. As expected the transmission coefficient vanishes at $q = \pi$ (plane wave band edge with zero group velocity), but also in this special case of an acoustic system it takes

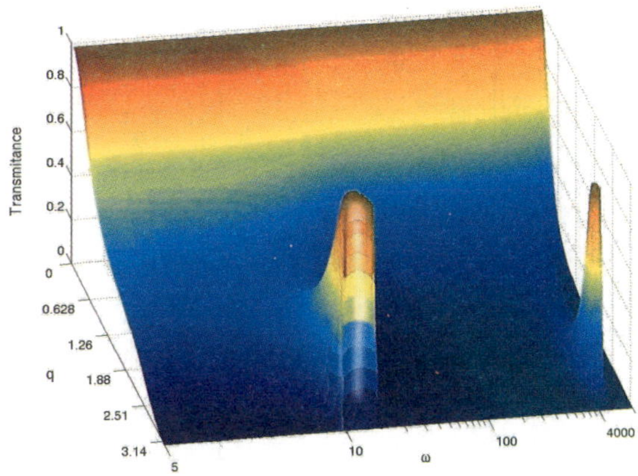

Fig. 23. Transmission coefficient versus wave number q and breather frequency Ω_b for an acoustic chain (see text for details).

value $t = 1$ at $q = 0$ due to mechanical momentum conservation. Note the two peaks in Fig. 23 where $t = 1$ again, due to bifurcations of localized Floquet states from the continuous part of the Floquet spectrum.[52,57]

In Fig. 24 the above case for $\Omega_b = 4.5$ is compared with the result for a chain with additional $w_3 = 1$.[55] Note the additional resonant perfect transmission peaks due to additional localized Floquet eigenstates and also the remarkable resonant perfect reflection minima due to Fano resonances.[54] Only recently these Fano resonances have been explained by localized modes of closed channels resonating with the open channel.[56] In some limiting cases these localized modes have been even computed numerically to predict and observe a Fano resonant reflection for other systems.[56]

7. Breathers in dissipative systems

So far we have been discussing computational methods of studying breathers in Hamiltonian lattices. Any experiment will however show up with some dissipation. When this dissipation is of fluctuating nature, it could be simulated using a heat bath. However it is possible to consider also simple deterministic extensions of the above problems. In Josephson junction systems (see the chapters by Mazo and Ustinov in this volume) this is actually even implemented experimentally. Here we will only mention some of the basic new features one is faced with when computing dissipative

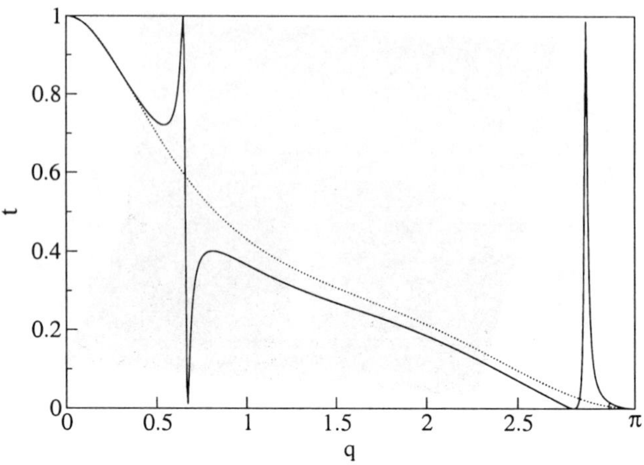

Fig. 24. Transmission coefficient versus wave number q for $\Omega_b = 4.5$ for an acoustic chain with $w_2 = w_4 = 1$ (dotted line) and additional $w_3 = 1$ (solid line), (see text for details).

breathers and their properties.[46,58]

7.1. *Obtaining dissipative breathers*

Consider the following set of equations of motion:

$$\ddot{x}_l = -\frac{\partial H}{\partial x_l} - \gamma \dot{x}_l - I \tag{90}$$

with

$$H = \sum_l [1 - \cos x_l - C(1 - \cos(x_l - x_{l-1}))] \ . \tag{91}$$

For $\gamma = I = 0$ this system is Hamiltonian and corresponds to the Takeno-Peyrard model of coupled pendula.[46,59] This model allows both for usual discrete breathers, but also for so-called roto-breathers. While for a usual breather $x_l(t+T_b) = x_l(t)$ for all l, for the simplest version of a roto-breather one pendulum is performing rotations

$$x_0(t + T_b) = x_0(t) + 2\pi m \ . \tag{92}$$

Here m is a winding number characterizing the roto-breather (again the simplest realization is $m = 1$). Note that at variance with a usual breather ($m = 0$), roto-breathers are not invariant under time reversal.

For nonzero γ and $I = 0$ the nonzero dissipation will lead to a decay of all breather and roto-breather solutions. But for nonzero time-independent I roto-breathers may still exist. The reason is that the rotating pendulum will both gain energy due to the nonzero torque I and dissipate energy due to the nonzero friction γ, so an energy balance is possible (whereas that is impossible for breathers with $m = 0$).

Instead of families of breather periodic orbits in Hamiltonian systems, dissipative roto-breathers will be attractors in the phase space. Attractors are characterized by a finite volume basin of attraction surrounding them. Any trajectory which starts inside this basin, will be ultimately attracted by the roto-breather. Thus dissipative breathers form a countable set of solutions.

To compute such a dissipative roto-breather, we can simply make a good guess in the initial conditions and then integrate the equations of motion until the roto-breather is reached. This method is very simple, but may suffer from long transient times, and also from complicated structures of the boundaries of the basin of attraction.

The Newton method can be applied here as well. Although we do not know the precise period of the roto-breather, we do not need it either. Instead of defining a map which integrates the phase space over a given time T_b, we may define a map which integrates the phase space of all but the rotating pendulum coordinate from its initial value $x_0(t = 0) = 0$ to $x_0(t_{map}) = 2\pi m$. Different trajectories will have different values of t_{map} which is not a problem. The only two things we have to worry about are: to find a trajectory which leads to a rotation of x_0 and as usual to be sufficiently close to the desired solution in order for the Newton map to converge. Once the solution is found, $T_b = t_{map}$.

7.2. *Perturbing dissipative breathers*

As long as a dissipative roto-breather is stable, the volume of its basin of attraction is finite, and small deviations will return the perturbed trajectory back to the breather. Upon the change of some control parameter the breather may still persist but get unstable. Consider the linearized phase space flow around a roto-breather of (90,91):

$$\ddot{\epsilon}_l = -\sum_m \frac{\partial^2 H}{\partial x_l \partial x_m}|_{\{x_{l'}(t)\}}\epsilon_m \; - \gamma\dot{\epsilon}_l \; . \tag{93}$$

In analogy with 6.1 we may introduce a (quasi-symplectic) matrix $\mathcal{R}$ which maps the phase space of the perturbations onto itself by integration of (93)

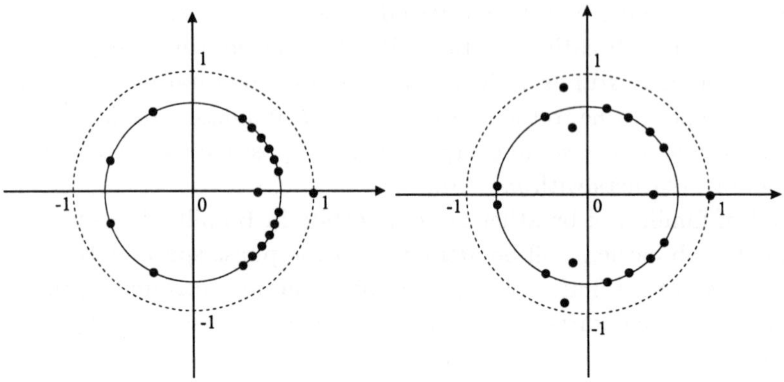

Fig. 25. Schematic view of an outcome of the Floquet analysis of a dissipative breather. Floquet eigenvalues (filled circles), the unit circle (large radius) and the inner circle of radius R (96) are plotted in the complex plane. Left picture: stable breather (all eigenvalues are located on the circle with radius R). Right picture: stable breather close to instability (two eigenvalues have collided on the inner circle, and one is departing outside towards the unit circle). Note that the group of closely nearby lying eigenvalues on the unit circle correspond to the plane wave continuum (extended Floquet eigenstates), while the separated eigenvalues on the inner circle correspond to localized Floquet eigenstates.

over one breather period.[58] By using the transformation

$$\epsilon_l(t) = \mathrm{e}^{-\frac{1}{2}\gamma t}\kappa_l(t) \tag{94}$$

we obtain

$$\ddot{\kappa}_l = -\sum_m \frac{\partial^2 H}{\partial x_l \partial x_m}|_{\{x_{l'}(t)\}}\kappa_m - \frac{1}{4}\gamma^2 \kappa_l \, . \tag{95}$$

Equations (95) define a Floquet problem with a symplectic matrix $\mathcal{F}$ with properties discussed above. By backtransforming to $\mathcal{R}$ we find that those eigenvalues which are located on the unit circle for $\mathcal{F}$ reside now on a circle with less radius

$$R(\gamma) = \mathrm{e}^{-\gamma T_b/2} \, . \tag{96}$$

If μ is an eigenvalue of $\mathcal{R}$, so are

$$\mu^* \, , \; \mathrm{e}^{-\gamma T_b}\frac{1}{\mu} \, , \; \mathrm{e}^{-\gamma T_b}\frac{1}{\mu^*} \, . \tag{97}$$

There is still one eigenvalue $\mu = 1$ which corresponds to perturbations tangent to the breather orbit. The related second eigenvalue is located at $\mathrm{e}^{-\gamma T_b}$, contrary to the Hamiltonian case. The schematic outcome of a Floquet analysis of a dissipative breather is shown in Fig. 25.

8. Computing quantum breathers

A natural question is what remains of discrete breathers if the corresponding quantum problem is considered.[60] Since the Schrödinger equation is linear and translationally invariant all eigenstates must obey the Bloch theorem. Thus we cannot expect eigenstates of the Hamiltonian to be spatially localized (on the lattice). On the other side the correspondence between the quantum eigenvalue problem and the classical dynamical evolution needs an answer.

The concept of tunneling is a possible answer to this puzzle. Naively speaking we quantize the family of periodic orbits associated with a discrete breather located somewhere on the lattice. Notice that there are as many such families as there are lattice sites. The quantization (e.g., Bohr-Sommerfeld) yields some eigenvalues. Since we can perform the same procedure with any family of discrete breather periodic orbits which differ only in their location on the lattice, we obtain N-fold degeneracy for every thus obtained eigenvalue, where N stands for the number of lattice sites. Unless we consider the trivial case of, say, uncoupled lattice sites, these degeneracies will be lifted. Consequently, we will instead obtain bands of states with finite band width which can even hybridize with other states. These bands will be called quantum breather bands. The inverse tunneling time of a semiclassical breather from one site to a neighboring one is a measure of the bandwidth.

We can then formulate the following expectation: if a classical nonlinear Hamiltonian lattice possesses discrete breathers, its quantum counterpart should show up with nearly degenerate bands of eigenstates, if the classical limit is considered. The number of states in such a band is N, and the eigenfunctions are given by Bloch-like superpositions of the semiclassical eigenfunctions obtained using the mentioned Bohr-Sommerfeld quantization of the classical periodic orbits. By nearly degenerate we mean that the bandwidth of a quantum breather band is much smaller than the spacing between different breather bands and the average level spacing in the given energy domain, and the classical limit implies large eigenvalues.

Another property of a quantum breather state is that such a state shows up with exponential localization in appropriate correlation functions.[61] This approach selects all particle-like states, no matter how deep one is in the quantum regime. In this sense quantum breather states belong to the class of particle-like bound states.

Intuitively it is evident that for large energies and N the density of states becomes large too. What will happen to the expected quantum breather bands then? Will the hybridization with other non-breather states destroy the particle-like nature of the quantum breather, or not? What is the impact of the nonintegrability of most systems allowing for classical breather solutions? Since the quantum case corresponds to a quantization of the classical phase space, we could expect that chaotic trajectories lying nearby classical breather solutions might affect the corresponding quantum eigenstates.

From a computational point of view we are very much restricted in our abilities to study quantum breathers. Ideally we would like to study quantum properties of a lattice problem in the large energy domain (to make contact with classical states) and for large lattices. This is typically impossible, since solving the quantum problem amounts to diagonalizing the Hamiltonian matrix with rank b^N where b is the number of states per site, which should be large to make contact with classical dynamics. Thus typically quantum breather states have been so far obtained numerically for small one-dimensional systems ($N \leq 8$).[61,62,63]

One of the few exceptions is the quantum discrete nonlinear Schrödinger equation with the Hamiltonian[64]

$$H = -\sum_{l=1}^{N} \left[\frac{1}{2}(a_l^\dagger a_l)^2 + C(a_l^\dagger a_{l+1} + h.c.) \right] \tag{98}$$

and the commutation relations

$$a_l a_m^\dagger - a_m^\dagger a_l = \delta_{lm} \tag{99}$$

with δ_{lm} being the standard Kronecker symbol. This Hamiltonian conserves the total number of particles

$$B = \sum_l n_l \ , \ n_l = a_l^\dagger a_l \ . \tag{100}$$

For b particles and N sites the number of basis states is

$$\frac{(b+N-1)!}{b!(N-1)!} \ . \tag{101}$$

For $b = 0$ there is just one trivial state of an empty lattice. For $b = 1$ there are N states which correspond to one-boson excitations. These states behave pretty much as classical extended wave states. For $b = 2$ the problem is still exactly solvable, because it corresponds to a two-body problem on a lattice. A corresponding numerical solution is sketched in Fig. 26.[64] Note the wide two-particle continuum, and a single band located below. This single band corresponds to quasiparticle states characterized by one single

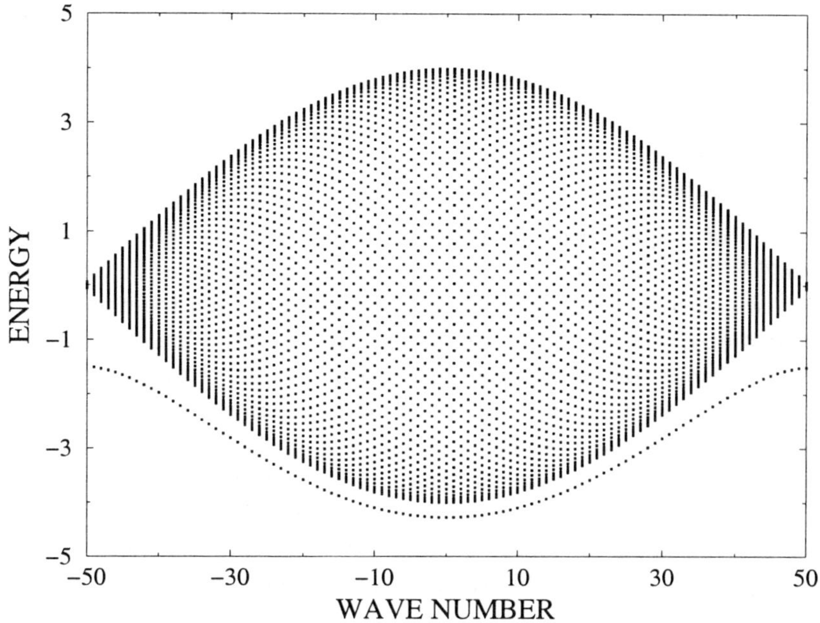

Fig. 26. Spectrum of the quantum DNLS with $b = 2$ and $N = 101$. The energy eigenvalues are plotted versus the wavenumber of the eigenstate.

quantum number (related to the wavenumber q). These states are two-particle bound states. The dispersion of this band is given[64] by

$$E = -\sqrt{1 + 16C^2 \cos^2\left(\frac{q}{2}\right)} . \tag{102}$$

Any eigenstate from this two-particle bound state band is characterized by exponential localization of correlations, i.e. when represented in some set of basis states, the amplitude or overlap with a basis state where the two particles are separated by some number of sites is exponentially decreasing with increasing separation distance. Note that a compact bound state is obtained for $q = \pm\pi$, i.e. for these wave numbers basis states with nonzero separation distance do not contribute to the eigenstate at all.

Increasing the number of particles to $b = 3$ or larger calls for computational tools. Eilbeck[65] has recently provided with updated codes in Maple in order to deal with systems with up to $b = 4$ and $N = 14$, implying a Hilbert space dimension of 2380 (there are $\binom{N+b-1}{b}$ ways to distribute b identical particles on N sites). While these studies revealed a lot of new structures of the corresponding spectra, we still have to wait for more systematic studies. Since the classical regime is still not easily reachable for these large systems, we will discuss in the next sections systematic studies of small systems, which allow to boost the energies into the semiclassical domain.

8.1. *The dimer*

A series of papers was devoted to the properties of the quantum dimer.[66,67,68] This system describes the dynamics of bosons fluctuating between two sites. The number of bosons is conserved, and together with the conservation of energy the system appears to be integrable. Of course, one cannot consider spatial localization in such a model. However, a reduced form of the discrete translational symmetry - namely the permutational symmetry of the two sites - can be imposed. Together with the addition of nonlinear terms in the classical equations of motion the dimer allows for classical trajectories which are not invariant under permutation. The phase space can be completely analyzed, all isolated periodic orbits can be found. There appears exactly one bifurcation on one family of isolated periodic orbits, which leads to the appearance of a separatrix in phase space. The separatrix separates three regions - one invariant and two non-invariant under permutations. The subsequent analysis of the quantum dimer demonstrated the existence of pairs of eigenstates with nearly equal eigenenergies.[66] The separatrix and the bifurcation in the classical phase space can be traced in the spectrum of the quantum dimer.[68]

The classical Hamiltonian may be written as

$$H = \Psi_1^*\Psi_1 + \Psi_2^*\Psi_2 + \frac{1}{2}\left((\Psi_1^*\Psi_1)^2 + (\Psi_2^*\Psi_2)^2\right) + C\left(\Psi_1^*\Psi_2 + \Psi_2^*\Psi_1\right) \quad . \tag{103}$$

with the equations of motion $\dot{\Psi}_{1,2} = i\partial H/\partial \Psi_{1,2}^*$. The model conserves the norm (or number of particles) $B = |\Psi_1|^2 + |\Psi_2|^2$.

Isolated periodic orbits (IPO) satisfy the relation grad$H$ $||$ grad$B$. Let us parameterize the phase space of (103) with $\Psi_{1,2} = A_{1,2}\mathrm{e}^{i\phi_{1,2}}$, $A_{1,2} \geq 0$. It follows that $A_{1,2}$ is time independent and $\phi_1 = \phi_2 + \Delta$ with $\Delta = 0, \pi$ and $\dot{\phi}_{1,2} = \omega$ being also time independent. Solving the algebraic equations for the amplitudes of the IPO's we obtain

$$\mathrm{I} : A_{1,2}^2 = \frac{1}{2}B \; , \; \Delta = 0 \; , \; \omega = 1 + C + \frac{1}{2}B \; , \tag{104}$$

$$\mathrm{II} : A_{1,2}^2 = \frac{1}{2}B \; , \; \Delta = \pi \; , \; \omega = 1 - C + \frac{1}{2}B \; , \tag{105}$$

$$\mathrm{III} : A_1^2 = \frac{1}{2}B\left(1 \pm \sqrt{1 - 4C^2/B^2}\right) \; , \; \Delta = 0 \; , \; \omega = 1 + B \; . \tag{106}$$

IPO III corresponds to two elliptic solutions which break the permutational symmetry. IPO III exist for $B \geq B_b$ with $B_b = 2C$ and occur through a bifurcation from IPO I. The corresponding separatrix manifold is uniquely defined by the energy of IPO I at a given value of $B \geq B_b$. This manifold separates three regions in phase space - two with symmetry broken solutions, each one containing one of the IPO's III, and one with symmetry conserving solutions containing the elliptic IPO II. The separatrix manifold itself contains the hyperbolic IPO I. For $B \leq B_b$ only two IPO's exist - IPO I and II, with both of them being of elliptic character. Remarkably there exist no other IPO's, and the mentioned bifurcation and separatrix manifolds are the only ones present in the classical phase space of (103).

To conclude the analysis of the classical part, we list the energy properties of the different phase space parts separated by the separatrix manifold. First it is straightforward to show that the IPO's (104)-(106) correspond to maxima, minima or saddle points of the energy in the allowed energy interval for a given value of B, with no other extrema or saddle points present. It follows

$$E_1 = H(\text{IPO I}) = B + \frac{1}{4}B^2 + CB \ , \tag{107}$$

$$E_2 = H(\text{IPO II}) = B + \frac{1}{4}B^2 - CB \ , \tag{108}$$

$$E_3 = H(\text{IPO III}) = B + \frac{1}{2}B^2 + C^2 \ . \tag{109}$$

For $B < B_b$ we have $E_1 > E_2$ (IPO I - maximum, IPO II - minimum). For $B \geq B_b$ it follows $E_3 > E_1 > E_2$ (IPO III - maxima, IPO I - saddle, IPO II - minimum). If $B < B_b$, then all trajectories are symmetry conserving. If $B \geq B_b$, then trajectories with energies $E_1 < E \leq E_3$ are symmetry breaking, and trajectories with $E_2 \leq E \leq E_1$ are symmetry conserving.

The quantum eigenvalue problem amounts to replacing the complex functions Ψ, Ψ^* in (7) by the boson annihilation and creation operators $a, a^\dagger$ with the standard commutation relations (to enforce the invariance under the exchange $\Psi \Leftrightarrow \Psi^*$ the substitution has to be done on rewriting $\Psi\Psi^* = 1/2(\Psi\Psi^* + \Psi^*\Psi)$):

$$H = \frac{5}{4} + \frac{3}{2}\left(a_1^\dagger a_1 + a_2^\dagger a_2\right) + \frac{1}{2}\left((a_1^\dagger a_1)^2 + (a_2^\dagger a_2)^2\right) + C\left(a_1^\dagger a_2 + a_2^\dagger a_1\right) \ . \tag{110}$$

Note that $\hbar = 1$ here, so the eigenvalues b of $B = a_1^\dagger a_1 + a_2^\dagger a_2$ are integers. Since B commutes with H we can diagonalize the Hamiltonian in the basis of eigenfunctions of B. Each value of b corresponds to a subspace of the dimension $(b+1)$ in the space of eigenfunctions of B. These eigenfunctions are products of the number states $|n\rangle$ of each degree of freedom and can be characterized by a symbol $|n,m\rangle$ with n bosons in the site 1 and m bosons in the site 2. For a given value of b it follows $m=b-n$. So we can actually label each state by just one number n: $|n,(b-n)\rangle \equiv |n\rangle$. Consequently the eigenvalue problem at fixed b amounts to diagonalizing the matrix

$$H_{nm} = \begin{cases} \frac{5}{4} + \frac{3}{2}b + \frac{1}{2}\left(n^2 + (b-n)^2\right) & n = m \\ C\sqrt{n(b+1-n)} & n = m+1 \\ C\sqrt{(n+1)(b-n)} & n = m-1 \\ 0 & \text{else} \end{cases} \quad (111)$$

where $n,m = 0,1,2,...,b$. Notice that the matrix H_{nm} is a symmetric band matrix. The additional symmetry $H_{nm} = H_{(b-n),(b-m)}$ is a consequence of the permutational symmetry of H. For $C=0$ the matrix H_{nm} is diagonal,

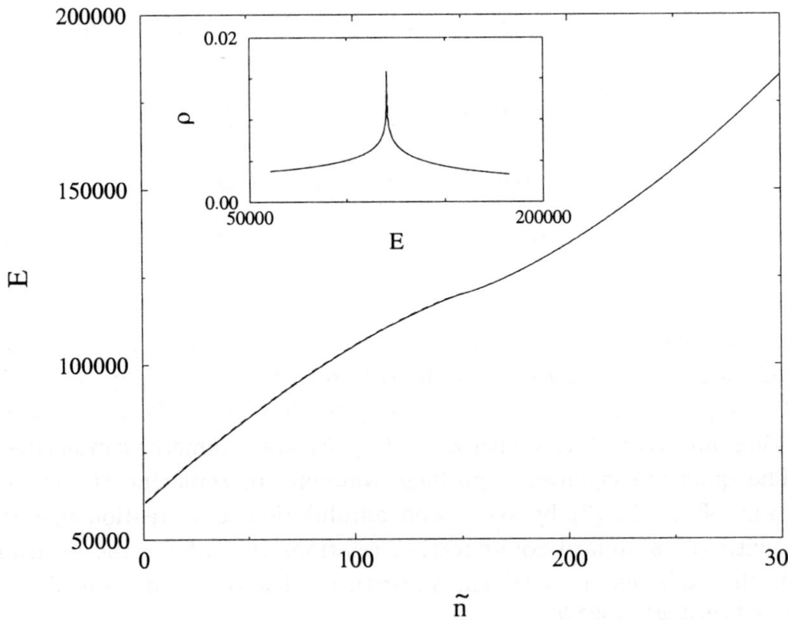

Fig. 27. Eigenvalues versus ordered state number $\tilde{n}$ for symmetric and antisymmetric states ($0 < \tilde{n} < b/2$ for both types of states). Parameters: $b = 600$ and $C = 50$. Inset: Density of states versus energy.

with the property that each eigenvalue is doubly degenerate (except for the state $|b/2\rangle$ for the even values of b). The classical phase space contains only symmetry broken trajectories, with the exception of IPO II and the separatrix with IPO I (in fact in this limit the separatrix manifold is nothing but a resonant torus containing both IPO's I and II). So with the exception of the separatrix manifold, all tori break permutational symmetry and come in two groups separated by the separatrix. Then quantizing each group will lead to pairs of degenerate eigenvalues - one from each group. There is a clear correspondence to the spectrum of the diagonal ($C = 0$) matrix H_{nm}. The eigenvalues $H_{00} = H_{bb}$ correspond to the quantized IPO's III. With increasing n the eigenvalues $H_{nn} = H_{(b-n),(b-n)}$ correspond to quantized tori further away from the IPO III. Finally the states with $n = b/2$ for even b or $n = (b-1)/2$ for odd b are tori most close to the separatrix. Switching the side diagonals on by increasing C will lead to a splitting of all pairs of eigenvalues. In the case of small values of b these splittings have no correspondence to classical system properties. However, in the limit of large b we enter the semiclassical regime, and due to the integrability of the system, eigenfunctions should correspond to tori in the classical phase space which satisfy the Einstein-Brillouin-Keller quantization rules. Increasing C from zero will lead to a splitting ΔE_n of the eigenvalue doublets of $C = 0$. In other words, we find pairs of eigenvalues, which are related to each other through the symmetry of their eigenvectors and (for small enough C) through the small value of the splitting. These splittings have been calculated numerically and using perturbation theory.[66,68] In the limit of large b the splittings are exponentially small for the energies above the classical separatrix energy (i.e. for classical trajectories which are not invariant under permutation). If the eigenenergies are lowered below the classical separatrix energy, the splittings grow rapidly up to the mean level spacing.

In Fig. 27 the results of a diagonalization of a system with 600 particles ($b = 600$) is shown.[68] The inset shows the density of states versus energy, which nicely confirms the predicted singularity at the energy of the separatrix of the classical counterpart. In order to compute the exponentially small splittings, we may use e.g. a Mathematica routine which allows to choose arbitrary values for the precision of computations. Here we chose precision 512. In Fig. 28 the numerically computed splittings are compared to perturbation theory results. As expected, the splittings become extremely small above the separatrix. Consequently these states will follow for long times the dynamics of a classical broken symmetry state.

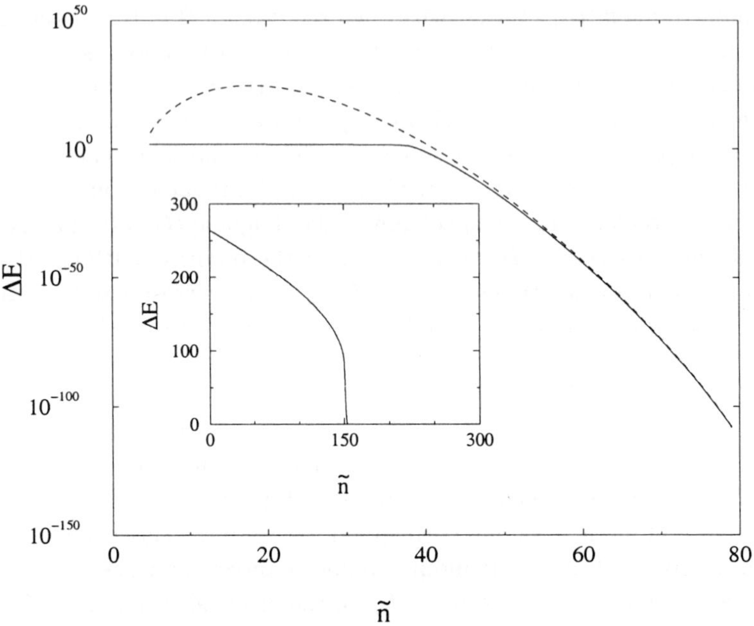

Fig. 28. Eigenvalue splittings versus $\tilde{n}$ for $b = 150$ and $C = 10$. Solid line - numerical result, dashed line - perturbation theory. Inset: Same for $b = 600$ and $C = 50$. Only numerical results are shown.

8.2. *The trimer*

The integrability of the dimer does not allow a study of the influence of chaos on the tunneling properties of the mentioned pairs of eigenstates. A natural extension of the dimer to a trimer adds a third degree of freedom without adding a new integral of motion. Consequently the trimer is nonintegrable. A still comparatively simple numerical quantization of the trimer allows to study the behavior of many tunneling states in the large-energy domain of the eigenvalue spectrum.[69]

Similarly to the dimer, the quantum trimer Hamiltonian is represented in the form

$$H = \frac{15}{8} + \frac{3}{2}(a_1^\dagger a_1 + a_2^\dagger a_2 + a_3^\dagger a_3) + \frac{1}{2}\left[(a_1^\dagger a_1)^2 + (a_2^\dagger a_2)^2\right] + C(a_1^\dagger a_2 + a_2^\dagger a_1) + \delta(a_1^\dagger a_3 + a_3^\dagger a_1 + a_2^\dagger a_3 + a_3^\dagger a_2) \ . \quad (112)$$

Again $B = a_1^\dagger a_1 + a_2^\dagger a_2 + a_3^\dagger a_3$ commutes with the Hamiltonian, thus we can diagonalize (112) in the basis of eigenfunctions of B. For any finite eigenvalue b of B the number of states is finite, namely $(b + 1)(b + 2)/2$.

Thus the infinite dimensional Hilbert space separates into an infinite set of finite dimensional subspaces, each subspace containing only vectors with a given eigenvalue b. These eigenfunctions are products of the number states $|n\rangle$ of each degree of freedom and can be characterized by a symbol $|n, m, l\rangle$ where we have n bosons on site 1, m bosons on site 2, and l bosons on site 3. For a given value b it follows that $l = b - m - n$. So we can actually label each state by just two numbers (n, m): $|n, m, (b - n - m)\rangle \equiv |n, m\rangle$. Note that the third site added to the dimer is different from the first two sites. There is no boson-boson interaction on this site. Thus site 3 serves simply as a boson reservoir for the dimer. Dimer bosons may now fluctuate from the dimer to the reservoir. The trimer has the same permutational symmetry as the dimer.

The matrix elements of (112) between states from different b subspaces vanish. Thus for any given b the task amounts to diagonalizing a finite dimensional matrix. The matrix has a tridiagonal block structure, with each diagonal block being a dimer matrix (111). The nonzero off-diagonal blocks contain interaction terms proportional to δ. Since H commutes with $\hat{P}_q$ we consider symmetric $|\Psi\rangle_s$ and antisymmetric $|\Psi\rangle_a$ states. The structure of the corresponding symmetric and antisymmetric decompositions of H is similar to H itself. In the following we will present results for $b = 40$. We will also drop the first two terms of the RHS in (112), because these only lead to a shift of the energy spectrum. Since we evaluate the matrix elements explicitly, we need only a few seconds to obtain all eigenvalues and eigenvectors with the help of standard Fortran routines. In Fig. 29 we plot a part of the energy spectrum as a function of δ for $C = 2$.[69] As discussed above, the Hamiltonian decomposes into noninteracting blocks for $\delta = 0$, each block corresponding to a dimer with a boson number between 0 and b. For $\delta \neq 0$ the nonzero block-block interaction leads to typical features in the spectrum, like, e.g., avoided crossings. The full quantum energy spectrum extends roughly over 10^3, which implies an averaged spacing of order 10^0. Also the upper third of the spectrum is diluted compared to the lower two thirds.

The correspondence to the classical model is obtained with the use of the transformation $E_{cl} = E_{qm}/b^2 + 1$ and for parameters C/b and δ/b (the classical value for B is $B = 1$).

The main result of this computation so far is that tunneling pairs of eigenstates of the dimer persist in the nonintegrable regime $\delta \neq 0$. However at certain pair-dependent values of δ a pair breaks up. From the plot in Fig. 29 we cannot judge how the pair splittings behave. In Fig.30 we plot

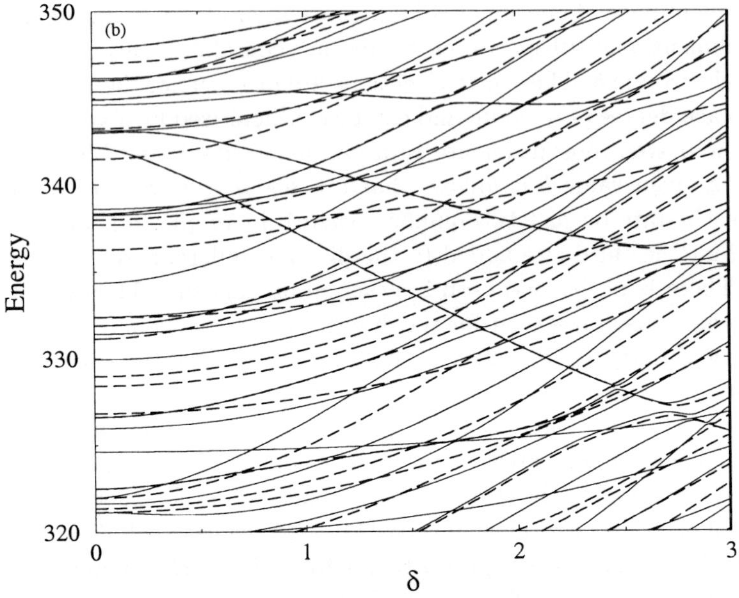

Fig. 29. A part of the eigenenergy spectrum of the quantum trimer as a function of δ with $b = 40$ and $C = 2$. Lines connect data points for a given state. Solid lines - symmetric eigenstates; thick dashed lines - antisymmetric eigenstates.

the pair splitting of the pair which has energy ≈ 342 at $\delta = 0$.[70] Denote with x, y, z the eigenvalues of the site number operators n_1, n_2, n_3. We may consider the quantum states of the trimer at $\delta = 0$ when z is a good quantum number and then follow the evolution of these states with increasing δ. The state for $\delta = 0$ can be traced back to $C = 0$ and be thus characterized in addition by x and y. The chosen pair states are then characterized by $x = 26(0)$, $y = 0(26)$ and $z = 14$ for $C = \delta = 0$. Note that this pair survives approximately 30 avoided crossings before it is finally destroyed at coupling strength $\delta \approx 2.67$ as seen in Fig. 29.

From Fig. 30 we find that the splitting rapidly increases gaining about eight orders of magnitude when δ changes from 0 to slightly above 0.5. Then this rapid but nevertheless smooth rise is interrupted by very sharp spikes when the splitting ΔE rises by several orders of magnitude with δ changing by mere percents and then abruptly changes in the opposite direction sometimes even overshooting its pre-spike value. Such spikes, some larger, some smaller, repeat with increasing δ until the splitting value approaches

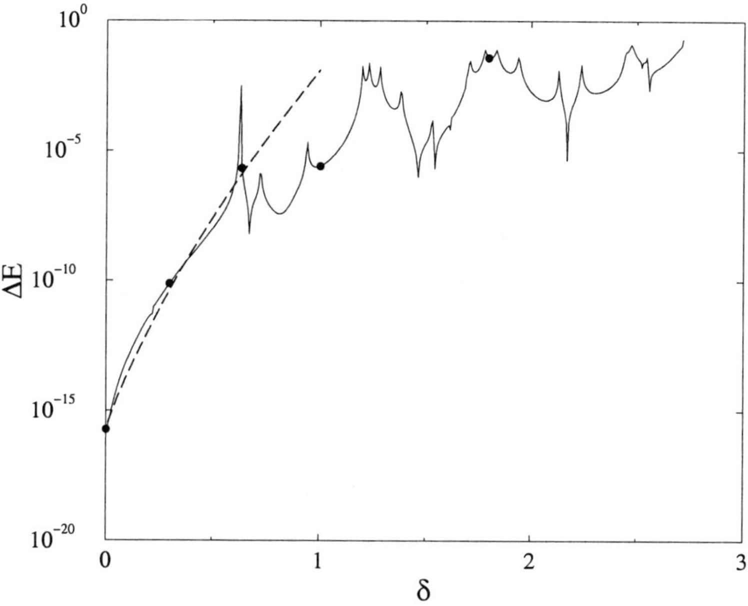

Fig. 30. Level splitting versus δ for a level pair as described in the text. Solid line - numerical result. Dashed line - semiclassical approximation. Filled circles - location of wave function analysis in Fig. 31.

the mean level spacing (of order one in the figure). Only then one may say that the pair is destroyed since it can be hardly distinguished among the other trimer levels.

Another observation is presented in Fig. 31.[70] We plot the intensity distribution of the logarithm of the squared symmetric wave function of our chosen pair for five different values of $\delta = 0$, 0.3 , 0.636 , 1.0 , 1.8 (their locations are indicated by filled circles in Fig. 30). We use the eigenstates of B as basis states. They can be represented as $|x, y, z>$ where x, y, z are the particle numbers on sites 1, 2, 3, respectively. Due to the commutation of B with H two site occupation numbers are enough if the total particle number is fixed. Thus the final encoding of states (for a given value of b) can be chosen as $|x, z)$ (see also discussion above for details). The abscissa in Fig. 31 is x and the ordinate is z. Thus the intensity plots provide us with information about the order of particle flow in the course the tunneling process. For $\delta = 0$ (Fig. 31(a)) the only possibility for the 26 particles on site 1 is to directly tunnel to site 2. Site 3 is decoupled with its 14

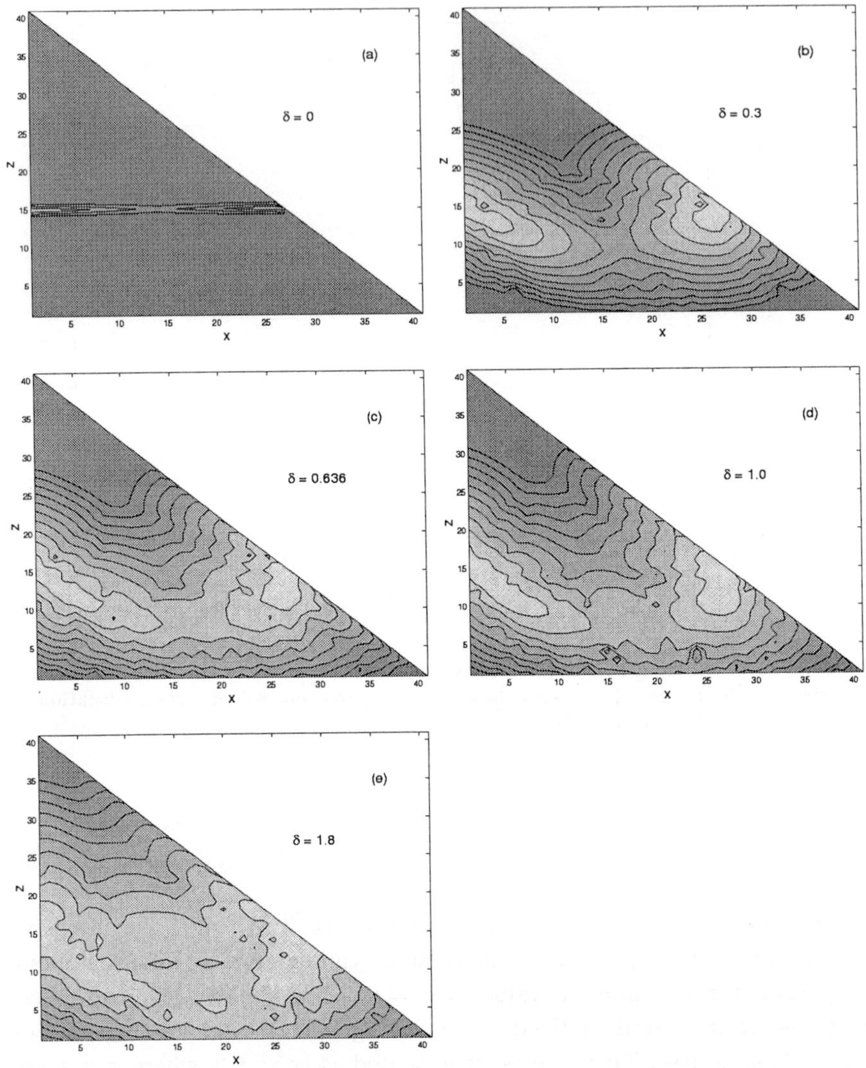

Fig. 31. Contour plot of the logarithm of the symmetric eigenstate of the chosen tunneling pair (cf. Fig. 30) for five different values of $\delta = 0,\ 0.3,\ 0.636,\ 1.0,\ 1.8$ (their location is indicated by filled circles in Fig. 30). (a): three equidistant grid lines are used; (b-e): ten grid lines are used. Minimum value of squared wave function is 10^{-30}, maximum value is about 1.

particles not participating in the process. The squared wave function takes the form of a compact rim in the (x, z) plane which is parallel to the x axis. Nonzero values of the wave function are observed only on the rim. This direct tunneling has been described in 8.1. When switching on some nonzero coupling to the third site, the particle number on the dimer (sites 1,2) is not conserved anymore. The third site serves as a particle reservoir which is able either to collect particles from or supply particles to the dimer. This coupling will allow for nonzero values of the wave function away from the rim. But most importantly, it will change the shape of the rim. We observe that the rim is bended down to smaller z values with increasing δ. That implies that the order of tunneling (when, e.g., going from large to small x values) is as follows: first, some particles tunnel from site 1 to site 2 and simultaneously from site 3 to site 2 (Fig. 32(a)). Afterwards particles flow from site 1 to both sites 2 and 3 (Fig. 32(b)). With increasing δ the structure of the wave function intensity becomes more and more complex, possibly revealing information about the classical phase space flow structure. Thus

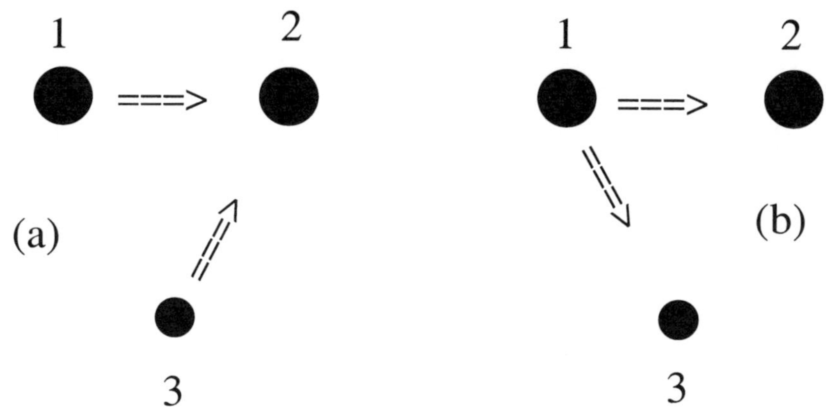

Fig. 32. Order of tunneling in the trimer. Filled large circles - sites 1 and 2, filled small circle - site 3. Arrows indicate direction of transfer of particles.

we observe three intriguing features. First, the tunneling splitting increases by eight orders of magnitude when δ increases from zero to 0.5. This seems to be unexpected, since at those values perturbation theory in δ should be applicable (at least Fig. 29 indicates that this should be true for the levels themselves). The semiclassical explanation of this result was obtained in Ref. 70.

The second observation is that the tunneling begins with a flow of particles from the bath (site 3) directly to the empty site which is to be filled (with simultaneous flow from the filled dimer site to the empty one). At the end of the tunneling process the initially filled dimer site is giving particles back to the bath site. Again this is an unexpected result, since it implies that the particle number on the dimer is increasing during the tunneling, which seems to decrease the tunneling probability, according to the results for an isolated dimer. These first two results are closely connected (see Ref. 70 for a detailed explanation). The third result concerns the resonant

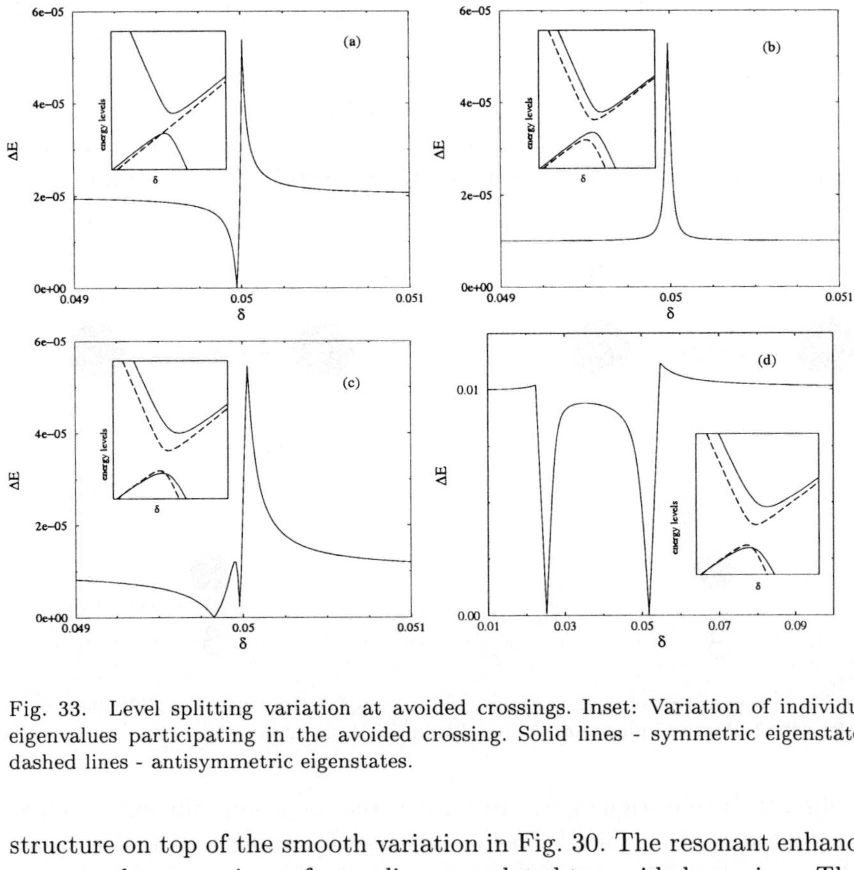

Fig. 33. Level splitting variation at avoided crossings. Inset: Variation of individual eigenvalues participating in the avoided crossing. Solid lines - symmetric eigenstates, dashed lines - antisymmetric eigenstates.

structure on top of the smooth variation in Fig. 30. The resonant enhancements and suppressions of tunneling are related to avoided crossings. Their presence implies that a fine tuning of the system parameters may strongly suppress or enhance tunneling which may be useful for spectroscopic devices. In Fig. 33 we show the four various possibilities of avoided crossings between a pair and a single level and between two pairs, and the schematic outcome for the tunneling splitting.[70]

8.3. *Quantum roto-breathers*

When discussing classical breather solutions we have been touching some aspects of roto-breathers, including their property of being not invariant under time reversal symmetry. In a recent study Dorignac et al have provided[71] with an analysis of the corresponding quantum roto-breather properties in a dimer with the Hamiltonian

$$H = \sum_{i=1}^{2} \left\{ \frac{p_i^2}{2} + \alpha(1 - \cos x_i) \right\} + \varepsilon(1 - \cos(x_1 - x_2)) \ . \tag{113}$$

The classical roto-breather solution consists of one pendulum rotating and the other oscillating with a given period T_b. Since the model has two symmetries - permutation of the indices and time-reversal symmetry, which may be both broken by classical trajectories, the irreducible representations of quantum eigenstates contain four symmetry sectors (with possible combinations of symmetric or antisymmetric states with respect to the two symmetry operations). Consequently, a quantum roto-breather state is belonging to a quadruplet of weakly split states rather than to a pair as discussed above. The schematic representation of the appearance of such a quadruplet is shown in Fig. 34.[71] The obtained quadruplet has an additional fine structure as compared to the tunneling pair of the above considered dimer and trimer. The four levels in the quadruplet define three characteristic tunneling processes. Two of them are energy or momentum transfer from one pendulum to the other one, while the third one corresponds to total momentum reversal (which restores time reversal symmetry). The dependence of the corresponding tunneling rates on the coupling ε is shown for a specific quadruplet from Ref. 71 in Fig. 35. For very weak coupling $\varepsilon \ll 1$ the fastest tunneling process will be momentum reversal, since tunneling between the pendula is blocked. However as soon as the coupling is increased, the momentum reversal turns into the slowest process, with breather tunneling from one pendulum to the other one being orders of magnitude larger. Note that again resonant features on these splitting curves are observed, which are related to avoided crossings.

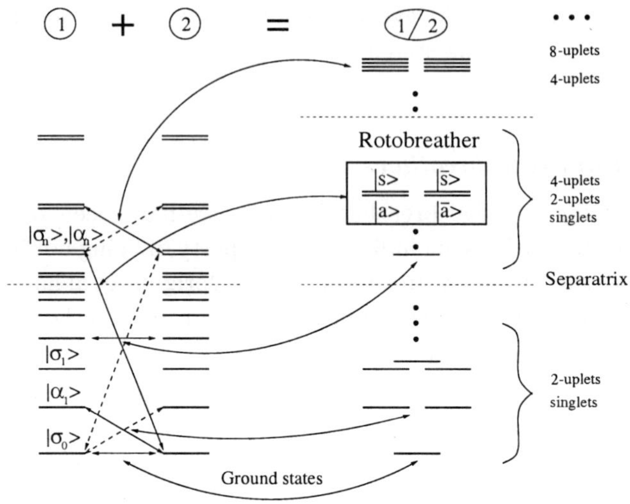

Fig. 34. Schematic representation of the sum of two pendula spectra. Straight solid arrows indicate the levels to be added and dashed arrows the symmetric (permutation) operation. The result is indicated in the global spectrum by a curved arrow. The construction of the quantum roto-breather state is explicitly represented.

9. Some applications instead of conclusions

Instead of providing with a standard conclusion, we will discuss in this last part some selected computational results of discrete breather studies, which have been boosting the understanding of various aspects of DBs or confirming analytical predictions.

Rather simple numerical observations of breathers showed that in one-dimensional acoustic chains a breather is usually accompanied by a kink-type static lattice distortion[72] - a fact later explained[73] and even used in analytical existence proofs.[17,19] Other numerical observations revealed that stable discrete breathers may be perturbed in an asymmetric way such that a separatrix may be crossed leading to possible movability (see discussion in Ref. 6).

While exact moving breather solutions in generic Hamiltonian lattices have not been observed, the understanding of some reasons[74,75] and their removal by considering dissipative breathers successfully allowed to obtain dissipative moving breathers.[58]

Traces of energy thresholds of discrete breathers[47] have been observed in the properties of correlation functions at thermal equilibrium.[77]

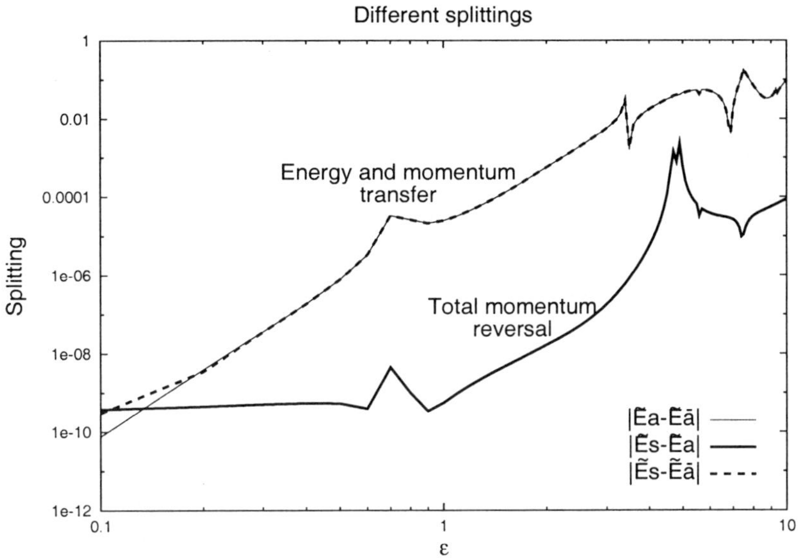

Fig. 35. Dependence of different splittings of a quadruplet on ε. Only three of them have been displayed, each being associated with a given tunneling process.[71]

Numerical studies of collisions between moving breathers showed that the energy exchange typically leads to the growth of the largest breather[34,76,31] - a fact which is not well explained yet.

The explained high precision numerical routines for obtaining discrete breathers have been used in order to obtain discrete breathers in acoustic two-dimensional lattices.[73] The predicted algebraic decay of the static lattice deformation and its dipole symmetry have been nicely observed prior to analytical proofs of existence.[19]

Another example concerns the case of algebraically decaying (long range) interactions on a lattice. While analytical proofs correctly stated that the asymptotic spatial decay of breathers will be also algebraic in such a case, numerical high precision computations showed that there is more to say.[78] The spatial breather profile in such systems shows an exponential decay on intermediate length scales with a crossover to algebraic decay on larger distances. Afterwards this crossover was explained analytically and estimates of the crossover distance well coincided with numerical results.

The tracing of bifurcations and instabilities explained an often observed puzzling exchange of stability of various breather types. The outcome of the numerical studies was that these different types of breather families are connected through unexpected asymmetric breather families.[50]

The understanding that two-channel scattering of plane waves by breathers is inelastic was used to perform numerical experiments which nicely showed the expected slow energy decrease of a breather in such a case.[52]

The appearance of local Floquet modes according to analytical predictions should lead to the appearance of perfect transmission of waves through breathers.[57,52] This fact has been nicely observed in various numerical studies.

The theoretical understanding of Fano resonances in wave scattering by breathers lead to a numerical scheme which allows to compute and thus predict the parameters of various models which should provide with resonant Fano backscattering. Direct numerical scattering computations have shown the correctness of these considerations and computations.[56]

The launching of a localized initial state in a quantum trimer showed up with unexpected echoes in the quantum evolution. These echoes have been explained with the help of the numerically obtained spectrum and eigenfunctions by relating it to the existence of quantum breather states .[69]

The interested user may consult the web page http://www.mpipks-dresden.mpg.de/~flach/html/dbreather.html for java applications written by A. E. Miroshnichenko, which allow for launching your favorite breather in your favorite system. There the interested reader may also find more references, related web addresses and links to related activities.

Acknowledgments

I would like to thank M. V. Ivanchenko, O. I. Kanakov and V. Shalfeev for providing with numerical results prior publication. Thanks are due to A. Miroshnichenko for useful discussions during the preparation of this work. I am indebted to all friends and colleagues with whom I had the chance to work and publish together and whose results have been used here, and from whom I benefited by discussing issues related to this work. Finally I am sincerely apologizing for any possible missing citations.

References

1. A. A. Ovchinnikov, *Sov. Phys. JETP* **30**, 147 (1970).
2. A. M. Kosevich and A. S. Kovalev, *Sov. Phys. JETP* **67**, 1793 (1974).
3. A. J. Sievers and S. Takeno, *Phys. Rev. Lett.* **61**, 970 (1988).

4. A. J. Sievers and J. B. Page, in *Dynamical Properties of Solids VII Phonon Physics The Cutting Edge*, Eds. G. K. Horton and A. A. Maradudin (Elsevier, Amsterdam, 1995).
5. S. Aubry, *Physica D* **103**, 201 (1997).
6. S. Flach and C. R. Willis, *Phys. Rep.* **295**, 181 (1998).
7. See also focus issues *Physica D* **113** Nr.2-4 (1998); *Physica D* **119** Nr.1-2 (1998); *CHAOS* **13** Nr.2. (2003).
8. H. S. Eisenberg, Y. Silberberg, R. Morandotti, A. R. Boyd, and J. S. Aitchison, *Phys. Rev. Lett.* **81**, 3383 (1998); A. A. Sukhorukov, Yu. S. Kivshar, H. S. Eisenberg, and Y. Silberberg, *IEEE J. Quantum Electron.* **39**, 31 (2003).
9. J. W. Fleischer, M. Segev, N. K. Efremidis, and D. N. Christodoulides, *Nature* **422**, 147 (2003).
10. B. I. Swanson, J. A. Brozik, S. P. Love, G. F. Strouse, A. P. Shreve, A. R. Bishop, W. Z. Wang, and M. I. Salkola, *Phys. Rev. Lett.* **82**, 3288 (1999).
11. E. Trias, J. J. Mazo, and T. P. Orlando, *Phys. Rev. Lett.* **84**, 741 (2000); P. Binder, D. Abraimov, A. V. Ustinov, S. Flach, and Y. Zolotaryuk, *Phys. Rev. Lett.* **84**, 745 (2000).
12. U. T. Schwarz, L. Q. English and A. J. Sievers, *Phys. Rev. Lett.* **83**, 223 (1999).
13. M. Sato, B. E. Hubbard, A. J. Sievers, B. Ilic, D. A. Czaplewski, and H. G. Craighead, *Phys. Rev. Lett.* **90**, 044102 (2003).
14. A. Trombettoni and A. Smerzi, *Phys. Rev. Lett.* **86**, 2353 (2001); E. A. Ostrovskaya and Yu. S. Kivshar, *Phys. Rev. Lett.* **90**, 160407 (2003).
15. R. S. MacKay and S. Aubry, *Nonlinearity* **7**, 1623 (1994).
16. S. Flach, *Phys. Rev. E* **51**, 1503 (1995).
17. R. Livi, M. Spicci, and R. S. MacKay, *Nonlinearity*, **10**, 1421 (1997).
18. J. A. Sepulchre and R. S. MacKay, *Nonlinearity* **10**, 679 (1997).
19. S. Aubry, *Ann. Institut Henri Poincare* **68**, 381 (1998).
20. S. Aubry, G. Kopidakis, and V. Kadelburg, *Discrete and Continuous Dynamical Systems - Series B* **1**, 271 (2001).
21. G. James, *J. Nonlinear Sci.* **13**, 27 (2003).
22. D. W. Heermann, *Computer Simulation Methods* (Springer, Berlin 1990).
23. W. H. Press, S. A. Teukolsky, W. T. Vetterling and B. P. Flannery, *Numerical Recipes in Fortran* (Cambridge University Press, Cambridge, 1992).
24. M. Abramowitz and I. A. Stegun, Eds., *Handbook of Mathematical Functions* (Dover Publications, Inc., New York 1965).
25. S. Flach and C. R. Willis, *Phys. Lett. A* **181**, 232 (1993).
26. S. Flach, C. R. Willis and E. Olbrich, *Phys. Rev. E* **49**, 836 (1994).
27. S. Flach, K. Kladko and C. R. Willis, *Phys. Rev. E* **50**, 2293 (1994).
28. V. M. Burlakov, S. A. Kisilev and V. I. Rupasov, *Phys. Lett. A* **147**, 130 (1990).
29. Y. S. Kivshar and M. Peyrard, *Phys. Rev. A* **46**, 3198 (1992).
30. I. Daumont, T. Dauxois and M. Peyrard, *Nonlinearity* **10**, 617 (1997).
31. T. Cretegny, T. Dauxois, S. Ruffo and A. Torcini, *Physica D* **121**, 109 (1998).
32. M. Peyrard, *Physica D* **119**, 184 (1998).

33. M. Ivanchenko, O. Kanakov, V. Shalfeev and S. Flach, to be published.
34. T. Dauxois and M. Peyrard, *Phys. Rev. Lett.* **70**, 3935 (1993); T. Dauxois, M. Peyrard and C. R. Willis, *Phys. Rev. E* **48**, 4768 (1993).
35. A. Bikai, N. K. Voulgarakis, S. Aubry and G. P. Tsironis, *Phys. Rev. E* **59**, 1234 (1999).
36. G. P. Tsironis and S. Aubry, *Phys. Rev. Lett.* **77**, 5225 (1996).
37. R. Reigada, A. Sarmiento and K. Lindenberg, *Phys. Rev. E* **64**, 066608 (2001).
38. V. M. Burlakov, S. Kisilev and V. N. Pyrkov, *Solid State Comm.* **74**, 327 (1990).
39. V. M. Burlakov, S. Kisilev and V. N. Pyrkov, *Phys. Rev. B* **42**, 4921 (1990).
40. T. Dauxois, M. Peyrard and A. R. Bishop, *Phys. Rev. E* **47**, 684 (1993).
41. S. Flach and G. Mutschke, *Phys. Rev. E* **49**, 5018 (1994).
42. S. Flach, *Phys. Rev. E* **50**, 3134 (1994).
43. S. Flach, *Phys. Rev. E* **51**, 3579 (1995).
44. Yu. S. Kivshar, *Phys. Rev. E* **48**, R43 (1993).
45. F. Fischer, *Ann. Physik* **2**, 296 (1993).
46. J. L. Marin and S. Aubry, *Nonlinearity* **9**, 1501 (1996).
47. S. Flach, K. Kladko and R. S. MacKay, *Phys. Rev. Lett.* **78**, 1207 (1997).
48. T. Cretegny and S. Aubry, *Physica D* **113**, 162 (1998).
49. J. L. Marin and S. Aubry, *Physica D* **119**, 163 (1998).
50. J. L. Marin, S. Aubry and L. M. Floria, *Physica D* **113**, 283 (1998).
51. S. Flach and C. R. Willis, in: *Nonlinear Excitations in Biomolecules*, Ed. M. Peyrard, (Springer, Berlin and Les Editions de Physicuqe, Les Ulis, 1995).
52. T. Cretegny, S. Aubry and S. Flach, *Physica D* **119**, 73 (1998).
53. S. W. Kim and S. Kim, *Physica D* **141**, 91 (2000).
54. S. W. Kim and S. Kim, *Phys. Rev. B* **63**, 212301 (2001).
55. S. Flach, A. E. Mirochnishenko and M. V. Fistul, *CHAOS* **13**, 596 (2003).
56. S. Flach, A. E. Mirochnishenko, V. Fleurov and M. V. Fistul, *Phys. Rev. Lett.* **90**, 084101 (2003).
57. S. Kim, C. Baesens and R. S. MacKay, *Phys. Rev. E* **56**, R4955 (1997).
58. J. L. Marin, F. Falo, P. J. Martinez, and L. M. Floria, *Phys. Rev. E* **63**, 066603 (2001).
59. S. Takeno and M. Peyrard, *Phys. Rev. E* **55**, 1922 (1997).
60. R. S. MacKay, *Physica A* **288**, 174 (2000).
61. W. Z. Wang, J. T. Gammel, A. R. Bishop and M. I. Salkola, *Phys. Rev. Lett.* **76**, 3598 (1996).
62. S. A. Schofield, R. E. Wyatt and P. G. Wolynes, *J. Chem. Phys.* **105**, 940 (1996).
63. P. D. Miller, A. C. Scott, J. Carr and J. C. Eilbeck, *Phys. Scr.* **44**, 509 (1991).
64. A. C. Scott, J. C. Eilbeck and H. Gilhoj, *Physica D* **78**, 194 (1994).
65. J. C. Eilbeck, in: *Localization and Energy Transfer in Nonlinear Systems*, Ed. L. Vazquez (World Scientific, Singapore, in press).
66. L. Bernstein, J. C. Eilbeck and A. C. Scott, *Nonlinearity* **3**, 293 (1990).
67. L. Bernstein, *Physica D* **68**, 174 (1993).

68. S. Aubry, S. Flach, K. Kladko and E. Olbrich, *Phys. Rev. Lett.* **76**, 1607 (1996).
69. S. Flach and V. Fleurov, *J. Phys: Condens. Matter* **9**, 7039 (1997).
70. S. Flach, V. Fleurov and A. A. Ovchinnikov, *Phys. Rev. B* **63**, 094304 (2001).
71. J. Dorignac and S. Flach, *Phys. Rev. B* **65**, 214305 (2002).
72. S. R. Bickham, S. A. Kisilev and A. J. Sievers, *Phys. Rev. B* **47**, 14206 (1993).
73. S. Flach, K. Kladko and S. Takeno, *Phys. Rev. Lett.* **79**, 4838 (1997).
74. S. Aubry and T. Cretegny, *Physica D* **119**, 34 (1998).
75. S. Flach and K. Kladko, *Physica D* **127**, 61 (1999).
76. O. Bang and M. Peyrard, *Phys. Rev. E* **53**, 4143 (1996).
77. M. Eleftheriou, S. Flach and G. P. Tsironis, *Physica D*, in print.
78. S. Flach, *Phys. Rev. E* **58**, R4116 (1998).

CHAPTER 2

VIBRATIONAL SPECTROSCOPY AND QUANTUM LOCALIZATION

François Fillaux

LADIR-CNRS, UMR 7075 Université P. et M. Curie,
2 rue Henry Dunant, 94320 Thiais, France
E-mail: fillaux@glvt-cnrs.fr

These lecture-notes are meant to provide newcomers with an overview of the impact of vibrational spectroscopy in the field of nonlinear dynamics of atoms and molecules, in the perspective of energy localization. In the introduction, the terminology of nonlinear excitations and tentative experimental evidences are briefly recalled in a brief historical perspective. The basic principles of vibrational spectroscopy are presented in section 11 for infrared, Raman and inelastic neutron scattering. The potentialities for each technique to probing energy localization are discussed. In section 12, nonlinear dynamics in isolated molecules are treated within the framework of normal versus local mode representations. It is shown that these complementary representations are not necessarily distinctive of weak versus strong anharmonicity, in the context of chemical complexity. It is emphasized that local modes and energy localization are totally independent concepts. In section 4, examples of nonlinear dynamics in crystals are reviewed: multiphonon bound states, strong coupling between phonons and electrons probed with resonance Raman, local modes and quantum rotation in one-dimension probed with inelastic neutron scattering, strong coupling in hydrogen-bonded crystals and self-trapping probed with time-resolved vibrational-spectroscopy. The extended character of eigenstates in crystals free of impurities and disorder, the nature of the interaction of periodic lattices with plane waves, the Franck-Condon principle and the particle-wave duality in the quantum regime are key factors preventing observation of energy localization. It is shown that free spatially-localized nondissipative classical waves give rise to free pseudoparticles that behave as planar waves in the quantum regime. In conclusion, a clear demonstration that energy localization corresponds to eigenstates is eagerly expected for further evidencing these states with vibrational spectroscopy.

1. Introduction

Nonlinear excitations giving rise to spatially localized non-dissipative waves in an extended lattice could be a source of phenomena and technological principles in advanced materials research.[4–1] They are also speculated key elements in complex events on the molecular level of life functioning.[5–7] Vibrational spectroscopy techniques have a great potential for observing nonlinear excitations of atoms and molecules. However, most of the spectra are convincingly rationalized within the framework of the (quasi-) harmonic approximation (namely, normal modes and phonons). Only spectra of highly anharmonic degrees of freedom (for example proton motions in hydrogen bonds, rotational tunneling, *etc.*), or high-excited vibrational states deserve tentative approaches in terms of nonlinear excitations. Apart from these pathological cases, there is no well-established fingerprint distinctive of nonlinear dynamics. The spectroscopic signature of quantum analogue to classical "spatially localized non-dissipative waves" is barely known, and still a matter of controversy because, in many cases, the confrontation of theoretical models with experiments is hampered by the complexity of real systems.

The purpose of these lecture-notes is to examine how nonlinear dynamics can be probed with vibrational spectroscopy techniques.

1.1. *Nonlinear dynamics and energy localization*

Since the pioneering work of Fermi, Pasta and Ulam,[8] evidencing energy localization in one-dimensional anharmonic lattices, numerical simulations and progress in the resolution of nonlinear equations have provided support to a wealth of nonlinear excitations with various properties: solitary waves, solitons, breathers, self-trapping states, discrete breathers (DBs), intrinsic localized modes (ILMs), *etc.* Nonlinear waves normally coexist with phonons that are solutions of the Hamiltonian in the harmonic approximation. In the classical regime nonlinear excitations may give rise to spontaneous energy localization in simple chains, free of impurity.

Solitary waves are spatially localized non-dispersive solutions, either exact or sufficiently accurate to be physically relevant, of Hamiltonians containing nonlinear potential terms. They have two special properties: first, dispersionless waves travel undistorted and, second, after any number of collisions, solitary waves recover asymptotically (as $t \to \infty$) their waveform and velocity. Among solitary waves, *breathers* are spatially localized non-dispersive waves with a time-periodic internal degree of freedom.[2,9,10] *Soli-*

tons are those solitary waves whose energy density profiles are asymptotically restored to their original shapes and velocities after collisions.[9] They behave like dimensionless particles. Only exactly integrable nonlinear equations possess exact many-soliton solutions.[2] Among them, the Korteweg-de Vries (KdV), the nonlinear Schrödinger (NLS) and the sine-Gordon (SG) equations play outstanding roles in physics.[2,4]

ILMs[1] or DBs,[3,11,10] are not solitons. They are spatially localized periodic vibrations which can occur naturally at finite temperature in impurity-free 1D discrete lattices with sufficient anharmonicity. For a given internal energy of the chain, one can intuitively distinguish planar-wave modes, with vanishing amplitudes of oscillation extending over the chain and localized DB modes with large amplitudes for a small number of contiguous sites. DBs may exist in a broader class of systems than solitons, for they are not restricted to integrable systems.

It is certainly needless to remind the reader that quantum effects are prevailing at the molecular level and in the condensed matter. However, the correspondence between classical solitons and pseudoparticle states of the quantized version is not trivial. It has been shown for the NLS and SG equations, which are also exactly integrable in the quantum regime, that one can associate not only a quantum soliton-particle state with a classical soliton solution, but a whole series of excited states as well, by quantizing fluctuation about the soliton.[12,13] As the quantum soliton-particle states are eigenstates they should be observable with spectroscopy techniques. However, this requires crystal lattices that can be modelled with an integrable Hamiltonian for which extended particle states of the quantized version are known. This is certainly exceptional and to the best of our knowledge, the 4-methylpyridine crystal is the only example ever reported of soliton dynamics represented with the quantized version of the sine-Gordon Hamiltonian (see below Sec. 4.5.1).

A major difficulty for spectroscopic studies is that the existence of nonlinear solutions analogous to classical DBs or ILMs in the quantum regime is not proved. Consequently, it is unknown whether these excitations are eigenstates observable with spectroscopy techniques in quantum nonintegrable systems. At first glance, spatially localized eigenstates cannot exist in a quantum lattice with translational invariance. However, it has been suggested that the translational invariance can be recovered if quantized DBs are allowed to "tunnel" from site to site.[10] Then, localization should not survive in the quantum regime and the insistent question remains: how to distinguish the band structure for DBs from those usually encountered for phonons? According to numerical calculations, multiphonon bound states are eigenstates for some quantum nonlinear lattice models that could be regarded as natural counterparts of DBs in classical lattices.[14] This apparent confusion of quantum DBs with multiphonon bound states is quite puzzling for spectroscopists (see Sec. 4.2). Nevertheless, a few experimental works have claimed evidences for energy localization, in molecules or crystal lattices.[4] These works deserve thorough examination as they could highlight guiding rules for further researches.

Self-trapping occurs in many-body systems for which strong interactions between a few degrees of freedom decreases the energy of some eigenstates. This is one of the well-known properties of hydrogen bonded systems. This is a purely quantum effect.

Some dynamical models, similar to those usually encountered for electronic excitations, have been tentatively applied to vibrational dynamics of nuclei in molecules or crystals. They give rise to localized excitations like polarons and excitons.

A *polaron* is a defect in an ionic crystal that is formed when an excess charge at a point polarizes the lattice in its vicinity. Thus, if an electron is captured by a halide ion in an alkali halide crystal the metal ions move toward it and the other negative ions shrink away. As the electron moves through the lattice it is accompanied by this distortion. Dragging this distortion around effectively makes the electron into a more massive particle.[15]

An *exciton* appears upon an electronic excitation of a molecule, or atom, or ion in a crystal. If the excitation corresponds to the removal of an electron from one orbital of a molecule and its transfer to an orbital of higher energy, the excited state of the molecule can be envisaged as the coexistence of an electron and a hole. The particle-like hopping of this electron-hole pair from molecule to molecule is the migration of the exciton through the crystal.[15]

1.2. *Nonlinear dynamics and vibrational spectroscopy*

Long before the first numerical evidences for energy localization in a lattice,[8] spectroscopic studies, even back to the early days, have emphasized the existence of *localized* vibrations in molecules and solids. It has been observed that many chemical groupings give rise to distinctive frequencies, which offer a powerful analytical tool for many purposes in fundamental and applied studies. Since localization arises from the intrinsic heterogeneity of systems containing different atoms linked by different chemical bonds, it could be termed intrinsic localization as well and this terminology is quite ambiguous. In order to avoid further confusion, we propose to call them chemically-ILMs, or CILMs. Similarly, localization may arise from local *impurities* (for example isotope substitution) in molecules or crystals. Although the main purpose of the presented here lecture-notes is to examine spectroscopic fingerprints of localization upon nonlinearity, as long as we deal with real experiments carried out on real samples, it is impossible to ignore the chemical complexity.

It is probably needless to recall that the *normal mode* representation, for atomic displacements with infinitely small amplitudes, is prevailing in the field of vibrational spectroscopy. However, back again to the early days of spectroscopic investigations, it has been observed that the *local mode* picture could be, in some instances, more adequate. At first, this was suspected for highly excited states of symmetrical molecules, close to the dissociation threshold of some chemical bonds, in a range where anharmonicity is important.[16] The considerable amount of work devoted to rationalize the apparently conflicting representations with normal or local modes is largely echoed in this chapter. Unfortunately, the concept of local mode has been confused by some authors with "energy localization".[4] Consequently, it has been speculated that vibrational spectroscopy should be relevant to observing energy localization in molecules. It will be shown that this is far from being as simple as it may seem.

The organization of this presentation is the following. Vibrational spectroscopy techniques are presented briefly in Section 2. In addition to optical techniques (infrared and Raman) recent developments of inelastic neutron scattering techniques open up new prospects for the characterization of nonlinear dynamics. For each technique, we emphasize the limitations imposed by basic laws of quantum physics to observing energy localization. Molecular vibrations are discussed in Section 3, in the perspective of normal versus local mode separation. The complementarity of the two representations is emphasized. The local mode representation and energy localization are clearly distinguished. Section 4 deals with vibrational spectra of collective dynamics in crystal lattices. The distinctive signatures of phonons, solitons and localized excitations are analyzed. It is shown that neither phonon bound-states, nor solitons, nor self-trapping states can give rise to energy localization.

The purpose of this chapter in the LOCNET-lecture-notes is to stimulate interaction between theorists and experimentalists in the field of vibrational spectroscopy applied to nonlinear dynamics. This presentation will certainly sound quite superficial for spectroscopists, rather frustrating for theorists, and extremely uncomfortable for the author who is very far from having well footed opinions on the different problems he has to deal with. Lets hope that strongly motivated young researchers, as it was a pleasure to meet so many of them during various LOCNET meetings, will find helpful the list of references.

2. Vibrational spectroscopy techniques

Vibrational spectroscopy provides information on forces between atoms, molecules and ions, in various states of the matter. Vibrational frequencies are related to electronic structures *via* multidimensional potentials that govern the dynamics. However, there are fundamental and technical limitations to full determination of potential hypersurfaces from vibrational spectra of complex systems. The development of mathematical models and quantum chemistry methods allows us to analyze vibrational dynamics of systems with increasing complexity. However, in spite of spectacular progresses, the confrontation of experiments with theory is far from being free of ambiguities and the interpretation of vibrational spectra remains largely based on experimentation.

Since the very beginning, almost a century ago, vibrational spectroscopy has revealed that dynamics are quantal in nature: the energy is quantized and dynamics are described in terms of eigenstates. The quantum theory has developed predominantly within the harmonic approximation, that is the simplest expansion, to quadratic terms, of the potential hypersurface. Vibrational dynamics can be thus represented with harmonic oscillators (normal modes) corresponding to coherent oscillations of all degrees of freedom at the same frequency. As the main source of information is interaction with light, spectra are largely related to symmetry. Group theory and symmetry-related selection rules have emerged as very efficient tools. From this long history, paved with remarkable successes, the interpretation of vibrational spectra with normal modes is largely prevailing.

2.1. *Some definitions*

Before presenting the basics of optical (Sec. 2.2) and neutron scattering techniques (Sec. 2.3), it is worthwhile to recall definitions of some physical parameters. Numerical values necessary for an estimation of the relevant orders of magnitude are gathered in Table 1.

For vibrational spectroscopists, energies of eigenstates are traditionally expressed in "wavenumber" units ($\tilde{\nu}$) or "cm^{-1}" that is the number of

Table 1. Some characteristics of vibrational spectroscopy techniques.

	E_i (cm^{-1})	λ_i	$\mathbf{k_i}$ (Å^{-1})	Spatial scale (Å)	Coherence length
Infrared	< 5000	$> 2\ \mu$	$< 3 \times 10^{-4}$	> 3000	$\sim 10^{-3}$ m
Raman	2×10^4	$\approx 0.5\mu$	10^{-3}	≈ 800	∞
Neutrons	5000–1	$\approx$ 0.4–9 Å	$\approx$ 16–1	< 1	≈ 10–10^3 Å

wavelengths per cm:

$$\tilde{\nu} = \frac{\nu}{c} = \frac{1}{\lambda}; \tag{1}$$

where ν is the frequency, c the velocity of light and λ the wavelength. It is very important to keep in mind that the wavenumber and frequency are different parameters, yet these two terms are often used interchangeably. Thus, an expression such as "frequency shift of 10 cm^{-1}" is used conventionally by infrared and Raman spectroscopists and this convention will be used hereafter.

If an electromagnetic field or neutrons interact with a molecule or a crystal, a transfer of energy can occur only when Bohr's frequency condition is satisfied:

$$\Delta E = h\nu = \hbar\omega = h\frac{c}{\lambda} = hc\tilde{\nu}. \tag{2}$$

ΔE is the difference in energy between two eigenstates, usually expressed in cm^{-1} units, h is Planck's constant, $\hbar = h/2\pi$, and $\omega = 2\pi\nu$. Thus, $\tilde{\nu}$ is proportional to the energy of transition. Most of the $|0\rangle \rightarrow |1\rangle$ transitions occur in the range between 1 and 5000 cm^{-1} and spectrometers commonly used for condensed matter studies operate with a resolution on the order of 1 cm^{-1}. (High-resolution spectroscopy in the gas phase can go very far beyond this value, by several orders of magnitude.)

Routine *infrared spectroscopy* measures the absorption of an incident radiation as a function of the energy. Therefore, the incident beam must span the whole frequency range and the wavelength is in the range from 2 to 1000 μm (10^{-6}m). *Raman spectroscopy* measures the light scattered by a sample irradiated with a sharp monochromatic laser beam, very often in the visible region ($\lambda_i \sim 0.5\ \mu$m).

The *neutron* is a dimensionless particle whose *kinetic momentum* $\mathbf{p}$ is related to the de Broglie wavelength λ as

$$|\mathbf{p}| = \frac{h}{\lambda} = \hbar|\mathbf{k}|, \tag{3}$$

where $\mathbf{k}$ is the *wavevector* parallel to the beam direction and such that $|\mathbf{k}| = k = 2\pi/\lambda$. The *kinetic energy* is

$$E = \frac{\hbar^2 k^2}{2m_n} \approx 2.08k^2, \tag{4}$$

where m_n is the neutron mass. From particle physics, neutron energy is traditionally given in meV or Terahertz units (1 THz $\equiv 10^{12}$ Hz $\equiv$ 33.356 cm^{-1}):

$$1\ \mathrm{meV}\ \equiv 10^{-3}\mathrm{eV} \equiv 10^{3}\mu\mathrm{eV} \equiv 0.24\ \mathrm{THz} \equiv 8.07\ \mathrm{cm}^{-1} \equiv 11.61\ \mathrm{K}; \tag{5}$$

Numerics in Eq. (4) were obtained with E and k in meV and Å $^{-1}$ units, respectively. Then, the neutron velocity is

$$v(\ \mathrm{ms}^{-1}) \approx 3.956\ 10^{3}k, \tag{6}$$

and the neutron wavelength is

$$\lambda(\ \text{Å}) = \sqrt{\frac{\hbar^2}{2m_n E}} \approx \frac{9.04}{\sqrt{E}}, \tag{7}$$

with E in meV units. Therefore, neutrons with energy ranging from 1 to 500 meV ($\approx$ 8 to 4000 cm^{-1}) have wavelengths in the range of about 9–0.4 Å . Wavevectors are in the range 16–0.7 Å $^{-1}$ and velocities from 60 to 3 km/s. With time resolution $\approx 10^{-6}$s, it is possible to estimate the neutron kinetic energy from the time-of-flight over a distance of a few meters.

2.1.1. *Spatial resolution*

The uncertainty principle, $\Delta x \Delta k \sim 1$, with Å and Å^{-1} units, respectively, gives an estimate of the shortest distances that can be resolved by each technique (see Table 1). With photons in the infrared or visible range, the best spatial resolution is $\sim 10^3$ Å. Therefore, the discrete structure of gas, liquids, and crystals cannot be resolved. As opposed to this, neutrons, like X-rays, are diffracted by crystals and can be used to determine structures. This spatial resolution is different in nature from the resolution limit $\sim \lambda$ for microscopes (focusing).

2.1.2. *Coherence length*

The coherence length, namely $l_c = \lambda^2/(2\Delta\lambda)$, where $\Delta\lambda$ is the resolution of the beam, determines the distance upon which coherent excitations can be probed. For optical techniques, when operated with standard resolution, this length is virtually infinity compared to bond lengths and crystal lattice parameters. Only coherent states extending over all indistinguishable molecules or sites are probed. With neutrons, for a standard resolution $\Delta\lambda/\lambda \approx 1\%$, the coherence length can be varied by orders of magnitudes. Therefore, dynamical correlation of indistinguishable atoms or molecules in crystals can be probed under favorable conditions.

2.1.3. *Energy localization*

In the context of energy localization and transport, it should be borne in mind that the spatial width and wavevector distribution of a wavepacket are correlated. At first glance, one can suppose that energy localization on the scale of, say, 1 Å can be probed with wavepackets having a spatial extension of that order of magnitude. However, the wavevector distribution width should be ~ 1 Å^{-1}. Quick examination of Table 1 shows that this is impossible with optical techniques. With neutrons at rather high energy, such wavepackets can be prepared (Compton effect).

Supposing one can prepare the desired wavepacket, the width irreversibly increases as time is passing. This dispersion is a consequence of the distribution in wavevectors. Therefore, in order to excite specifically a particular site or chemical bond, it should be necessary to prepare a wavepacket at time t_0 in such a way that it focuses on that site at time $t_0 + t$.

However, quantum indistinguishability imposes further restrictions, for it is impossible to constrain wavepackets to interact only with *the* site they are due to focus on. The concept of trajectory must be abandoned. Only eigenstates are relevant. Excitations resulting from spatially-localized wavepackets correspond to a chaotic regime due to superposition of a virtually infinite number of eigenstates. Therefore, the concept of energy localization and experimental methods able to evidence this effect must be examined critically within the framework of quantum mechanics.

2.1.4. *The Franck-Condon principle*

The Franck-Condon principle states that: because nuclei are much more massive than electrons, an electronic transition takes place while the nuclei in a molecule are effectively stationary.[15] Originally, the principle governs probabilities of transitions between the vibrational levels of different molecular electronic states. As it is directly related to the existence of adiabatic potentials in different electronic states (Born-Oppenheimer), the principle can be extended to vibrational transitions in hydrogen bonds for which the vibrational states of the fast stretching proton mode are analogous to the electronic states with respect to the slow motion of heavy atoms (see Sec. 4.5.2). The relevancy of the adiabatic approximation depends on the frequency ratio for the fast and slow motions. For electronic transitions $\sim 1 - 10$ eV and nuclear vibrations $\sim 10 - 100$ meV, the ratio is ~ 100. For hydrogen bonds, the ratio is ~ 10 ($\approx 3000/100$). For the Amide-I band in acetanilide, the ratio is certainly much less and nonadiabatic corrections may be necessary (see Sec. 4.5.3).

2.2. *Optical techniques*

Infrared and Raman spectroscopy are based on the interaction of light and matter. With infrared spectroscopy we measure the absorption/transmission of a sample. The Raman effect is the inelastic scattering process of an electromagnetic wave (see Fig. 1). Detailed presentations of the theoretical framework for infrared and Raman spectroscopy can be found in many textbooks.[17–21] Here, we give only a few definitions to guide the reader.

Let us recall that light is an electromagnetic wave bearing electric and magnetic field components. Hereafter, only interaction of the electric field and matter is considered and we shall ignore the magnetic component. Nuclei and electrons determine the distribution of electric charges in a sample. The barycenters of positive and negative charges define the dipole moment vector $\mathbf{M}$. In the cases of interest for the present lecture, infrared spectroscopy measures the interaction of an incident light beam with the derivatives of the dipole moment with respect to the various vibrational degrees of freedom:

$$\left.\begin{aligned} \mathbf{M} &= \mathbf{M}_0 + \sum_i \frac{\partial \mathbf{M}}{\partial x_i} x_i + \frac{1}{2} \sum_{ij} \frac{\partial^2 \mathbf{M}}{\partial x_i \partial x_j} x_i x_j + \cdots \\ \text{or } \mathbf{M} &= \mathbf{M}_0 + \sum_i \mathbf{M}_i x_i + \frac{1}{2} \sum_{ij} \mathbf{M}_{ij} x_i x_j + \cdots \end{aligned}\right\} \tag{8}$$

The transition between an initial state $|\varphi_0\rangle$ at energy E_0 and a final state $|\varphi_f\rangle$ at E_f occurs at frequency $\hbar\omega_{0f} = E_f - E_0$ and the intensity is proportional to the square amplitude of the transition matrix element: $|\langle\varphi_0|\mathbf{M}|\varphi_f\rangle|^2$. Only vibrations generating a change of the dipole moment are infrared active.

The origin of Raman[22] spectra is markedly different from that of infrared spectra. In Raman, the sample is irradiated with an intense laser

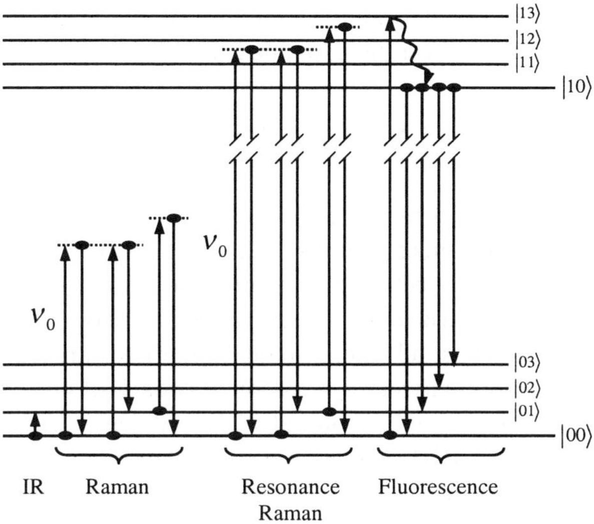

Fig. 1. Schematic view of the energy level schemes for infrared (IR), "normal" Raman, resonance Raman and fluorescence. The horizontal lines represent vibronic states of a harmonic oscillator in one-dimension at frequency ν_i. Within the Born-Oppenheimer approximation, the quantum numbers $|ij\rangle$ refer to electronic and vibrational states, respectively. Dash lines represent "virtual" states.

beam at frequency ν_0, usually in ranges of energy corresponding to ultra-violet, visible or near infrared radiations. In "normal" Raman, ν_0 does not correspond to any eigenstate (see Fig. 1). The interaction of the incident radiation with the sample is represented as a "*virtual*" state that can be represented as a linear combination of eigenstates. The life-time of virtual states is supposed to be vanishingly small. The scattered light consists of *Rayleigh* scattering, that is very intense elastic scattering at ν_0, and Raman scattering, that is very weak inelastic scattering, normally $\sim 10^{-5}$ of the incident beam intensity. The *Stokes* and *anti-Stokes* lines are observed at $\nu_0 + \nu_i$ and $\nu_0 - \nu_i$, respectively.

Normal Raman spectroscopy probes the variations of the polarizability tensor with respect to the degrees of freedom, in the ground electronic state. When an electrical field is applied to a system the electron distribution is modified and the geometry is distorted: the nuclei are attracted toward the negative pole and the electron toward the positive one. The sample acquires an induced dipole moment as the barycenters of the charges are displaced. The polarizability tensor $[\alpha]$ defines the correspondence between the incident electrical field $\mathbf{E}$ and the induced dipole moment $\mathbf{M}' = [\alpha]\mathbf{E}$. The polarizability tensor can be expanded in a Taylor series analogous to Eq. (8),

$$[\alpha] = [\alpha]_0 + \sum_i [\alpha]_i x_i + \frac{1}{2}\sum_{ij} [\alpha]_{ij} x_i x_j + \cdots \tag{9}$$

and the intensity is proportional to $|\langle\varphi_0|[\alpha]|\varphi_f\rangle|^2$.

Because the interaction of photons with matter is quite complicated and depends strongly on the electronic structure, the various derivatives of the dipole moment and polarizability tensor cannot be estimated easily. Even calculations with the most advanced quantum methods are far from providing reliable estimation. Therefore, a major drawback of optical techniques is that intensities cannot be interpreted with confidence.

Resonance Raman (RR) scattering occurs when the exciting line intercepts the manifold of an electronic excited state (see Fig. 1). In the solid state, upper vibrational levels are normally broad and form a continuum. Excitation in this continuum produces resonance Raman spectra that show extremely strong enhancement of some Raman bands strongly coupled to this particular electronic transition. This technique is ideally suited to single out distinctive vibrations of a *chromophore* embedded in a complex environment (see below Sec. 4.3).

Resonance fluorescence occurs when the sample is excited to an electronic eigenstate and then relaxes spontaneously to lower excited states (see Fig. 1). The distinction from resonance Raman is largely a matter of time-scale: the life-time of the excited state in resonance Raman is still very short ($\sim 10^{-14}$s) compared to fluorescence ($\sim 10^{-8}$–10^{-5}s). However, there is no clear cut borderline and the two contributions are usually superimposed. In principle, only time-resolved spectroscopy techniques are able to distinguish Raman and fluorescence effects.

2.3. *Neutron scattering techniques*

Vibrations in solids can be probed also with neutrons. An incident monochromatic beam is scattered by a sample and analyzed with a detector at a general position in space (see Fig. 2). The incident and scattered neutrons are represented with plane waves whose wavevectors are $\mathbf{k_i}$ and $\mathbf{k_f}$, respectively. The *momentum transfer* vector is $\mathbf{Q} = \mathbf{k_i} - \mathbf{k_f}$. One can distinguish *elastic-inelastic* and *coherent-incoherent* scattering processes.[23] Indeed, neutron sources are very expensive and quite rare compared to optical sources like black bodies or lasers. However neutrons give information

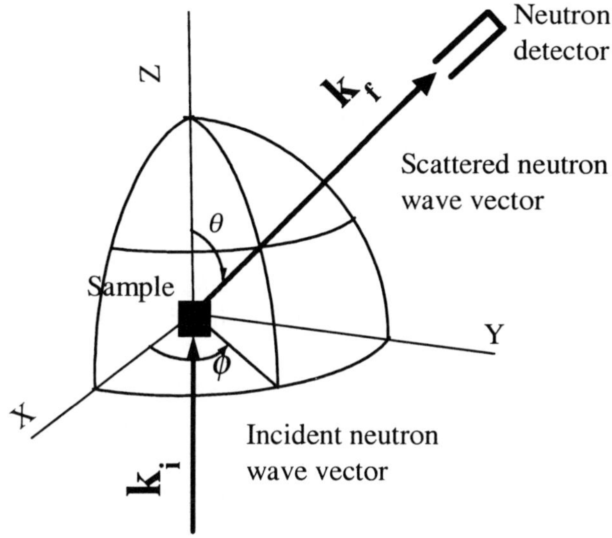

Fig. 2. Schematic view of a neutron scattering experiment.

that cannot be obtained with other techniques. Some interesting properties are featured below.

2.3.1. *Nuclear cross-sections*

The interaction of neutrons and matter is extremely weak. It is dominated by *spin-spin* interaction. Interaction with electron-spins is negligible. Nuclei can be treated as dimensionless scattering centers (Fermi potential). The *nuclear cross-sections* are strictly independent of the electronic surrounding (ionic or neutral, chemical bonding, *etc*). Therefore, the scattering cross-section of any sample can be calculated exactly, from the known nuclear cross-section of each constituent. Compared to optical techniques, INS intensities can be fully exploited and the spectra can be interpreted with more confidence. They are related to nuclear displacements involved in each vibrational eigenstate.

2.3.2. *Coherent versus incoherent scattering*

For each nucleus, one can distinguish coherent and incoherent scattering cross-sections. *Elastic coherent scattering* by a crystal gives rise to *Bragg diffraction*, analogous to what is more routinely measured with X-rays. A great advantage of neutron diffraction is the possibility to locate protons that are normally barely seen with X-rays. *Inelastic coherent scattering* performed on single crystals probes the frequency dispersion of *phonons* in momentum space. *Inelastic incoherent scattering* probes the phonon *density-of-states* over the entire *Brillouin-zone* (see below Sec. 4).

2.3.3. *Contrast*

The incoherent scattering cross-section of the *hydrogen* nucleus (or *proton*) is more than one order of magnitude greater than for any other atom. Therefore, protons largely dominate neutron scattering intensities for hydrogenous samples (see below Sec. 2.4). In many systems the non-hydrogenous matrix environment can be regarded as virtually transparent. This great sensitivity to proton motions can be further exploited because the incoherent cross-section for deuterium atoms (or deuterons, ^{2}H) is about 40 times weaker. Therefore, isotope substitution at specific sites greatly simplifies the spectra. Skill chemists can thus assign bands with great confidence. Isotope labelling is also extremely useful to interpreting infrared and Raman spectra. However, whereas some band frequencies are shifted quite significantly, the spectra of deuterated samples are not greatly simplified, regarding the number of bands and their intensities.

2.3.4. *Penetration depth*

As a consequence of the weak interaction with matter, neutrons can penetrate many samples over rather long distances and probe the bulk material. As opposed to that, the incident light can hardly penetrate samples with high refractive indices and optical techniques probe only a very thin layer at the surface. On the other hand, INS requires much greater amounts of samples in the beam and longer measuring times. Nowadays, INS measurements of molecules in the gas phase are not yet possible.

2.3.5. *Wavelength*

With optical techniques, vibrational dynamics are probed on spatial scales much greater than molecular sizes, or unit cell dimensions in crystals, commonly encountered (see Table 1). For molecules, only overall variations of the dipole moment or polarizability tensor can be probed. In crystals, only a very thin slice of reciprocal space about the center of the Brillouin-zone ($\mathbf{k} \approx 0$) can be probed. Owing to the coherent length for conventional steady sources, this corresponds to in-phase vibrations of a virtually infinite number of unit cells. With optical techniques, band intensities are largely determined by symmetry related selection rules. However, it should be borne in mind that transitions allowed by symmetry are not necessarily intense and can be apparently missing. Conversely, symmetry related selection rules hold only in the *harmonic approximation*. Therefore, supposedly "forbidden" transitions can be observed in some cases.

With the INS technique, wavelengths compare to the sizes of chemical bonds and momentum transfer values can probe in details Brillouin-zones in many crystals. Symmetry related selection rules are irrelevant. It is clear that in the context of localization/dispersion, INS should provide more information than optical techniques.

2.3.6. *Scattering function*

In the context of energy localization, we shall avoid any further consideration on coherent scattering, either elastic (diffraction) or inelastic (phonon dispersion). On the other hand, to measure the scattering function of hydrogen atoms with incoherent scattering can be extremely useful to determine the localization degree of particular excitations. As a tutorial exercise, we present below the textbook example of the isolated harmonic oscillator in one dimension with mass m. The Hamiltonian

$$H = -\frac{\hbar^2}{2m}\frac{d^2}{dx^2} + \frac{1}{2}m\omega_0^2 x^2 \tag{10}$$

can be solved analytically.[24,25] Energy levels and wave functions are

$$\left.\begin{aligned} E &= \left(n + \frac{1}{2}\right)\hbar\omega_0 \\ \Psi_n(x) &= \frac{1}{\sqrt{2^n n!}} H_n\left(\sqrt{\frac{m\omega_0}{\hbar}}x\right)\exp\left(-\frac{m\omega_0}{2\hbar}x^2\right) \end{aligned}\right\} \tag{11}$$

where H_n is the Hermite polynomial of order n. The frequency is $\omega_0/2\pi$. Eigenstates are equidistant in energy (see Fig. 3). The neutron scattering function accounts for the coupling of initial and final states via the neutron plane wave:

$$S(Q_x, \omega) = \left|\langle\Psi_i(x)|\exp(iQ_x x)|\Psi_f(x)\rangle\right|^2 \delta(\omega - \omega_{if}). \tag{12}$$

It can be calculated analytically for transitions arising from the ground state as:

$$S_{0\to n}(Q_x, \omega) = \frac{(Q_x u_x)^{2n}}{n!}\exp\left(-Q_x{}^2\, u_x{}^2\right)\delta(\omega - n\omega_0), \tag{13}$$

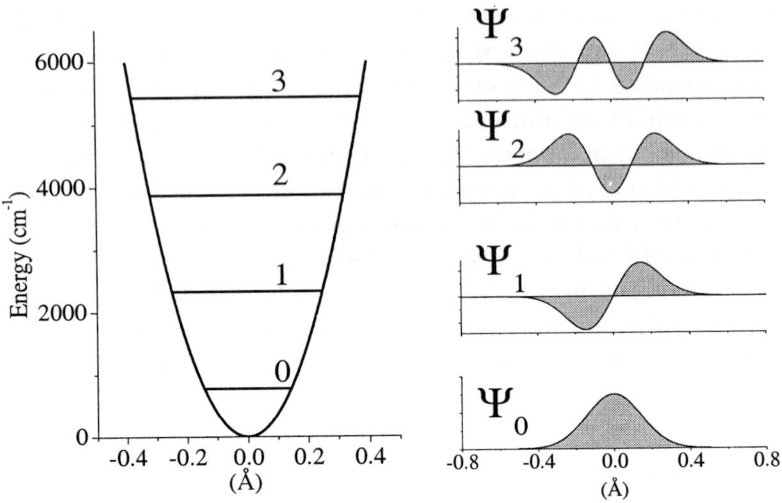

Fig. 38. The harmonic oscillator: energy levels and wavefunctions. $\hbar\omega_0 = 1600$ cm^{-1}, $m = 1$ amu.

where

$$u_x^2 = \langle \Psi_0(x)|x^2|\Psi_0(x)\rangle = \frac{\hbar}{2m\omega_0} \tag{14}$$

is the mean square amplitude of the oscillator in the ground state.

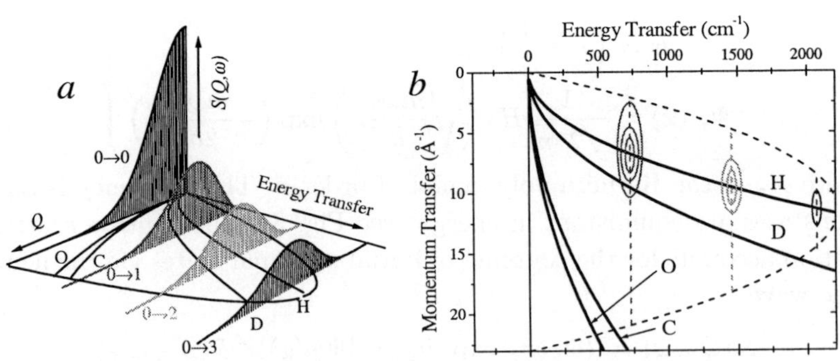

Fig. 39. Landscape (a) and isocontour map (b) representations of the incoherent neutron scattering function for the proton harmonic oscillator. The intensity is a maximum along the recoil line labelled H. Recoil lines for oscillators with masses corresponding to D, C and O atoms are shown. For a fixed incident energy, only momentum transfer values inside the parabolic area (dashed lines) can be measured.

It is thus possible to observe with neutrons all transitions $|0\rangle \rightarrow |n\rangle$ arising from the ground state (see Fig. 4). The intensity is a maximum at $Q_x^2 u_x^2 = n$. In the $S(Q_x, \omega)$ map of intensity, each transition appears as an island of intensity, thanks to some broadening in energy. Profiles along Q_x are directly related to the effective oscillator mass, *via* u_x^2. The maxima of intensity occur along the *recoil line* for the corresponding oscillator mass:

$$E \approx \frac{\hbar^2 Q^2}{2m}. \tag{15}$$

Equations (12) and (13) can be generalized for a set of harmonic oscillators. Then, combination bands can be observed, in addition to transitions for each oscillator. As opposed to this, intensities measured for higher transitions with optical techniques are proportional to

$$|\langle \Psi_m | \, x^p \, | \Psi_n \rangle|^2 \neq 0 \text{ if } m = n \pm p. \tag{16}$$

For most of the active transitions, the high order derivatives of the dipole moment or polarizability tensor are very weak and optical spectra are largely dominated by $|0\rangle \rightarrow |1\rangle$ transitions. Overtones and combination bands are usually very weak.

2.4. *A (not so) simple example*

This section is a tutorial for readers with little experience in vibrational spectroscopy. The purpose is to illustrate at a very qualitative level the complementarity of the various techniques and what information can be sorted out of each of them.

Unfortunately, we have to start with a terrifying name: *potassium hydrogen bistrifluoroacetate*, whose chemical formula is $KH(CF_3COO)_2$. This salt can be obtained quite easily by mixing trifluoroacetic acid (CF_3COOH) and potass (KOH), in water. In the crystal the hydrogen bistrifluoroacetate ions, namely $H(CF_3COO)_2^-$ (see Fig. 5), are surrounded by potassium ions.[26] Two trifluoroacetate entities, CF_3COO^-, are linked by a very short (strong) and centrosymmetric *hydrogen bond* with O$\cdots$O distance of 2.435 Å (see below Sec. 4.5.2 for further presentation of hydrogen bonding). This sample is extremely well-suited to INS studies as there is only one hydrogen atom with large incoherent cross-section. The contribution from other atoms is largely negligible. Proton dynamics studies of such strong symmetrical hydrogen bonds have an impact on many fields.

The infrared, Raman and INS spectra compared in Fig. 5 reveal different aspects of the vibrational dynamics of the $H(CF_3COO)_2^-$ dimers.[27]

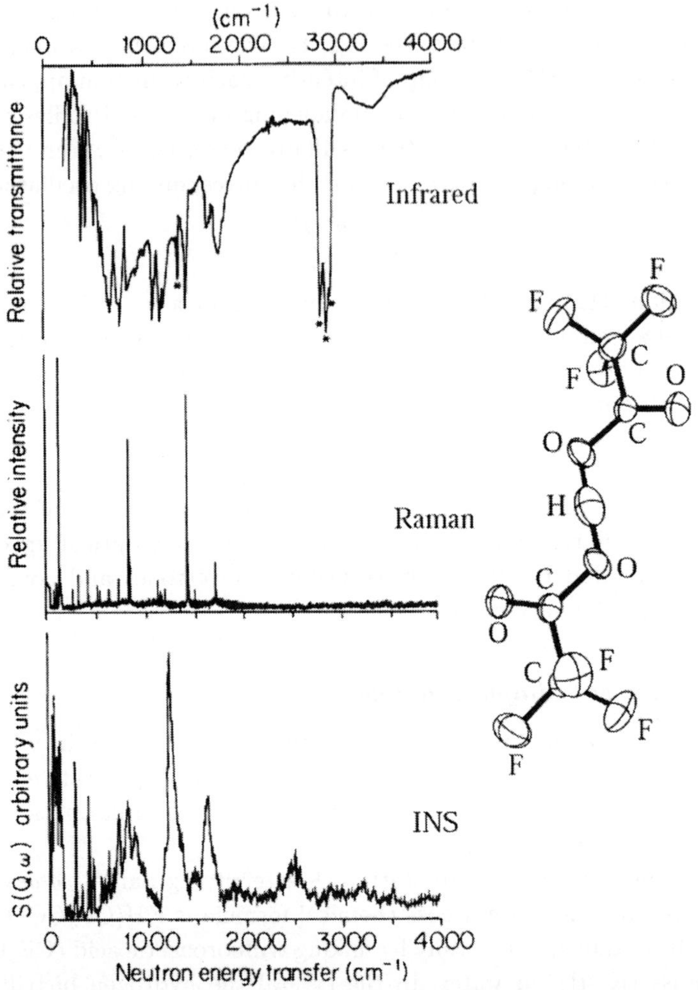

Fig. 5. Schematic view of the hydrogen bistrifluoroacetate complex after Ref. 26 and infrared (Nujol® mull), Raman and INS spectra (from top to bottom) at 20 K, after Ref. 27. *: Nujol® band. C: carbon. H: hydrogen. O: oxygen. F: fluorine.

The infrared spectrum is dominated by a very broad and intense band whose absorption is a maximum (the transmittance is a minimum) at ≈ 800 cm^{-1}. This band extends itself almost continuously from ≈ 300 to 2000 cm^{-1}. This spectacular response in the infrared is distinctive of the centrosymmetric hydrogen bond which can be regarded as essentially ionic in nature. The two oxygens share the electron of the H-atom. This is conventionally written as: $O^{0.5-} \cdots H^{+} \cdots O^{0.5-}$. By symmetry, there is no permanent electric dipole. However, the derivative of the dipole moment with respect to the displacements of the neat charge of the proton is very large along the H-bond (stretching or ν_a OHO) and very weak perpendicular to the bond, either parallel to the plane defined by the COO groups (in-plane bending or δ OHO) or perpendicular (out-of-plane bending or γ OHO). The broad infrared band can be thus assigned to the stretching mode, whereas the bending modes are barely visible.

The Raman spectrum is dominated by very sharp bands due to the COO and CF_3 entities. These entities contain many electrons, and the variation of the polarizability tensor with atomic displacements is a maximum. On the other hand, there is no visible counterpart to the huge infrared band. Because of its pronounced ionic character, there is virtually no deformation of the electronic orbitals for proton displacements and the derivatives of the polarizability tensor are very small.

The INS profile represents a cut of the scattering function $S(Q, \omega)$ along the recoil line for the proton mass (see Fig. 4). Along this trajectory, the intensity arises from vibrational modes involving proton displacements. The band intensity for H-free vibrations is much weaker for two reasons. First, the scattering cross-sections of C, O and F atoms are small. Second, the maximum of the scattering function for these heavy oscillators occurs along recoil lines located at much greater Q-values, beyond the accessible Q-range. Consequently, most of the observed INS intensity arises from proton motions and the assignment scheme is straightforward, for there is only one proton in the dimer and all dimers are indistinguishable. The three bands at $\approx$ 800, 1200 and 1600 cm^{-1}, can be safely attributed to the three degrees of freedom of the proton. The weaker band at $\approx$ 2500 cm^{-1} is the overtone of the sharp band at 1200 cm^{-1}. Then, back to the infrared spectrum, there is a clear correspondence between the INS band at $\approx$ 800 cm^{-1} and the broad proton stretching band. Furthermore, it is easy to identify the weak infrared bands due to the proton bending modes, at about the same frequencies as in INS.

The strong INS intensity of the proton modes indicates that the proton vibrations are largely isolated from the other internal degrees of freedom and there is clearly a pronounced localization. This is primarily due to the very light mass of the proton compared to the other atoms (see below Sec. 3), rather than to anharmonicity. Furthermore, the very sharp INS bands in the 200-500 cm^{-1} range reveal weak but significant coupling of the proton displacements with the CF_3 modes. Finally, below 200 cm^{-1}, the

INS spectral profile arises from the density-of-states of the crystal lattice. In this low frequency range, the dynamics can be represented, to a good level of approximation, with translational and librational (rotational) motions of rigid CF_3COO^- entities.

The decoupled dynamics of the proton with respect to the bistrifluoroacetate framework is by no means equivalent to energy localization in the crystal. The infrared and INS bands for proton modes correspond to coherent oscillations of a large number of entities throughout crystal domains.

3. Molecular vibrations

3.1. *The harmonic approximation: Normal modes*

The problem of the vibrational motions of polyatomic molecules is a very old one and the basic principles were well understood since the early times of vibrational spectroscopy.[17,18,28] The standard treatment begins with the assumption that the vibrational amplitudes are infinitesimal, which implies that the potential energy is a purely quadratic function of the vibrational coordinates. Then the problem for a nonlinear molecule made of $\mathcal{N}$ atoms can be reduced to that of a set of $N = 3\mathcal{N} - 6$ uncoupled harmonic oscillators, or normal modes. The total vibrational energy equals the sum of the energies of the harmonic oscillations,

$$E = \sum_{i}^{N} \left(n_i + \frac{1}{2} \right) \hbar\omega_i, \tag{17}$$

where n_i is the vibrational quantum number and $\omega_i/2\pi$ is the harmonic frequency associated with mode i.

In traditional molecular vibrational spectroscopy important subjects have been the determination of harmonic frequencies for vibrational modes and the assignment of a set of quantum numbers to observed bands on the basis of normal-modes.[17–21] Since this model is based on small-amplitude vibrations, it describes well low-lying energy levels and small deviations has been treated in most cases by perturbative expansion, as proposed in early times by Dunham.[28]

Although this model provides a credible description of "rigid" molecules for low vibrational quantum numbers, it is bound to be inaccurate when the quantum numbers are not small, and to fail completely when they are large. Then, the so called "harmonic" approximation breaks down and higher order terms in the potential expansion must be taken into account. There is no general analytical solution, and the choice of the best representation (or approximation) depends on the system under investigation and of the experimental techniques that are used.

3.2. *Anharmonicity*

Beyond the harmonic scheme, we can distinguish two types of anharmonic terms: diagonal terms, which depend on the coordinate of a single oscillator, and off-diagonal terms, which depend on the coordinates of two or more oscillators. The latter terms, which couple the motions of different zeroth-order harmonic oscillators, are by far the hardest to deal with. As a result of this coupling, the total vibrational energy cannot be written as a sum of the contributions of individual oscillators, since it contains, in addition, terms depending on quantum numbers associated to two or more oscillators:

$$E = \sum_{i}^{N} \left(n_i + \frac{1}{2}\right) \hbar\omega_i + \sum_{i}^{N} \sum_{j \geq i}^{N} \left(n_i + \frac{1}{2}\right) \left(n_j + \frac{1}{2}\right) hX_{ij} + \cdots \tag{18}$$

High-order potential terms necessary for an accurate description of large amplitude displacements arise from two main causes: repulsion and dissociation. For very small internuclear distances, repulsive forces dominate the potential and cause it to vary much faster than quadratically with distances. For very large internuclear distances, the potential varies more slowly than quadratically, and ultimately reaches a constant value which equals the dissociation energy of the oscillator.

As a rule, available spectroscopic data fall far short of what is needed for a complete determination of the $N(N+1)/2$ anharmonicity constants X_{ij} that characterize a molecule, except for a few simple triatomic molecules.[17] Anharmonic constants of higher orders are virtually inaccessible to experimental determination. Hence, it is necessary to use quantum chemistry methods to estimate multidimensional potential surfaces for molecular systems.

3.3. *Local modes*

Experimental studies of highly excited vibrational transitions in the electronic ground state offer a way of exploring the multidimensional energy potential surface along coordinates leading to dissociation. This is a region of chemical interest where the potential energy function begins to reflect intermolecular interactions. It is also the energy region where the zeroth-order Born-Oppenheimer approximation can begin to fail. Therefore, the high order vibrational states to be found in the visible and near ultraviolet spectral regions are of importance to molecular structure determination and chemical reaction mechanisms.

3.3.1. *Diatomic molecules*

For diatomic molecules early spectroscopic studies have demonstrated that very large vibrational amplitudes correspond to dissociation. Birge and Sponer[16] observed that the vibrational levels follow a two parameter relation, namely

$$E_n - E_0 = nA - n^2 B, \tag{19}$$

suggesting the Morse function[29]

$$V(r) = D_e \left\{ \exp\left[-a\left(r - r_e\right)\right] - 1 \right\}^2 \tag{20}$$

as a reasonable approximation to the actual vibrational potential (see Fig. 6). D_e is the dissociation energy, r_e is the equilibrium bond length (the calculated minimum in the potential well) and a is the Morse constant. This

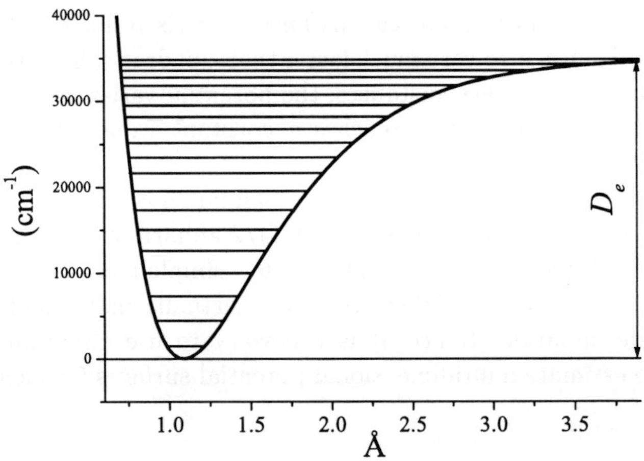

Fig. 6. Representation of the Morse function and energy levels for a diatomic molecule. $D_e = 35000$ cm^{-1}, $a = 1.795$ Å^{-1}, $r_e = 1.086$ Å, $\mu = 1$ amu.

function has the invaluable advantage of analytical formulae for eigenstates (and eigenfunctions):

$$E_n - E_0 = \frac{-D_e}{K^2}[n^2 + (1 - 2K)n], \text{ with } K = \frac{2}{a}\left[\frac{\pi c \mu D_e}{\hbar}\right]^{1/2}, \tag{21}$$

where μ is the effective oscillator mass. D_e and a can be thus determined from Eqs. (19) and (21).

3.3.2. *Polyatomic molecules*

In polyatomic molecules vibrational dynamics for large atomic displacements become a nearly intractable problem when all degrees of freedom have to be treated simultaneously. The number of potential terms increases dramatically and, as the total vibrational energy is raised, the number of states, or ways each state can be partitioned among the vibrational degrees of freedom, grows rapidly. The states may become so closely spaced that they overlap and form a quasicontinuum. In general, each vibrational state is then a complex mixture of many zeroth-order states. The vibrational motion is quite complicated and energy is distributed throughout the molecule.

However, the literature on radiationless transitions in polyatomic molecules has recognized the importance of highly excited stretching vibrations involving the hydrogen atom as the principal accepting modes in transitions between electronic states having large energy gaps. These accepting modes are also the most active in overtone absorption spectroscopy within the ground electronic state, and this fact has prompted interest in such spectroscopy. In many molecules, the hydrogen-atom-based stretching overtone spectrum turns out to be remarkably simple, or diatomic-like according to Eq. (19). This observation has led to contrast the conventional overtone and combination description of excited vibrational states based on normal modes with that obtained with a "local mode" model.

The local mode description of vibrations was already in use by 1929.[30,31] Since those early times, the study of local modes has concentrated on overtones of stretching vibrations of O–H and C–H bonds.[32–58] Their high fundamental stretching frequencies (above 3000 cm^{-1}) cause them to be far out of resonance with fundamentals involving more massive parts of a molecule, so that local behavior can stand out. It was supposed that in a local mode all quanta should be in a single bond and the pure local mode spectrum of a polyatomic molecule with identical oscillators should mimic that of a diatomic molecule. Local modes should have long (hundreds of vibrational periods) lifetimes in which the motion (and energy) are largely confined. They could exist even when they are embedded in a dense manifold of other vibrational modes, sometime referred to as a "bath". Therefore, local modes provide a counterintuitive and fascinating contrast to the normal mode behavior. They are by far the best candidates to studying energy localization arising from anharmonicity.[4] In the next subsection we examine more thoroughly the dynamical regimes for which normal or local mode representations should be preferred.

3.4. *Local versus normal mode separability*

In Polyatomic molecules the relation between normal-mode anharmonicity constants X_{ij} in (18), bond-dissociation energies D_e and normal-mode dissociation energies D_i, is not immediately obvious. For those normal modes in which the vibrational energy is evenly distributed among a number of equivalent chemical bonds, the dissociation energy D_i refers to the sum of the bond energies $\sum D_e$ of all of these bonds. This implies a large value of D_i and thus a small value of X_{ii}. However, the physically important process is, of course, the rupture of a single chemical bond and, for large E, the molecule will oscillate according to a pattern which is closer to a local mode rather than to a normal mode. The dissociation energy is $D_e \ll D_i$ and the anharmonicity constant for the local mode (X_{LL}) is much larger than X_{ii}. Therefore, the normal and local mode representations give totally different views of the energy potential function for the molecule. It is then necessary to determine which key factors make one representation more convenient, or accurate, than the other.

From the adiabatic Born-Oppenheimer separation of nuclear and electronic motions, the vibrational Hamiltonian for a polyatomic molecule can be written as[18,43]

$$H = -\frac{\hbar^2}{2} g^{1/4} \sum_{i,j}^{N} \frac{\partial}{\partial S_i} g^{-1/2} G_{ij} \frac{\partial}{\partial S_j} g^{1/4} + V(\{S_k\}) \tag{22}$$

$V(\{S_k\})$ is a multidimensional function, characteristic of the ground-electronic-state charge distribution, describing the average potential felt by the nuclei. It depends in general on a set of N internal coordinates $\{S_k\}$. The first term represents the kinetic energy of the nuclei. $g^{1/2}$ is the Jacobian for the transformation from the Cartesian coordinates to $\{S_k\}$; the G_{ij} are the elements of the $\mathbf{G}$ matrix in the $\{S_k\}$ system. If we define the momentum operators conjugate to $\{S_k\}$ by

$$\left.\begin{aligned} p_k &= i\hbar g^{-1/4} \frac{\partial}{\partial S_k} g^{1/4} \\ p_k^{\dagger} &= i\hbar g^{1/4} \frac{\partial}{\partial S_k} g^{-1/4} \end{aligned}\right\} \tag{23}$$

then (22) can be expressed as

$$H(\{S_k\}) = \sum_{i,j}^{N} p_i^\dagger G_{ij} p_j + V(\{S_k\}). \tag{24}$$

The solutions to the eigenvalue equation are of course complicated multidimensional functions offering little insight into molecular vibrations. We seek a zeroth-order approach that will provide both a simpler mathematical analysis and a clearer physical picture. For example, if we could reduce the Hamiltonian to a sum of terms describing independent one-dimension oscillators,

$$H(\{S_k\}) = \sum_{i=1}^{N} h(S_i), \tag{25}$$

the eigenstates would be simple products of 1D eigenstates,

$$\psi_{n_1 n_2 n_3 \cdots}(S_1, S_2, S_3, \cdots) = \prod_{i=1}^{N} \phi_{n_i}(S_i), \tag{26}$$

and the overall molecular solutions would reduce to superpositions of N uncoupled motions. In general, of course, such a separation is not possible and we are forced to make the following approximations. First, we neglect the dependence of both g and G_{ij} on S_k and evaluate them at the equilibrium configuration S_k^0. This allows us to write the kinetic energy in the form

$$\mathrm{KE} = -\frac{\hbar^2}{2} \sum_{i,j}^{N} G_{ij} \frac{\partial^2}{\partial S_i \partial S_j}. \tag{27}$$

Similarly, we may expand the potential energy surface in a Taylor series about S_k^0:

$$V(\{a_k\}) = \sum_{i,j}^{N} k_{ij} a_i a_j + \sum_{i,j,l}^{N} k_{ijk} a_i a_j a_l + \cdots, \tag{28}$$

where $a_k \equiv S_k - S_k^0$. The full Hamiltonian has terms to all orders in a_k:

$$H(\{a_k\}) = \sum_{i,j}^{N} \left(-\frac{\hbar^2}{2} G_{ij} \frac{\partial^2}{\partial a_i \partial a_j} + k_{ij} a_i a_j \right) + \sum_{i,j,l}^{N} k_{ijl} a_i a_j a_l + \cdots \tag{29}$$

3.4.1. *Zeroth-order descriptions of the nuclear Hamiltonian*

The Hamiltonian (29) can separate *via* at least two different routes.

Normal modes

We can neglect all cubic and higher order terms, leaving a quadratic or normal mode Hamiltonian

$$H^{NM}\left(\{a_k\}\right) = \sum_{i,j}^{N}\left(-\frac{\hbar^2}{2}G_{ij}\frac{\partial^2}{\partial a_i \partial a_j} + k_{ij}a_i a_j\right). \tag{30}$$

It is then always possible to find a set of coordinates $\{\xi_i\}$ that separate H^{NM} into a sum of independent 1D harmonic-oscillator Hamiltonians

$$H^{NM}\left(\{\xi_k\}\right) = \sum_{i}^{N}\left(-\frac{\hbar^2}{2}G_{ii}\frac{\partial^2}{\partial \xi_i^2} + k'_{ii}\xi_i^2\right). \tag{31}$$

The set $\{\xi_i\}$ are the molecular normal coordinates, and the eigenstates are products of the corresponding harmonic oscillator solutions

$$\psi^{NM}_{n_1 n_2 n_2 \cdots} = \phi^{(1)}_{n_1}(\xi_1)\phi^{(2)}_{n_2}(\xi_2)\phi^{(3)}_{n_3}(\xi_3)\cdots. \tag{32}$$

These results are exact in the limit of infinitesimal displacements and provide a reasonable approximation to the vibrational motions as long as the nuclear configuration does not stray too far from its equilibrium value.

Local modes

In seeking an alternative separation of (29) which might be more appropriate to large displacements, it is natural to consider the following approximation: we drop all the cross terms, even those at the quadratic level, and define a local mode Hamiltonian by

$$H^{LM}\left(\{a_k\}\right) = \sum_{i}^{N}\left(-\frac{\hbar^2}{2}G_{ii}\frac{\partial^2}{\partial a_i^2} + k_{ii}a_i^2 + k_{iii}a_i^3 + \cdots\right). \tag{33}$$

The eigenfunctions of H^{LM} are products of the anharmonic oscillator eigenfunctions associated with each term in (33):

$$\psi^{LM}_{n_1 n_2 n_3 \cdots} = \chi^{(1)}_{n_1}(a_1)\chi^{(2)}_{n_2}(a_2)\chi^{(3)}_{n_3}(a_3)\cdots \tag{34}$$

As we discuss in details below, the preference of (33) over (30) as a zeroth-order representation for (29) depends on the molecule in question, the particular choice of internal coordinates $\{a_k\}$, and the energy range of interest.

There are many cases where adding "pure" $(\xi_i^3, \xi_i^4, \cdots)$ terms to (31) still provides a poorer approximation to the full nuclear Hamiltonian than does (33), if in (33) the $\{a_k\}$ are chosen to be valence (local) coordinates; this is because of the large mixing of normal modes arising from cubic and higher-order coupling terms ($\xi_i \xi_j \xi_l$ and $\xi_i^2 \xi_j^2 \cdots$).

3.4.2. *Breakdown of the zeroth-order descriptions*

Table 2. Density of vibrational states per cm^{-1} for simple molecules ρ_{vib}, after Ref. 57

Molecule	Energy (cm^{-1})	Closest ν_{X-H} quantum number	ρ_{vib} Direct count
H_2O	10 600	3	0.0050
	13 830	4	0.0075
	16 900	5	0.0095
C_2H_2	10 000	3	4.1
	13 000	4	4.1
C_6H_6	3 000	1	0.48
(21 in-plane)	6 000	2	21.7
	9 000	3	438
	12 000	4	5 460
C_6H_6	3 000	1	8.8
(30 modes)	6 000	2	1 680
	9 000	3	10^5

The part of the full Hamiltonian (22) not included in H^0 (H^{NM} or H^{LM}) can be considered as a perturbation $H^{(1)}$ coupling the zeroth-order solutions ψ^0 (ψ^{NM} or ψ^{LM}) and causing the breakdown of the separability:

$$\left.\begin{array}{l} H = H^0 + H^{(1)} \\ H^0 \psi_n^0 = E_n^0 \psi_n^0 \end{array}\right\} . \tag{35}$$

A diagonalization of H in the ψ_n^0 basis involves significant coupling and breakdown of the zeroth-order picture whenever

$$\left| \langle \psi_n^0 | H^{(1)} | \psi_m^0 \rangle \right| > \left| E_n^0 - E_m^0 \right| \tag{36}$$

That is, for comparable zeroth-order energies the best choice for H^0 will be that whose $H^1 \equiv H - H^0$ is the most diagonal. Furthermore, for a given choice of $H^0(\{\psi^0\})$ - and barring accidental degeneracies - there will be little coupling between states whose excitations involve modes of widely different frequencies. This leads to an approximate block diagonalization of H.[18] Finally, Eq. (36) implies that, no matter how small the coupling energy

$\langle \psi_n^0 | H^{(1)} | \psi_m^0 \rangle$, the zeroth-order description must break down as soon as the vibrational density-of-states $|E_n^0 - E_m^0|^{-1}$ becomes sufficiently large. This occurs as the energy in relative nuclear motion becomes high enough, regardless of how small the molecule and of how separable its modes. The densities of vibrational states given in Table 2 for some molecules illustrate the dramatic increase of the number of states for large quantum numbers.[57] For the benzene molecule, vibrational states are so closely spaced that they merge into a quasicontinuum. In the particular case of the water molecule, for example, the local mode separability survives for much higher excitation. Nevertheless, the separability breaks down dramatically at some "critical" energy as couplings begin to exceed the vibrational level spacing.

3.5. *The water molecule*

The water molecule shown schematically in Fig. 7, is a rather simple case of a nonlinear triatomic molecule for which both experiments and theoretical models have been thoroughly investigated. Displacements in the valence coordinates r, r', θ provide a simple starting choice for $\{a_k\}$. The nuclear Hamiltonian is symmetric in r and r' and (29) becomes

$$\left.\begin{aligned} H(r, r', \theta) = {} & G_{rr} p_r^2 + G_{r'r'} p_{r'}^2 + G_{rr'} p_r p_{r'} + G_{\theta\theta} p_\theta^2 + G_{r\theta} p_r p_\theta \\ & + k_{rr}\left(r^2 + r'^2\right) + k_{\theta\theta}\theta^2 + k_{rr'} r r' + k_{r\theta} r\theta \\ & + k_{rrr}\left(r^3 + r'^3\right) + k_{rrr'}\left(r + r'\right) r r' + k_{rr\theta}\left(r^2 + r'^2\right)\theta \\ & + k_{rr'\theta} r r' \theta + + k_{\theta\theta\theta}\theta^3 + k_{r\theta\theta}\left(r + r'\right)\theta^2 + \cdots \end{aligned}\right\} \quad (37)$$

In Fig. 7, the schematic view of the potential surface for the O$-$H stretch suggests that the normal mode model should apply for small amplitude displacements (i.e., small quantum numbers) whereas the local mode coordinates r, r' along the potential valleys are better adapted for large amplitude motions.

3.5.1. *The normal mode model*

To simplify our starting discussion of local versus normal-mode descriptions, it is convenient to suppress for the moment the bend. Then, the normal-mode Hamiltonian reduces to the following quadratic form:

$$H^{NM}(r, r') = G_{rr}\left(p_r^2 + p_{r'}^2\right) + G_{rr'} p_r p_{r'} + k_{rr}\left(r^2 + r'^2\right) + k_{rr'} r r'. \quad (38)$$

When expressed with respect to the normal coordinates

$$S = \frac{1}{\sqrt{2}}\left(r + r'\right) \quad \text{and} \quad A = \frac{1}{\sqrt{2}}\left(r - r'\right), \quad (39)$$

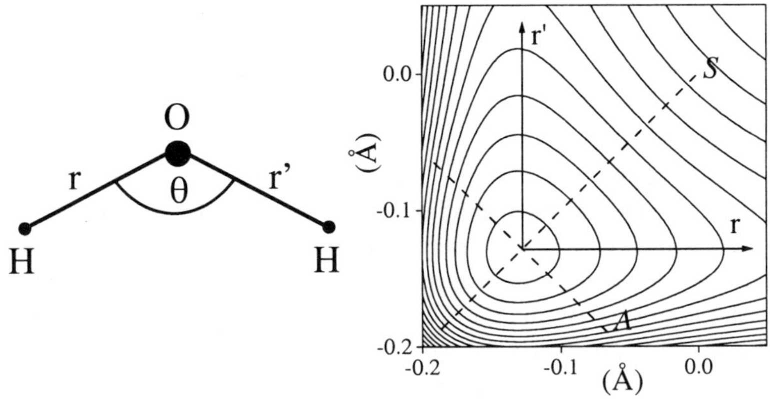

Fig. 7. The water molecule. Left: valence coordinates. Right: schematic view of the potential surface, of the normal symmetry coordinates S, A and of the valence coordinates r, r'.

H^{NM} becomes separable according to

$$H^{NM}(S,A) = G_{SS}p_S^2 + G_{AA}p_A^2 + k_{SS}S^2 + k_{AA}A^2 \tag{40}$$

with

$$\left.\begin{array}{lll} G_{SS} = G_{rr} + G_{rr'}, & & G_{AA} = G_{rr} - G_{rr'} \\ k_{SS} = k_{rr} + k_{rr'}, & \text{and} & k_{AA} = k_{rr} - k_{rr'} \end{array}\right\} \tag{41}$$

The frequencies associated with the normal modes of stretching are $\omega_{SS} = (G_{SS}k_{SS})^{1/2}$ and $\omega_{AA} = (G_{AA}k_{AA})^{1/2}$; the quadratic level coupling has split the degeneracy of the stretches. The extend of this splitting depends directly on the magnitudes of $G_{rr'}$ and $k_{rr'}$ relative to $G_{rr} = G_{r'r'}$ and $k_{rr} = k_{r'r'}$.

3.5.2. *The local mode model*

The local-mode Hamiltonian for the stretches includes cubic and higher order terms in r and r' separately, but no coupling terms:

$$\left.\begin{array}{l} H^{LM}(r,r') = G_{rr}\left(p_r^2 + p_{r'}^2\right) + k_{rr}\left(r^2 + r'^2\right) + k_{rrr}\left(r^3 + r'^3\right) \\ \qquad + k_{rrrr}\left(r^4 + r'^4\right) + \cdots \end{array}\right\} \tag{42}$$

Substituting for r and r' from (39), H^{LM} can be re-expanded in terms of the normal coordinates:

$$\left.\begin{array}{l} H^{LM}(S,A) = G_{SS}p_S^2 + G_{AA}p_A^2 + k_{SS}S^2 + k_{AA}A^2 \\ \qquad + k_{SSS}S^3 + k_{SAA}SA^2 + k_{SSSS}S^4 \\ \qquad + k_{AAAA}A^4 + k_{SSAA}S^2A^2 + \cdots, \end{array}\right\} \tag{43}$$

with

$$\left.\begin{aligned} k_{SS} = \; k_{AA} \; = k_{rr}, \qquad k_{SSS} = \frac{1}{\sqrt{2}}k_{rrr}, \quad k_{SAA} = \frac{3}{\sqrt{2}}k_{rrr}, \\ k_{SSSS} = k_{AAAA} = k_{rrrr}/2, \qquad \text{and} \qquad k_{SSAA} = 3k_{rrrr}. \end{aligned}\right\} \tag{44}$$

Thus we see that small coupling of the bond stretches implies large interaction between the local modes, and *vice versa.* In this sense the two representations are *complementary*, and for each given molecule we must decide which is the most appropriate.

In the case of H_2O all the quantities $k_{rr'}/k_{rr}, k_{rrr'}/k_{rrr}, k_{rrrr'}/k_{rrrr}$, and $k_{rrr'r'}/k_{rrrr}$ are very small compared with unity. Thus (42) holds and we expect the local-mode approximation to work well for the description of vibrationally excited states. This is because the potential anharmonicity is directed mainly along the individual bonds (see Fig. 7). For a linear molecule like CO_2, on the other hand, the anharmonicity resides primarily along the normal coordinates and the normal-mode description is more appropriate. These conclusions are further substantiated upon consideration of the kinetic energy couplings.

Kinetic energy coupling arises from the off-diagonal elements in the **G** matrix. In the normal-coordinate basis, **G** is of course diagonal; but in the valence-coordinate basis off-diagonal terms depend on the atomic masses and equilibrium bond lengths and angles. For example, the **G** matrix for XYX molecules[18] can be written as

$$\mathbf{G} = \begin{bmatrix} \mu_X + \mu_Y & \mu_Y \cos\theta_0 & -\frac{\mu_Y \sin\theta_0}{r_0} \\ \mu_Y \cos\theta_0 & \mu_X + \mu_Y & -\frac{\mu_Y \sin\theta_0}{r_0} \\ -\frac{\mu_Y \sin\theta_0}{r_0} & -\frac{\mu_Y \sin\theta_0}{r_0} & (\mu_X + \mu_Y)\left(\frac{1}{r_0^2} + \frac{1}{r'^2_0}\right) - 3\mu_Y \frac{\cos\theta_0}{r_0 r'_0} \end{bmatrix} \tag{45}$$

where $\mu_i = 1/m_i$. Then, $G_{i,j\neq i}/G_{i,i} \approx m_X/m_Y$. If $m_X \ll m_Y$, as for the water molecule, **G** is virtually diagonal and the kinetic energy contribution to the observed splitting will be small. This approximate diagonality (separability) of the nuclear kinetic energy persists even as the vibrational displacements become so large that the displacement dependence of **G** can no longer be neglected. It turns out that the heavy atom m_Y "*insulates*" the motion of one m_X atom from the other, *via* its large inertia.

3.5.3. *Vibrational wave functions and spectrum*

The above discussion suggests that we expect a good separation of the vibrational Hamiltonian for the water molecule with respect to the local coordinates r, r' and θ:

$$H(r, r', \theta) \approx \sum_{i=r,r',\theta} h(i), \tag{46}$$

where

$$h(r) = -\frac{\hbar^2}{2} G_{rr} \frac{\partial^2}{\partial r} + V(r' = \theta = 0; r), \tag{47}$$

and

$$h(\theta) = -\frac{\hbar^2}{2} G_{\theta\theta} \frac{\partial^2}{\partial \theta} + V(r = r' = 0; \theta) \tag{48}$$

and similarly, by symmetry for $h(r')$. That is the one dimensional potentials are defined by taking the appropriate slice through the full potential energy surface (see Fig. 7).

To examine the separability of the full Hamiltonian we diagonalize it using $h(i)$ eigenfunctions as our basis. These latter (local modes) are written in *symmetrized* form as

$$\psi_{lmn}^{LM} = \chi_l(\theta)\phi_{mn}(r, r') \tag{49}$$

with

$$\phi_{mn}(r, r') = \frac{1}{\sqrt{2}}[\chi_m(r)\chi_n(r') \pm \chi_n(r)\chi_m(r')], \tag{50}$$

where χ_l and χ_m are the eigenfunctions of $h(\theta)$ and $h(r)$, respectively. These basis states are coupled by the off-diagonal terms in $\mathbf{G}$ and the nonseparable part of V. To determine these couplings, then, we require a realistic potential energy surface...

Eq. (49) emphasizes that local modes, like normal modes, are consistent with the molecular symmetry, which is independent from the choice of the preferred representation. There is obviously no energy localization on a single coordinate. In the next section we summarize the salient conclusions arising from vibrational models of the water molecule.

3.5.4. *Eigenstates and eigenfunctions*

Using the quadratic approximation for $H^{(1)}$, the full Hamiltonian is expressed in the separable basis defined by slices of the empirical potential energy and diagonalization gives the overall vibrational eigenstates to be compared to the observed spectra. These data lead to a number of conclusions:

(1) The local mode basis is very accurate for the low lying vibrational states, e.g., the symmetric stretch is 98.4% "pure" superposition of states $|100\rangle$ and $|010\rangle$ (the quantum numbers refer to coordinates r, r' and θ, respectively).
(2) At higher energies the local mode description begins to break down. At the lowest energies where this occurs it is due mostly to strong coupling between two or three states which happen to be close in energy, e.g., the nearly degenerate levels $|200\rangle$ or $|020\rangle$ and $|110\rangle$. At higher energies larger numbers of zeroth-order levels are involved.
(3) The breakdown is irregular because some states are constrained by selection rules to interact most strongly with local mode levels whose energies are far removed.
(4) The strong coupling of local-mode levels, to form the molecular eigenstates, does not necessarily imply that the infrared absorption spectrum will be particularly irregular and complicated.
(5) Finally, the breakdown of the local-mode description has direct consequences on the dynamics of intramolecular vibrational energy redistribution.

3.6. *The algebraic force-field Hamiltonian*

Analysis of energy-level structures obtained with advanced spectroscopic techniques show that anharmonic oscillators, such as the Morse oscillator, are particularly well-adapted for zeroth-order states in the range of energy where anharmonic coupling terms prevail.[59–62] The algebraic approach,[63–66] borrowed from nuclear physics, is a powerful method to determine a potential function from the observed vibrational states. The Hamiltonian expanded with the Lie algebraic operators gives a matrix with rather simple block diagonal structure. Furthermore, an accurate representation of the eigenstates can be thus obtained with a rather small basis-size. The expansion coefficients can be determined *via* a least-square fitting to the observed energy levels and diagonalization can be performed for each block. Therefore, the assignment scheme is straightforward. This approach may become irrelevant in the chaotic regime where multiplet quantum numbers are no longer conserved quantities and interaction between multiplet manifolds may take place.[67–69]

For the water molecule, a total of 20 experimental energy levels of the stretching modes, with no quantum on the bending mode, were fitted with an expansion of the algebraic Hamiltonian.[59–62] Multiplets corresponding

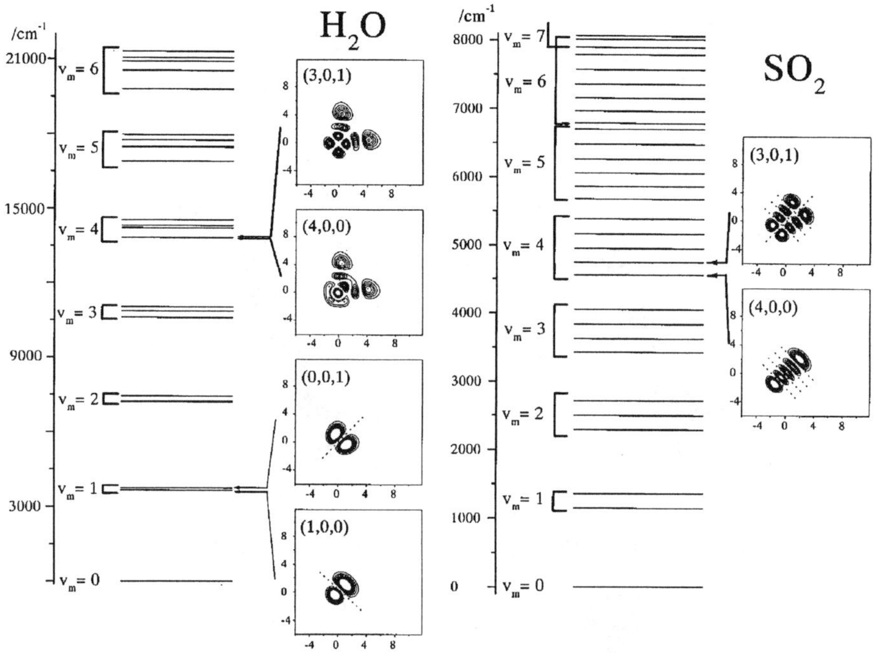

Fig. 8. Energy levels and probability densities for H_2O (left) and SO_2 (right) molecules, after Ref. 61.

to quantum numbers $\nu_m = \nu_r + \nu_{r'} + \nu_\theta$ are well separated and the components for each multiplet are distinguished on the energy-level diagram in Fig. 8 (left). The splitting within multiplets arises from the weak coupling between the two stretching motions and from various combinations of different frequencies for symmetric and antisymmetric modes. The isocontour maps of probability density (namely the squared wavefunctions) are graphic views of the difference between normal and local mode dynamics. For $\nu_m = 1$ nodal lines along the symmetric and antisymmetric coordinates are distinctive of normal modes. The local mode behavior appears clearly for $\nu_m = 4$ and the probability densities confirm that there is no energy localization.

The energy level diagram for the nonlinear triatomic SO_2 molecules was obtained *via* a similar fitting procedure to 53 energy levels.[69] The coupling between the two stretching coordinates gives a rather large splitting for the symmetric and antisymmetric modes. Consequently, overlapping of the multiplet structures occurs for $\nu_m > 4$ and the analysis is more complex. Nevertheless, the normal mode dynamics is still observed for $\nu_m = 4$ and persists clearly for $\nu_m \leq 11$. For increasing quantum numbers, the wavefunctions spread progressively along the antisymmetric direction and for ν_m as large as 23 a clear bifurcation into local modes appears (see Fig. 44a and b).

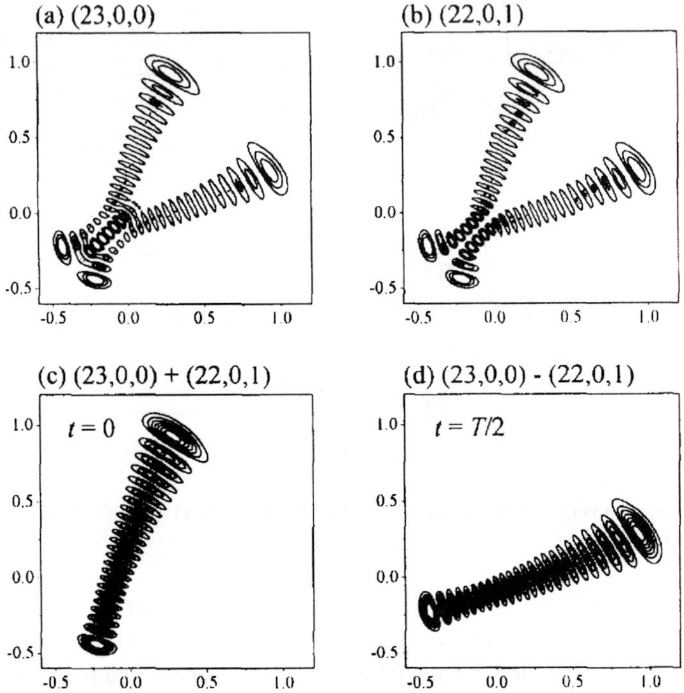

Fig. 9. Probability density of highly excited vibrational states $|23,0,0\rangle$ (a) and $|22,0,1\rangle$ (b) (at $\approx$ 24600 cm^{-1}) of the SO_2 molecule, after Ref. 69. In the time dependent representation, (c) and (d) represent the wavefunctions, where $T = 11$ ps characterizes the energy transfer between the two local modes.

For such large quantum numbers, an interaction takes place between the almost degenerate states $|23,0,0\rangle$ and $|22,0,1\rangle$. The wavefunctions for these states are almost identical along the local mode directions but they have different symmetry along the antisymmetric coordinate. Superpositions of the wave functions such as $(|23,0,0\rangle + |22,0,1\rangle)/\sqrt{2}$ and $(|23,0,0\rangle - |22,0,1\rangle)/\sqrt{2}$ are graphic representations of the local mode behavior (see Fig. 9c and d). They can be regarded as snapshots of the time evolution of the wavepacket corresponding to the superposition of the two states. Numerical calculations give a period of 11 ps for energy transfer between the local modes. However, this nice view is somewhat ideal. In reality, there is a manifold of states and chaotic dynamics is more likely to dominate.

3.7. *Other molecules*

The above presentation of normal and local modes for triatomic molecules is one of the simplest cases in molecular spectroscopy. For more complicated molecules the choice of coordinates providing the best representation of the vibrational dynamics is not so simple and might be not unique. Thorough examination of the chemical complexity, of the spatial arrangement of the nuclei and of the electronic structure (potential surface) is of great significance in many cases.

The internal coordinates of molecules often fall into smaller groups of similar motions having comparable frequencies. For different enough frequencies the oscillators in one group will be only weakly coupled to those in another. This partial decoupling gives so called chemical group frequencies. As an example, consider the *formaldehyde* molecule shown in Fig. 10.

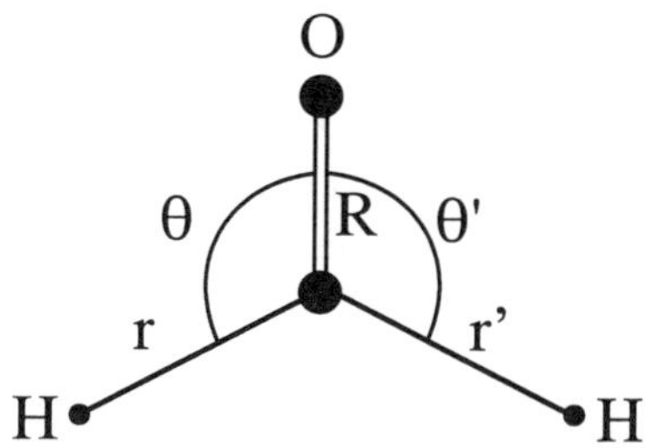

Fig. 10. Valence coordinates for the formaldehyde molecule.

Its six internal coordinates can be assigned as follows: two equivalent CH stretches r and r'; the CO stretch R; two equivalent HCO angle bends θ and θ'; and the out-of-plane bending γ. Table 3 shows that the CH stretches form a nearly degenerate pair of high frequency. The three bending motions have comparable frequencies, but separate by symmetry into the out-of-plane bending (which transforms according to the B_2 representation of the C_{3v} point group) and the pair of HCO angle bends (fairly well isolated from those of the other modes, and it forms a group of its own.[17] Thus the out-of-plane bend and the CO stretch constitute approximate normal coordinates as well. For

Table 3. Frequencies in cm^{-1} units of formaldehyde vibrations, after Ref. 43.

Coordinate	Description		Frequency
r, r'	Symmetric C–H stretch	ν_1	2780
r, r'	Asymmetric C–H stretch	ν_4	2874
R	C=O stretch	ν_2	1744
θ, θ'	Symmetric O=C–H in-plane bend	ν_3	1503
θ, θ'	Asymmetric O=C–H in-plane bend	ν_5	1280
γ	out-of-plane bend	ν_6	1167

the two groups comprised of pairs of equivalent modes, on the other hand, we must decide whether to leave them as local modes or linearly superpose them into their normal or coupled form.

Table 4 shows splitting ratios for equivalent stretches in a few different molecules. The ν CH and ν OH in formaldehyde and water have far lower

Table 4. Frequencies in cm^{-1} units and splitting ratios for equivalent stretches, after Ref. 43.

Molecule	$\nu_{asymmetric}$	$\nu_{symmetric}$	Splitting ratio
H_2CO	2780	2874	0.033
H_2O	3656	3759	0.027
CO_2	1388	2349	0.482
NO_2	1320	1621	0.203

splitting ratios than the CO and NO in CO_2 and NO_2. We have already discussed the nature of the kinetic energy coupling for H_2O and similar conclusions hold for the C–H stretches of formaldehyde. On the other hand, in linear molecules like CO_2 and NO_2, the off-diagonal terms in the **G** matrix (45) vanish. Moreover, the O, C and N atoms have similar masses and rather strong kinetic coupling between the two stretching coordinates arise. Therefore, the normal mode representation is straightforward.

For more and more complex molecules with rapidly growing numbers of degrees of freedom, detailed analysis of the **G** matrix, of the multidimensional potential surface and of the wavefunctions in highly excited vibrational states is impossible. It is then necessary either to make drastic simplifications, if one wishes to pursue some analytical treatment, or to utilize quantum chemistry methods for molecular dynamics simulations.

An example of oversimplified model is the local mode representation of the benzene molecule.[70] The vibrational dynamics is represented with an hexagonal arrangement of the C–H oscillators, with all the C atoms at the center. It is clear that this model cannot be used to analyze the normal versus local mode representations.

3.8. *Local modes and energy localization*

As the normal and local mode basis sets, Eqs. (30) and (33) are complementary representations of the Hamiltonian Eq. (29), they should give the same final results for eigenstates and eigenfunctions. By definition, both are able to account for the observed spectra, although calculations can be greatly simplified with one basis sets, compared to the other. The molecular symmetry should remain the same with both representations. From the analysis of water an formaldehyde molecules, it is clear that an eigenstate corresponding to energy localization in a small number of internal coordinates is better represented with local modes built as linear combinations of these internal coordinates. However, the isodensity maps for the H_2O and SO_2 molecules show that energy localization/delocalization does not depend of the local or normal mode representation for eigenstates.

The interpretation of the high resolution spectra of the stannane molecules SnH_4 and SnD_4 emphasizes the necessary distinction to be drawn between local modes and energy localization.[71–73] The Sn–H and Sn–D stretching overtones reveal that the dynamic symmetry of the molecules changes from that of a spherical top, in the low lying states, to a prolate symmetric top in high excited states. This is a clear transition from normal to local mode dynamics and it has been concluded that energy localization takes place in a single bond stretch oscillator, as represented schematically in Fig. 11.

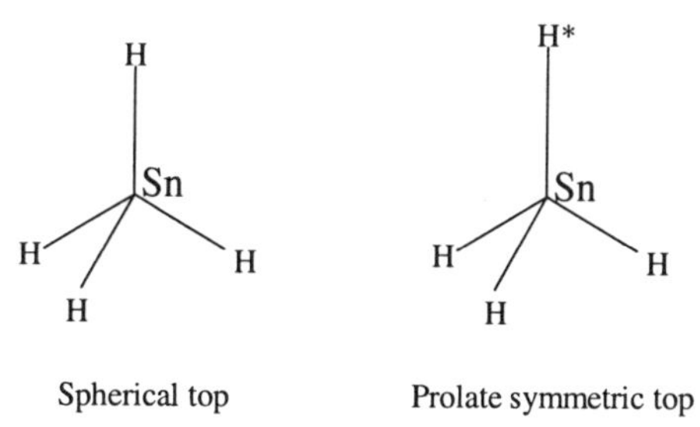

Fig. 11. Change of the dynamic symmetry for the stannane molecule, after Ref. 73.

However, this dynamic transition does not correspond to energy localization because it is impossible to excite specifically a particular bond of a single molecule, and not the others. In the ground state all protons or deuterium atoms of the spherical top are indistinguishable. When illuminated, molecules are excited in a superposition of indistinguishable prolate symmetric tops with equal lengthening probabilities for the four bonds. Consequently, there is no energy localization, although the local mode dynamics is clearly characterized.

In a thought experiment, one could isolate a single molecule with a well-defined orientation and illuminate with a polarized beam properly oriented to excite a single bond. However, to "isolate a single molecule in a well-defined orientation" is an implicit breakdown of the spherical top symmetry and proton or deuterium indistinguishability. This breakdown cancels the local mode degeneracy and energy localization becomes possible.

Along the same line of reasoning, molecular dynamics simulations of a series of hydrocarbon molecules suggest that vibrational energy initially localized on a unique C–H stretching oscillator remains localized on the ps timescale. [74] In these simulation, the vibrational dynamics are treated within classical mechanics and, therefore, the existence of localized eigenstates is not addressed. There is no straightforward contact with spectroscopic studies. Even if the local mode representation may be quite convenient for highly excited vibrational states, excitation with light of benzene molecules from the ground state with sixfold symmetry yields necessarily a superposition of states with energy evenly distributed over all local modes, according to the symmetry related selection rules.

4. Crystals

Vibrational spectroscopy of ideal crystals measures the interaction of an incident plane wave radiation with an infinite spatially periodic lattice *via*, for example, transmission/absorption or scattering process. This interaction between two ideally periodic systems probes primarily the extended eigenstates of the lattice. In the harmonic approximation, lattice dynamics are usually represented with an ideal gas of noninteracting phonons, see Sec. 4.1. As for molecules, anharmonicity can be treated either as a perturbation, like phonon-phonon interaction, or as a genuine dynamical problem giving rise to nonlinear excitations (see Sec. 4.2). In the same spirit as in the previous section, we will try to understand the difference between localized excitations and energy localization, with particular emphasis on the quantal nature of crystal-lattice dynamics. All along this section, close contact will be established with previous works referring to observation of ILMs or DBs (Secs. 4.2 and 4.3), solitons (Sec. 4.5.1), trapping-states (Sec. 4.5.2) and Davydov's model (Sec. 4.5.3).

4.1. *The harmonic approximation: Phonons*

Lattice dynamics within the harmonic approximation are treated in many textbook.[75–77] Within the adiabatic approximation, electron motion is ignored. The electron system is replaced by a spatially uniform distribution of negative charges. Forces correlate the individual motions of the lattice particles about their equilibrium positions. The lattice is limited to a volume V_g composed of N unit cells with cyclic boundary condition. The equilibrium positions are $\mathbf{R}_{n\alpha} = \mathbf{R}_n + \mathbf{R}_\alpha$, where $\mathbf{R}_n$ is a suitable reference point inside the unit cell and $\mathbf{R}_\alpha$ is the vector from this point to the αth basis atom. The index α will run from 1 to r for a basis made up of r particles. The time-dependent vector $s_{n\alpha}(t)$ is the instantaneous displacement of the $n\alpha$th ion from its equilibrium position.

The kinetic energy is

$$T = \sum_{n\alpha i} \frac{M_\alpha}{2} \dot{s}_{n\alpha i}^2 \qquad n = 1, \cdots, N, \qquad \alpha = 1, \cdots, r, \qquad i = 1, 2, 3. \tag{51}$$

M_α is the mass of the αth basis atom. The index i distinguishes the three Cartesian coordinates of the vector $s_{n\alpha}$ whose time derivative is $\dot{s}_{n\alpha i}$.

The potential energy is expanded in increasing powers of the displacement. The first (constant) term does not contribute to the dynamics. The second term, linear in the $s_{n\alpha i}$, is cancelled for particle oscillating about an equilibrium position. The third term is quadratic in the displacement and has the form

$$\frac{1}{2} \sum_{n\alpha i, n'\alpha' i'} \frac{\partial^2 V}{\partial R_{n\alpha i} \partial R_{n'\alpha' i'}} s_{n\alpha i} s_{n'\alpha' i'} = \frac{1}{2} \sum_{n\alpha i, n'\alpha' i'} \Phi_{n\alpha i}^{n'\alpha' i'} s_{n\alpha i} s_{n'\alpha' i'}. \tag{52}$$

The matrix $\Phi_{n\alpha i}^{n'\alpha' i'}$, with dimensions $(3rN \times 3rN)$, represents force constants in the i-direction on the αth particle in the nth cell when the α'th particle in the n'th cell is displaced by unit distance in the i'-direction. This matrix is real and symmetric. It is also invariant with respect to any translation or rotation of the lattice (the crystal symmetry may give rise to further interesting properties). The lattice periodicity reduces the system of $3rN$ equations of motion to a system of $3r$ equations. Then, special solutions written as

$$s_{n\alpha}^{(j)}(\mathbf{q}, t) = \frac{1}{\sqrt{M_\alpha}} \mathbf{e}_\alpha^{(j)}(\mathbf{q}) \exp\{i [\mathbf{q} \cdot \mathbf{R}_n - \omega_j(\mathbf{q}) t]\}, \quad j = 1, \cdots, 3r \tag{53}$$

can be used to construct general solutions. The function $\omega_j(\mathbf{q})$ is periodic in $\mathbf{q}$-space (reciprocal or momentum space) and can be totally characterized within a single Brillouin-zone. In principle, the sets of $\mathbf{q}$-values (N) and of $\omega_j(\mathbf{q})$-values $(3rN)$ are finite. However, discreteness can be ignored even for microscopic crystal domains. Then, the analytical function $\omega_j(\mathbf{q})$ splits into $3r$ branches and, according to time-reversal symmetry, $\omega_j(\mathbf{q}) = \omega_j(-\mathbf{q})$.

The $s_{n\alpha i}(t)$ can be represented as linear combinations of particular solutions such as

$$s_{n\alpha}(t) = \frac{1}{\sqrt{NM_\alpha}} \sum_{j\mathbf{q}} Q_j(\mathbf{q}, t)\, e_\alpha^{(j)}(\mathbf{q}) \exp(i\mathbf{q} \cdot \mathbf{R}_n) \tag{54}$$

where the time-dependent exponential factor in (53) has been included in $Q_j(\mathbf{q}, t)$ and the factor $1/\sqrt{N}$ has been sorted out. The Hamiltonian can be rewritten with normal coordinates as

$$H = \frac{1}{2} \sum_{j\mathbf{q}} \dot{Q}_j^*(\mathbf{q}, t)\, \dot{Q}_j(\mathbf{q}, t) + \omega_j^2 Q_j^*(\mathbf{q}, t)\, Q_j(\mathbf{q}, t). \tag{55}$$

The coupled individual oscillations of particles are thus represented with decoupled collective oscillations. It is then straightforward to introduced the conjugate momentum $P_j(\mathbf{q}, t) = \dot{Q}_j(\mathbf{q}, t)$ and the quantum Hamiltonian is

$$H = \frac{1}{2} \sum_{j\mathbf{q}} P_j^*(\mathbf{q}, t)\, P_j(\mathbf{q}, t) + \omega_j^2 Q_j^*(\mathbf{q}, t)\, Q_j(\mathbf{q}, t). \tag{56}$$

The quantized collective oscillations can be regarded as elementary excitations called phonons. Each state $(\mathbf{q}, j)$ can be occupied by $n_j(\mathbf{q})$ phonons of energy $\hbar\omega_j(\mathbf{q})$ and the total energy is

$$E = \sum_{j\mathbf{q}} \hbar\omega_j(\mathbf{q}) \left[n_j(\mathbf{q}) + \frac{1}{2} \right] \tag{57}$$

Phonons obey the Bose statistics: each state can be occupied by any number of such excitations. For a given lattice, the occupation number of oscillators at frequency $\hbar\omega$ depends only on the temperature. The probability is

$$P_n = \exp(-n\hbar\omega/k_B T)\, [1 - \exp(-\hbar\omega/k_B T)]^{-1} \tag{58}$$

and the mean number of phonons in state $j, \mathbf{q}$ is

$$\bar{n}_j(\mathbf{q}) = \frac{1}{\exp[\hbar\omega_j(\mathbf{q})/k_B T] - 1} \tag{59}$$

Phonons obey the statistic law of Bose-Einstein. They are bosons. In contrast to isolated molecules, the mean occupation number of the ground state is always infinity, at any temperature. It is impossible to depopulate this state by any means. Conversely, any number of phonons can be created in any state, *via* thermal excitation or irradiation.

The representation of phonon dispersion is largely recognized as a major success of the harmonic approximation in solids. The curves measured in many different systems are in rather good agreement with the harmonic theory, although significant deviation and singularities can be observed. In the next subsections, the textbook examples for the single and two atom chains are proposed as tutorials.

4.1.1. *The linear single-particle chain*

For a chain of identical particles (mass M) with no on-site potential and with nearest-neighbor quadratic-interaction (force constant f), Eq. (53) can be rewritten as

$$s_n \propto \frac{1}{\sqrt{M}} \exp\left\{i\left[qan - \omega\left(q\right)t\right]\right\}, \text{with } \omega = 2\sqrt{\frac{f}{M}}\left|\sin\frac{qa}{2}\right|. \tag{60}$$

a is the lattice parameter. The frequency is a periodic function of q and the first Brillouin-zone corresponds to $-\pi/a \leq q \leq \pi/a$ (see Fig. 12*a*). The wavelength in direct space is $\lambda = 2\pi/q$.

4.1.2. *The linear di-atom chain*

For a chain composed of diatomic oscillators with masses M_1 and M_2 and with interaction *via* a single force-constant,

$$\omega_{\pm}^2\left(q\right) = f\left(\frac{1}{M_1} + \frac{1}{M_2}\right) \pm f\sqrt{\left(\frac{1}{M_1} + \frac{1}{M_2}\right)^2 - \frac{4}{M_1 M_2}\sin^2\frac{qa}{2}}. \tag{61}$$

There are two branches, $\omega_+(q)$ and $\omega_-(q)$ (see Fig. 12*b*). At $q = 0$ the spatial wavelength of oscillations is infinity. All particles move synchronously in the same direction. At $q = \pi/a$ adjacent particles move antiphase in the single particle chain. For the di-atom chain, the two particles move in-phase at ω_- and antiphase at ω_+. The lower and upper curves are traditionally referred to as the acoustic and optical branches, respectively.

4.2. *Phonon-phonon interaction*

The concept of an ideal gas of noninteracting phonons does not survive if higher order terms are included in the expansion of the potential function. However, if the perturbation is weak, dynamics can be still represented with

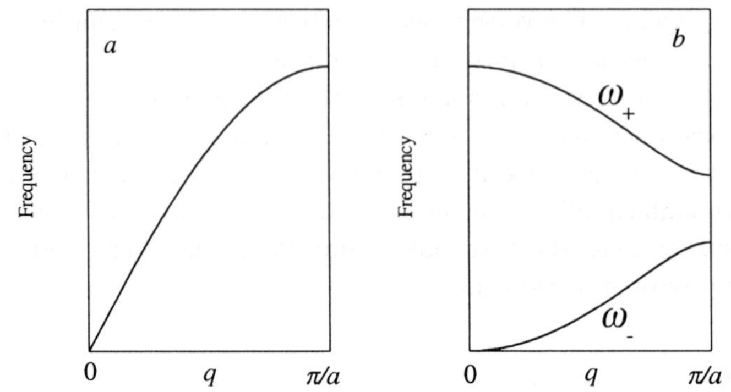

Fig. 12. Dispersion curves $\omega(q)$ for linear chains. a: single particle. b: di-atom chain.

phonon-like entities interacting *via* anharmonic coupling. Consequently, the energy of collective oscillations is shifted, their life-times are finite and multiphonon processes appear. Anharmonicity plays an important role in thermodynamics and heat conduction of crystals. This section is devoted to manifestations of two-phonon processes that have been sometimes confused with energy localization.[78]

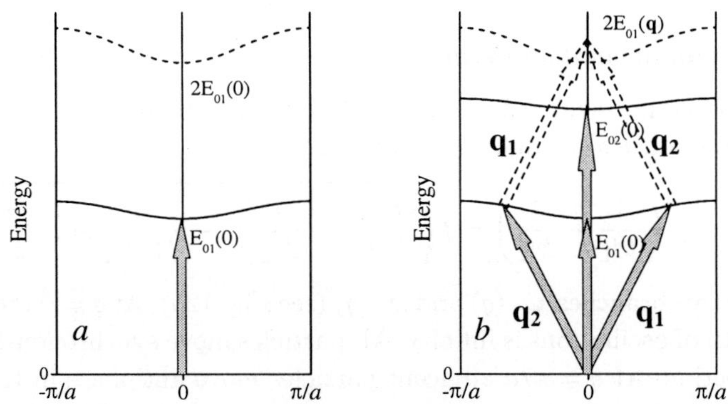

Fig. 13. Schematic representation of transitions that can be observed with optical techniques. a: harmonic scheme. b: anharmonic scheme.

In the harmonic approximation only transitions corresponding to the transfer of one quantum at $\mathbf{q} \approx 0$ can be observed at $E_{01}(0)$ with optical techniques (Fig. 13a), supposing there is no significant electrical anharmonicity. Transitions to overtones at $2 \times E_{01}(0)$ are forbidden for any $\mathbf{q}$-value.

For anharmonic potentials, the energy of the overtone $E_{02}(\mathbf{q})$ is different from $2 \times E_{01}(\mathbf{q})$ (see Fig. 13*b*). As a rule of thumb for molecular crystals, $E_{02}(\mathbf{q}) < 2 \times E_{01}(\mathbf{q})$, but there are exceptions. The one-photon $0 \rightarrow 2$ transition can be observed at $\mathbf{q} \approx 0$. This is generally referred to as a two-phonon bound state, or two-vibron for molecular internal vibrations.

In addition, two-photon transitions, with wavevectors (say) $\mathbf{q}_1$ and $\mathbf{q}_2$, can be excited with electromagnetic waves provided $\mathbf{q}_1 + \mathbf{q}_2 \approx 0$ (Fig. 13*b*). These transitions give rise to a continuum of intensity whose profile is determined by the dispersion of $2 \times E_{01}(\mathbf{q})$. Even in the absence of potential anharmonicity, two-phonon states can be observed thanks to electrical anharmonicity. In that case, the bound state is at the edge of the continuum. In real systems both mechanical and electrical anharmonicity may occur simultaneously.

Two-phonon (or even multiphonon) excitations have been reported in several molecular crystals, for example solid H_2,[79] D_2,[80] HCl,[81] CO_2 or NO_2,[82] *etc.* Two-phonon bound states have been reported also for CO adsorbed on Ruthenium (Ru).[83,84] In these solids the dispersion (bandwidth) of the intramolecular vibration is narrow, anharmonic effects dominate and dynamics are those of an ensemble of weakly coupled molecular oscillators. The bivibron bound states are observed below the bivibron continuum. When either the anharmonicity is decreased, or the dispersion is increased continuously, a transition may take place from the weakly to the strongly coupled case.[85]

It has been shown, and this can be inferred intuitively, that two-vibron bound states can be regarded as molecular in nature inasmuch as they can be represented with combinations of local oscillators.[85] This is indeed very similar in spirit to the local mode representation for isolated molecules. It means that if by appropriate means, or thanks to Maxwell's daemon, a single molecular site could be excited, the energy should remain localized for a long time, barring coupling terms with other degrees of freedom. However, excitation of a single molecule in a crystal cannot be realized *via* irradiation with plane waves, electromagnetic or neutrons. The $\mathbf{q} \approx 0$ selection rule for infrared and Raman and the coherence lengths (see Table 1) impose spatial delocalization of the excitation, practically throughout crystal domains. Moreover, all equivalent molecules are indistinguishable. Therefore, the proposal that the observed bivibron bound states for CO on Ru[83,84] and for D_2[78] correspond to energy localization (ILMs or DBs) is quite in conflict with the well-established foundations of vibrational spectroscopy.

4.3. *Phonon-electron interaction*

The resonance Raman technique has been used to probe anharmonicity in the chloride bridged mixed-valence platinate complex $[Pt(en)_2][Pt(en)_2Cl_2](ClO_4)_4$, where en = ethylenediamine. The crystal structure is composed of infinite chains of PtCl entities (see Fig. 14).[86] Each Pt atom is coordinated to two ethylenediamine units. The chains are well separated and interchain dynamical correlation is negligible. In this quasi-one-dimensional semiconductor the chains are Peierls distorted with the Cl atoms shifted off the central position between the metals. They can be represented as $Cl{-}Pt^{IV}{-}Cl$ entities separated Pt^{II} atoms.

Resonance Raman occurs upon irradiation with an exciting line whose energy corresponds the intervalence charge transfer band near 2.5 eV (see Sec. 2.2).[87] The technique probes dynamics in the electronic ground state and the intensities of Raman bands arising from the totally symmetric stretch $Cl{-}Pt^{IV}{-}Cl$ strongly coupled to the electronic transition is enhanced by several orders of magnitude. Then, the contrast is so high that the other modes are invisible.

Displacement of the equilibrium position along the symmetric stretch in the electronic excited state gives rise to series of overtone transitions. Frequencies give information on the effective potential anharmonicity in the ground state. Relative intensities are determined by the Franck-Condon factors for the vibrational states in the ground and excited electronic states. Intensities decrease exponentially as the order of the overtone increases.

For the compound with natural concentration of the two chlorine isotopes ^{35}Cl ($\approx$ 75.5%) and ^{37}Cl ($\approx$ 24.5%), the $|00\rangle \rightarrow |01\rangle$ transition shows a complicated spectrum that can be decomposed into at least six components between 305 and 315 cm^{-1}.[87] By enriching progressively the sample with the ^{35}Cl isotope, it has been shown that these bands correspond to modes localized by different isotopes. Moreover, because the masses are only slightly different, the impurity modes are not localized on a single site. The various transitions involve about five Pt atoms and the complexity of the spectrum is related to the statistics of the different sequences of isotopes. For the most enriched sample (close to 100%) only one band at 311 cm^{-1} survives.

For the $|00\rangle \rightarrow |02\rangle$ transition, the spectrum of the sample with natural abundance is simplified.[88] It reduces to only 3 components with relative intensities corresponding to those anticipated for the statistical distribution of isotopes in decoupled $Cl{-}Pt^{IV}{-}Cl$ entities. It can be concluded safely

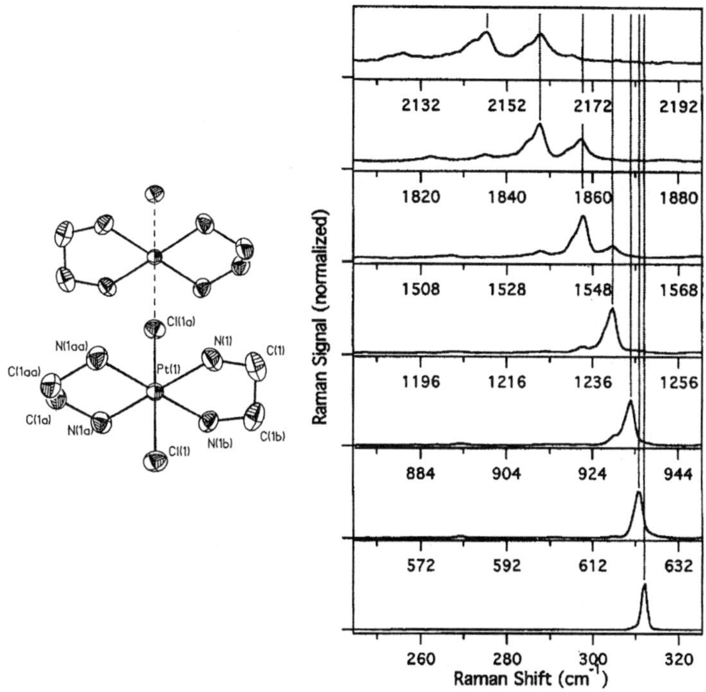

Fig. 14. Left: schematic view of the structure of the $[Pt(en)_2][Pt(en)_2Cl_2](ClO_4)_4$ crystal showing Peierls distortion, after Ref. 86. Right: Resonance Raman spectra of the isotopically pure ^{35}Cl complex, after Ref. 88. Moving upward in each panel, each x axis is offset by the appropriate integral multiple of the 312 cm^{-1} fundamental frequency. All spectra have been scaled vertically to equal peak intensities.

that the modes are more localized in state $|02\rangle$ than in state $|01\rangle$ and it was conjectured that this "intrinsic localization" is a distinctive consequence of anharmonicity. However, the enhanced localization may arise, at least partially, from the increased frequency difference between modes for different isotopes, because the mass effect in state $|02\rangle$ is twice that in state $|01\rangle$. Consequently, the spectra do not prove *per se* that intrinsic localization occurs in the absence of impurity. The mass effect should be thoroughly examined before further consideration be given to ILMs.

The resonance Raman spectra of the pure isotopic samples were presented as stack plots in which each successive spectrum is offset along the frequency axis by multiples of the fundamental frequency (see Fig. 14 for the pure ^{35}Cl derivative).[88] As anticipated, an increasing anharmonic redshift is observed for the dominant features (marked with vertical lines in Fig. 14) that are attributed quite safely to $|00\rangle \rightarrow |0n\rangle$ transitions, with $n \leq 7$. Furthermore, for higher overtones, each dominant peak has a marked counterpart offset by the fundamental frequency in the spectrum just above. Similar spectra were obtained for the ^{37}Cl analogue. These side bands were attributed to intrinsic localization.[88] The lowest energy dominant peaks were attributed to excitations with all quanta of vibrational energy localized in approximately one $PtCl_2$ unit, while the recurrent side peaks correspond to combination of all quanta but one in a single unit and one quantum in the first excited state. Dynamical models have been proposed along these lines. [89–91]

These conclusions are quite at variance from more conventional interpretations. First, as already mentioned in Sec. 2.2 and according to Table 1, resonance Raman corresponds to coherent excitation of the crystal at the center of the Brillouin zone. Therefore, ILMs cannot be probed with this technique. Second, according to the standard theory, the Raman scattering process is extremely fast, such that there is no time for structural or conformational rearrangement. Even if further evolution with time of extended excited states into localized vibrations would occur, this should give rise to fluorescence rather than Raman (see Fig. 1). Third, the most straightforward interpretation of the spectra is that combination bands arise from multiple scattering. As the absorption of the incident beam is quite high, and the electronic band quite broad, multiple resonance Raman scattering events have high probability. For example, a scattering event $|00\rangle \rightarrow |01\rangle$ can be followed by re-absorption, for the scattered light is still in the appropriate frequency range. Then, a second scattering event $|00\rangle \rightarrow |0n\rangle$ gives a combination band at $\nu_n + \nu_1$, exactly as observed in Fig. 14. The probability of multiple scattering events increases dramatically with the absorbance of the sample and with the power of the incident beam. Although there is no information on the power of the incident laser beam in Refs. 87 and 88, and multiple scattering was ignored, it can be suspected that the incident power was increased to observe high overtones with weaker and weaker intensities. Then, the probability of multiple scattering and the relative intensities of the combination bands were enhanced. A flaw of the stack presentation is that it does not give information on the relative intensities of the overtones. Fourth, the broadening of the high overtone bands on the low frequency side suggests contribution of "hot" transitions arising from the first excited vibrational state. At first glance, this may sound impossible since the sample was at a very low temperature. However, resonance Raman can give rise to optical pumping and the population of the resonant mode excited states can deviate significantly from that corresponding to thermal equilibrium, at the temperature of the cryostat. It is quite clear that wings on the low frequency sides of each dominant feature in Fig. 14 are counterparts to the dominant feature just above, offset negatively by the fundamental frequency. They can be safely attributed to $|01\rangle \rightarrow |0n\rangle$ transitions. The population of the first excited state should be further evaluated with resonance anti-Stockes Raman spectra.

In conclusion, resonance Raman of the $PtCl_2$ complex with natural abundance provides a graphic view of localization by natural isotopes of the chlorine atom. The spectra of the pure isotopic samples can be rationalized in terms of slightly anharmonic phonons and combinations with mainly the $|00\rangle \rightarrow |01\rangle$ and $|01\rangle \rightarrow |00\rangle$ transitions due to strong phonon-electron coupling. There is no straightforward experimental evidence for intrinsic localization.

4.4. *Local modes*

In the solid state, it is difficult to distinguishing normal and local modes with infrared and Raman. As overtones and combinations are normally weak, numerous, and broadened by the density-of-states, there is no straightforward assignment scheme. Isotope substitution is a source of information but it is rarely unambiguous, owing to the large number of matrix elements to be determined. Overtones and combinations are best observed with the INS technique. Moreover, effective oscillator masses for different eigenstates can be estimated (see Sec. 2.3). From direct information on anharmonicity and degree of mixing of internal coordinates, normal and local modes are distinguished in a more quantitative way. As already illustrated in Sec. 2.3.6, this technique can be fully exploited for hydrogenous systems.

The potassiumhydrogencarbonate ($KHCO_3$) crystal is an ideal system for INS experiments. It is composed of centrosymmetric dimer entities $(HCO_3^-)_2$ linked by moderately strong OH$\cdots$O hydrogen bonds with length $R_{O\cdots O} = 2.587$ Å at 14 K (see Fig. 15).[92]

Incoherent scattering by protons is much more intense than scattering by other atoms.[93] All protons are equivalent and band splitting arises from dynamical correlation (Davydov splitting). Intra- and interdimer coupling can be distinguished with optical and INS techniques. The intradimer coupling gives B_u and A_g symmetry species observed in the infrared and with Raman, respectively.[94–96] The INS technique on the other hand, probes

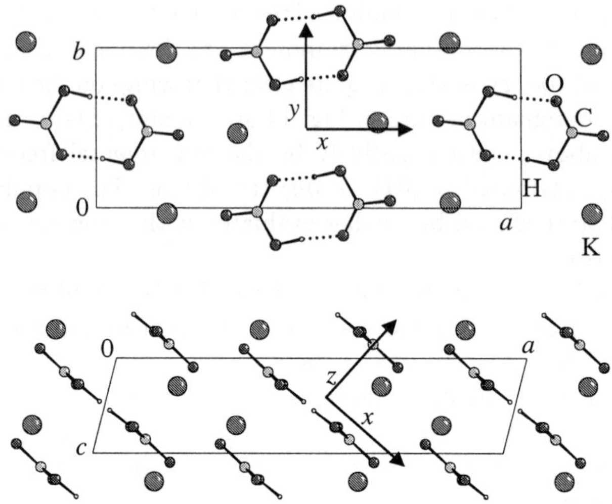

Fig. 15. Schematic view of the crystalline structure of $KHCO_3$ at 14 K, after Ref. 92.

the density-of-states. As the observed bandwidths for the proton modes are similar with the three techniques, interdimer coupling can be neglected. This is in accordance with the rather large distances between dimers in the crystal structure.

The $S(Q, \omega)$ maps of intensity measured with the MARI spectrometer at the ISIS pulsed neutron source (Rutherford Appleton Laboratory, Chilton, UK) show islands of intensity corresponding to the eigenstates involving large proton displacements (see Fig. 16). In these experiments, one has to compromise with the energy E_i of the incident neutron beam, which determines both energy and momentum transfer-ranges to be probed, and the resolution proportional to E_i ($\Delta E_i/E_i \approx 2\%$).

With $E_i = 500$ meV, the maxima of intensity for all observed eigenstates are very close to the recoil line for the proton mass, with only one exception for the in-plane bending δ CO_3 at ≈ 650 cm^{-1}. For the other eigenstates, a more quantitative analysis of cuts along the Q axis confirm that the effective oscillator masses are virtually equal to 1 amu. It can be thus concluded that coupling of the proton modes with the heavy atoms is marginal. This is quite at variance from previous normal mode representations used to account for infrared and Raman spectra.[95] However, this is consistent with the difference of masses for protons and carbonate entities, respectively. This local mode character is quite similar to those already discussed for water or formaldehyde (see Secs. 3.5 and 3.7). Therefore, it is not totally

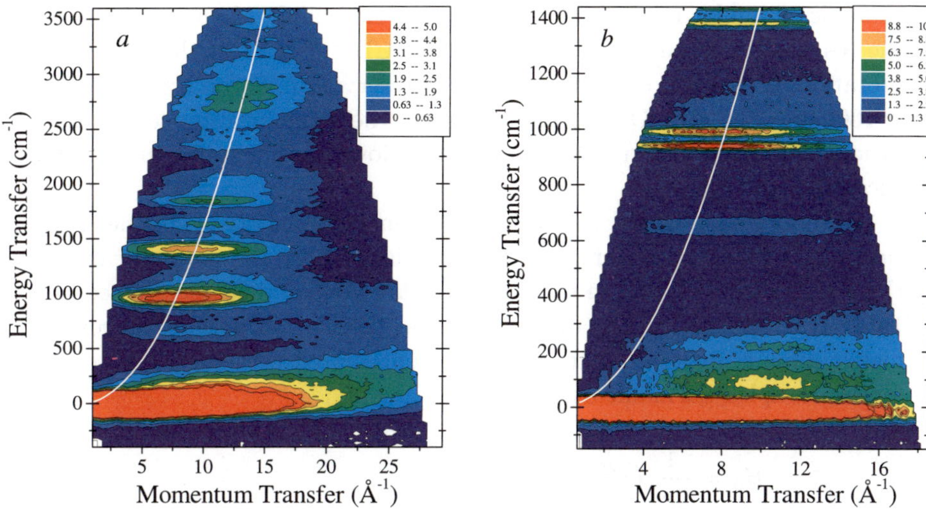

Fig. 16. $S(Q,\omega)$ for powdered $KHCO_3$ at 20 K measured with a fixed incident energy of 500 meV (*a*) and 200 meV (*b*). A qualitative band assignment scheme is: elastic scattering and lattice density-of-state at ≈ 0 cm^{-1}; in-plane bending δ CO_3 at ≈ 650 cm^{-1}; out-of-plane bending γ H at ≈ 960 cm^{-1}; in-plane bending δ H at ≈ 1400 cm^{-1}; stretching ν CO_3 at ≈ 1650 cm^{-1}; overtone $2 \times \gamma$ H at ≈ 1850 cm^{-1}; combination γ H + δ H at ≈ 2400 cm^{-1}; stretching ν H centered at ≈ 2800 cm^{-1} + $2 \times \delta$ H.

surprising to observe an effective mass of 1 amu for the out-of-plane and in-plane bending modes at about 1000 and 1400 cm^{-1}, respectively, for the overtone at ≈ 1850 and for the broad stretching mode at ≈ 2800 cm^{-1}. All these eigenstates can be regarded as CILMs. More surprising is the band at ≈ 1600 cm^{-1} that is traditionally attributed to the CO stretching mode: this band is quite intense in the infrared and shows very weak frequency shift upon deuteration. However, this eigenstate appears as a virtually pure proton mode on the map of INS intensity.

For a more detailed understanding of the coupling between the δ H and ν CO_3 coordinates, it is necessary to measure the scattering function of properly oriented single crystals to probe the orientation of the eigenvectors with respect to the crystal referential.[97,98] Then, it transpires that the local mode representation for the proton modes fully applies to the ground state, but, for symmetry reasons, the anharmonic coupling mixes the first excited states of the δ H and ν CO_3 coordinates. This graphic case emphasizes two important points. First, different techniques may give support to different representations of a particular eigenstate. Straightforward assignment schemes from infrared and Raman can be misleading. For example, a similar mixing of proton modes and Amide-I band reported for the N-methylacetamide crystal ($CD_3CONHCD_3$),[99] suggest that the interpretation of Amide-I band deserves some reservation. Second, the band at ≈ 1600 cm^{-1} in $KHCO_3$ is a nice example where the local mode representation holding in the ground state brakes down, accidentally, in the first excited state.

With increased energy resolution for an incident energy of 200 meV, lattice modes below ≈ 200 cm^{-1} are distinguished from the elastic peak and the γ H mode is resolved in two components at 940 and 990 cm^{-1} (see Fig. 16*b*). This splitting arises from dynamical correlation within centrosymmetric dimers. Proton dynamics can be represented with the Hamiltonian for two coupled harmonic oscillators

$$\begin{aligned} H_\alpha = & \frac{1}{2m}\left(P_{z1}^2 + P_{z2}^2\right) \\ & + \frac{1}{2} m\omega_{0z}^2 \left[\left(z_1 - z_0\right)^2 + \left(z_2 + z_0\right)^2 + 2\lambda_z\left(z_1 - z_2\right)^2\right]. \end{aligned} \tag{62}$$

P_{z1} and P_{z2} are kinetic momenta. The coordinates z_1 and z_2 are projections onto the z direction of the proton positions with respect to the projection of the dimer center of symmetry (see Fig. 15). The harmonic frequency of the uncoupled oscillators at equilibrium positions $\pm z_0$ is $\hbar\omega_{z0}$. The coupling potential proportional to λ_z depends only on the distance between the particles. The equilibrium positions of the coupled oscillators are at $\pm z_0' = \pm z_0/\left(1 + 4\lambda_z\right)$.

With normal coordinates corresponding to symmetric and antisymmetric displacements of the particles, such as

$$\left.\begin{aligned} z_s &= \frac{1}{\sqrt{2}}\left(z_1 - z_2\right), \; P_{zs} = \frac{1}{\sqrt{2}}\left(P_{z1} - P_{z2}\right), \\ z_a &= \frac{1}{\sqrt{2}}\left(z_1 + z_2\right), \; P_{za} = \frac{1}{\sqrt{2}}\left(P_{z1} + P_{z2}\right), \end{aligned}\right\} \tag{63}$$

the Hamiltonian splits into two harmonic oscillators at frequencies $\hbar\omega_{zs} = \hbar\omega_{z0}\sqrt{1 + 4\lambda_z}$ and $\hbar\omega_{za} = \hbar\omega_{z0}$, respectively. Wavefunctions and energy levels can be written as

$$\left.\begin{aligned} \Psi_{n_a n_s} &= \Psi_{n_a}\left(z_a\right)\Psi_{n_s}\left(z_s - \sqrt{2}z_0'\right), \\ E_{n_a n_s} &= \left(n_a + \frac{1}{2}\right)\hbar\omega_{za} + \left(n_s + \frac{1}{2}\right)\hbar\omega_{zs}. \end{aligned}\right\} \tag{64}$$

The scattering functions analogous to Eq. (13) are then

$$\left.\begin{aligned} S_{01}\left(Q_z,\omega_{za}\right) &= \frac{Q_z^2 u_{z0}^2}{2}\exp-\left(\frac{Q_z^2 u_{z0}^2}{2}\right)\delta\left(\omega_{za}-\omega\right), \\ S_{01}\left(Q_z,\omega_{zs}\right) &= \frac{Q_z^2 u_{z0}^2}{4\sqrt{1+4\lambda_z}}\exp-\left(\frac{Q_z^2 u_{z0}^2}{2\sqrt{1+4\lambda_z}}\right)\delta\left(\omega_{zs}-\omega\right). \end{aligned}\right\} \quad (65)$$

The symmetric and antisymmetric modes in Eq. (63) correspond to effective oscillator masses for single protons, as observed. Whereas in the classical regime effective masses associated to normal coordinates are arbitrary,[25] INS data demonstrate that masses are perfectly defined in the quantum regime. During the scattering process, half a quantum is transferred coherently to each proton in a dimer. For the γ H mode, this is possible for neutron plane-waves propagating along the z direction, perpendicular to the dimer planes (see Fig. 15). Further experiments have demonstrated that all protons are correlated in the ground state and must be regarded as a macroscopic quantum state.[92] Therefore, the marked local mode character of the proton eigenstates does not give rise to energy localization.

4.5. *Nonlinear dynamics*

4.5.1. *Quantum rotational dynamics for infinite chains of coupled rotors*

Spectroscopic studies of nonlinear dynamics in the 4-methylpyridine crystal highlight the correspondence between classical nonlinear excitations and eigenstates for quantum pseudoparticles. The confrontation of theory and experiments benefits from the remarkable adequacy of the dynamical model (the sine-Gordon equation) to the crystal structure and dynamics (collective rotational tunneling of methyl groups in one dimension), along with the great specificity of the inelastic neutron scattering (INS) technique. These advantages are documented below.

The 4-methylpyridine crystal (4MP or γ-picoline, C_6H_7N) is an ideal system for experimental studies of nonlinear dynamics arising from collective rotation of methyl groups.[102–105] For the isolated molecule the methyl group bound to the pyridine ring rotates almost freely around the C–C single bond. In the crystalline state, nearly free rotation survives and rotors are organized in infinite chains with rotational axes parallel to the c crystal axis (see Fig. 17).[100,101,106,107] The shortest intermolecular distances of 3.430(2) Å occur between face-to-face methyl groups. According to Ohms and coworkers,[106] paired methyl groups should be twisted by ±60° with

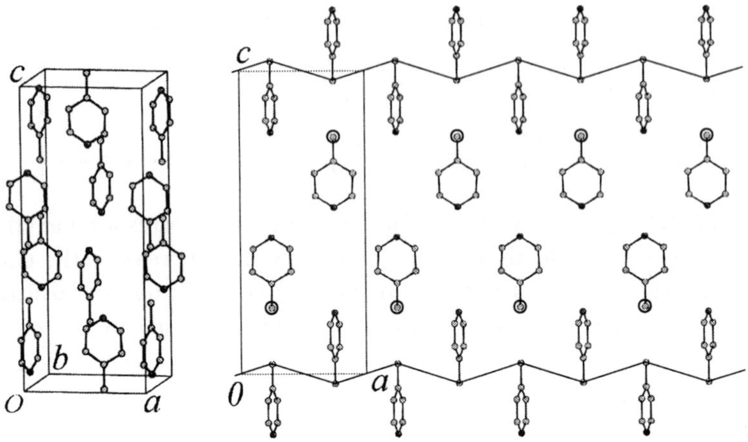

Fig. 17. Schematic view of the structure of the 4-methylpyridine crystal at 10 K. The symmetry is tetragonal ($I4_1/a$, Z=8). Left: view of the unit cell. Right: projection onto the (a, c) plane showing the infinite chains parallel to a (along the zigzag lines) or parallel to b (circles). For the sake of clarity, all H-atoms are hidden. After Refs. 100 and 101.

respect to each other and should perform combined rotation. However, this is not consistent with the C_2 site symmetry: to any particular orientation of one group corresponds four indistinguishable orientations obtained by symmetry with respect to the molecular plane and by $\pm\pi/2$ rotation. The twelve equivalent proton sites are indistinguishable in the probability density obtained with the Fourier difference method applied to neutron diffraction data (see Fig. 18). The effective intra-pair potential arising from the H...H pair potential averaged over all orientations is virtually a constant. Correlated rotation is cancelled.

The next shortest methyl-methyl distances of 3.956(1) Å occur parallel to a and b axes. One can distinguish two equivalent sets of orthogonal infinite chains of methyl groups. The zigzag lines in Fig. 17 correspond to chains parallel to one of the crystal axes (say a) and circles represent intersections with the (a, c) plane of chains parallel to the other axis (say b). As there is only one close-contact pair in common for two orthogonal chains, there is no coupling between collective excitations along a or b. Rotational dynamics are largely one dimensional in nature.[102–105] To the best of our knowledge, this structure is unique to observing collective quantum rotation in one-dimension.

The theoretical model. The rotational Hamiltonian for an infinite chain

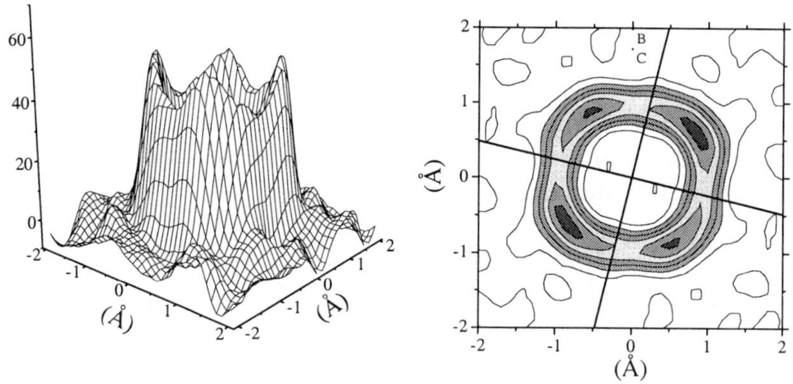

Fig. 18. 4-methylpyridine crystal at 10 K. Landscape view (left) and map of isodensity contours (right) of the H-atom distribution in the rotational (a, b) plane of the methyl groups, obtained with the Fourier difference method. The solid lines represent the orientations of the molecular planes. The ideal isotropic distribution anticipated for disordered rotors with fixed axes is only slightly modified by convolution with the probability density arising from molecular librations. After Refs. 108 and 109.

of coupled methyl-groups with the C_{3v} symmetry can be written as:

$$H = \sum_j -\frac{\hbar^2}{2I_r}\frac{\partial^2}{\partial\theta_j^2} + \frac{V_0}{2}\left(1 - \cos 3\theta_j\right) + \frac{V_c}{2}\left[1 - \cos 3\left(\theta_{j+1} - \theta_j\right)\right], \quad (66)$$

where θ_j is the angular coordinate of the jth rotor in the one-dimensional chain with parameter L. V_0 is the on-site potential which does not depend on lattice position, and V_c is the coupling ("strain" energy) between neighboring rotors.

At first glance, rotational tunneling was treated as a one-dimensional band-structure problem.[102,103,105] Then, eigenstates were plane waves with longitudinal wavevector $k_{\|}$. Apart from a phase factor, wavefunctions can be represented with the basis set for free rotors as

$$\left.\begin{aligned}
\varphi_{0A}\left(k_{\|},\theta\right) &= (2\pi)^{-1/2} a_{0A0}\left(k_{\|}\right) \\
&\quad + \pi^{-1/2}\sum_{n=1}^{\infty} a_{0A3n}\left(k_{\|}\right)\cos\left(3n\theta\right) \\
\varphi_{0E+}\left(k_{\|},\theta\right) &= \pi^{-1/2}\sum_{n=0}^{\infty}\left[a_{0E+(3n+1)}\left(k_{\|}\right)\cos(3n+1)\theta\right. \\
&\quad \left. + a_{0E+(3n+2)}\left(k_{\|}\right)\cos(3n+2)\theta\right] \\
\varphi_{0E-}\left(k_{\|},\theta\right) &= \pi^{-1/2}\sum_{n=0}^{\infty}\left[a_{0E-(3n+1)}\left(k_{\|}\right)\sin\left(3n+1\right)\theta\right. \\
&\quad \left. + a_{0E-(3n+2)}\left(k_{\|}\right)\sin\left(3n+2\right)\theta\right].
\end{aligned}\right\} \quad (67)$$

The lowest state is $|0_A\rangle$. The degenerate tunneling states with opposite angular momentums are $|0_{E+}\rangle$ (symmetrical) and $|0_{E-}\rangle$ (antisymmetrical). The tunnel splitting, $E_{0E\pm}(k_\parallel) - E_{0A}(k_\parallel)$, varies continuously between two extremes: E_{ip}, for in-phase tunneling ($k_\parallel = 0$), and E_{op}, for out-of-phase tunneling ($k_\parallel L = \pi/3$, that is $k_\parallel \approx 0.26$ Å^{-1} in the case of 4MP). The corresponding Hamiltonians are

$$\left.\begin{aligned} H_{ip} &= -\frac{\hbar^2}{2I_r}\frac{\partial^2}{\partial\theta_j^2} + \frac{V_0}{2}(1-\cos 3\theta) \\ H_{op} &= -\frac{\hbar^2}{2I_r}\frac{\partial^2}{\partial\theta_j^2} + \frac{V_0}{2}(1-\cos 3\theta) + \frac{V_c}{2}(1-\cos 6\theta)\,. \end{aligned}\right\} \tag{68}$$

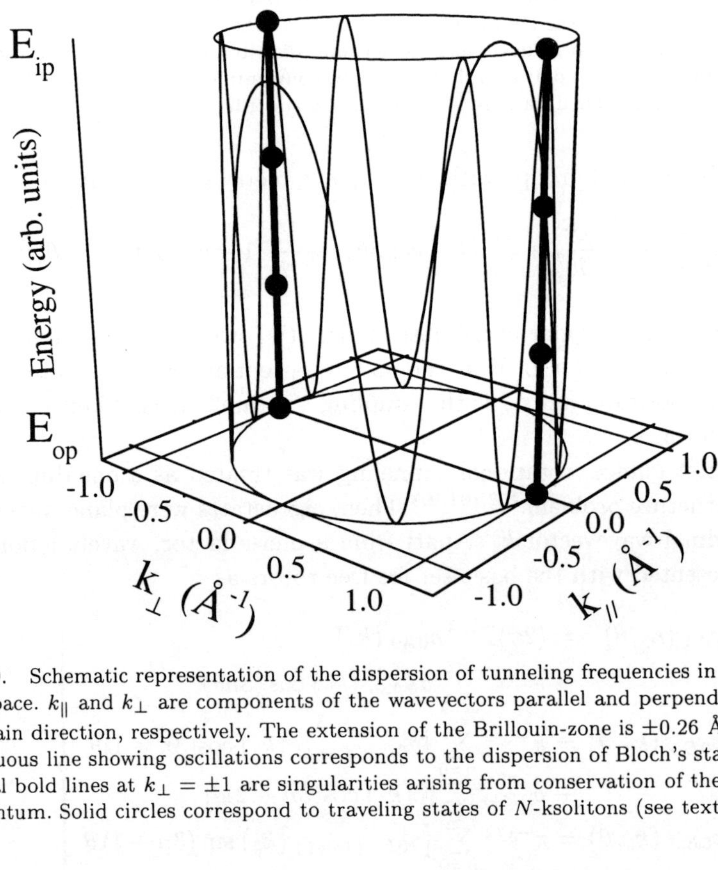

Fig. 19. Schematic representation of the dispersion of tunneling frequencies in momentum space. $k_\parallel$ and $k_\perp$ are components of the wavevectors parallel and perpendicular to the chain direction, respectively. The extension of the Brillouin-zone is ± 0.26 Å^{-1}. The continuous line showing oscillations corresponds to the dispersion of Bloch's states. The vertical bold lines at $k_\perp = \pm 1$ are singularities arising from conservation of the angular momentum. Solid circles correspond to traveling states of N-ksolitons (see text).

In momentum space, the Fourier transform of the wavefunctions is composed of circles at $|\mathbf{k}| = (3n+1)/r$ and $(3n+2)/r$. For free rotors, there is only one circle with a radius of $|\mathbf{k}| = 1/r \approx 1$ Å^{-1}. Amplitudes of high order rings are increasing functions of the potential barrier. For weak potential values, higher order rings can be ignored. The dispersion curve at first order shows periodic oscillations between E_{ip} and E_{op} on the cylinder with radius $|\mathbf{k}| = 1/r$ (see Fig. 19).

However, experimental observation, such as the anisotropy of the scattering function (see below) and the vanishing intensity of the "tunneling" transitions,[104] oppose to the existence of Bloch's states. With hindsight it turns out that proton permutation *via* tunneling arising from topological degeneracy cannot be represented with small amplitude displacements in the potential expanded to second order around the equilibrium position because the cyclic boundary condition does not hold. Therefore, collective tunneling must be represented with nonlinear excitations.

To the best of our knowledge, the existence of solitonic solutions for Eq. (66) is not proved. However, this equation is equivalent to the sine-Gordon equation in the strong coupling (or displacive) limit:

$$H \approx \sum_j -\frac{\hbar^2}{2I_r}\frac{\partial^2}{\partial\theta_j^2} + \frac{V_0}{2}(1-\cos 3\theta_j) + \frac{9V_c}{4}(\theta_{j+1}-\theta_j)^2. \qquad (69)$$

It is thus possible to start with the well-known solutions of this equation for an approximate, hopefully accurate enough, representation of collective rotational dynamics.

In the continuous limit, kinks, anti-kinks and breathers are exact solutions. These solitons do not interact with rotons (equivalent to phonons), that are harmonic oscillations of the chain about the equilibrium configuration. A kink or an anti-kink traveling along a chain rotates the methyl groups by $\pm 2\pi/3$. This is analogous to the propagation of classical jumps over the on-site potential barriers. The breather can be regarded as a bound pair in which a kink and an anti-kink oscillate harmonically with respect to the center of mass. All possible amplitudes of oscillation give a continuum of internal energy, below the dissociation threshold of the bound pair.

In the quantum regime,[13] the renormalized energy at rest for kinks and anti-kinks remains finite, $E_{0\mathrm{K}} = 4\sqrt{V_0 V_c}(1 - 9/8\pi) \approx 11.5$ meV, and the population density vanishes at low temperatures. On the other hand, the quantized internal oscillation of the breather gives a discrete spectrum of renormalized energies at rest. In the particular case where the on-site potential has threefold periodicity, there is only one internal state:

$$E_{0B} = 2E_{0K} \sin \left[\frac{9}{16\,(1 - 9/8\pi)} \right]. \tag{70}$$

As the breather at rest belongs to the ground state of the chain, the translational degree of freedom can be excited, *via* momentum transfer along the chain, even at a very low temperature. If the width of the breather waveform is much greater than L there is no pinning potential arising from the chain discreteness. Breathers can move along the chain like free dimensionless pseudoparticles with kinetic momentum $p_B = h/\lambda_B$, where λ_B is the de Broglie wavelength. The chain lattice diffracts the planar wave and stationary states occur for the Bragg's condition:

$$\lambda_B = L/n_B, \; p_B = n_B h/L; \quad n_B = 0, \pm 1, \pm 2, \cdots \tag{71}$$

This quantum effect is totally different in nature from the pinning effect, although they are both a consequence of the chain discreteness. The kinetic energy spectrum is then[102]

$$E_B(n_B) = \sqrt{E_{0B}^2 + n_B^2 \hbar^2 \omega_c^2}. \tag{72}$$

Within the framework of the quantum sine-Gordon theory, extended to Eq. (66), collective tunneling can be represented with long-lived metastable pseudoparticles, composed of kinks or anti-kinks, traveling along the chain. Calling on the integrability of the sine-Gordon equation, we consider pseudoparticles composed of an integer number "N" of kinks or anti-kinks moving together at the same velocity. In order to avoid tedious repetitions in the remainder of this paper, "***ksoliton***" is used to designate indifferently kinks or anti-kinks, whenever it is needless to distinguish these excitations. "N-ksolitons" are solitons,[81,110,111] and we suppose that they are also soliton-like solutions of the fully periodical Hamiltonian (66), at least to a level of accuracy compatible with experimental observations. On the one hand, N-ksolitons are dimensionless and stationary states arising from diffraction, hence should give a discrete energy spectrum analogous to the breather. On the other hand, as the translation of ksolitons is equivalent to rotation of methyl groups, these pseudoparticles should obey an additional conservation rule related to the angular momentum.

In the absence of many-particle effects, the energy at rest of a N-ksoliton is NE_{0K}, the de Broglie's wavelength associated to this pseudoparticle is $\lambda_{NK} = L/n_{NK}$. Furthermore, only states within the tunneling energy band,

Eq. (68), are stationary. Then, the kinetic energy spectrum is

$$\left.\begin{array}{c} E_{op} \leq E\left(N, n_{\mathrm{NK}}\right) = \sqrt{N^2 E_{0\mathrm{K}}^2 + n_{\mathrm{NK}}^2 \hbar^2 \omega_c^2} \leq E_{ip}, \\ n_{\mathrm{NK}} = 0; \pm 1; \pm 2; \cdots \end{array}\right\} \tag{73}$$

In addition, translational dynamics of N-ksolitons are related to the symmetry of the tunneling states $|0_{E\pm}\rangle$. These states have opposite angular momentums and, as a consequence of proton indistinguishability, only states whose total angular-momentum is zero are allowed. A topological index $\mathcal{I}$ can be associated to each pseudoparticle such that $\mathcal{I} = N^+ - N^-$ where N^+ and N^- are the numbers of kinks and anti-kinks, respectively ($N = N^+ + N^-$). Therefore, except for pseudoparticles composed of equal numbers of kinks and anti-kinks ($\mathcal{I} = 0$), only states with $k_{\|} = 0$ are allowed.[b] They may correspond either to two identical pseudoparticles with opposite $k_{\|}$-values or to multi-quantum states of single pseudoparticles. In reciprocal space, the conservation of the angular-momentum and kinetic energy, Eq. (73), give rise to singularities in energy at $k_{\|} = 0$ and $|k_{\perp}| = 1/r$ (see Fig. 19).

Tunneling spectroscopy is distinctive of quantum rotational dynamics of methyl groups. First, in the frequency range of rotational tunneling ($\approx \pm 500$ μeV for 4MP) there is no phonons, other than the acoustic ones, arising from the crystal lattice. Consequently, interaction of rotational and lattice dynamics can be largely ignored. Second, the specificity of tunneling transitions to methyl rotation can be fully exploited with INS because the incoherent cross-section is much greater for the hydrogen atom than for any other atom. The singling out of H-atoms, further enhanced by the very large angular amplitude associated with methyl tunneling, makes the assignment of spectra quite easy. Third, regarding the comparatively modest cross-section of the deuterium (^{2}H) atom, partial deuteration and isotope mixtures provide information on methyl rotation. Needless to say, the many advantages of INS are complementary to those of infrared and Raman spectroscopy.[112]

The assignment scheme of the tunneling spectra of 4MP was based on INS spectra of isotope mixtures,[102,105] partially deuterated analogues, [103] high resolution in energy,[104] infrared and Raman spectra,[112] and, more recently, INS measurements with single crystals.[101] The potential terms of

[b]Breather traveling states are also degenerate with respect to the propagation direction but the internal harmonic oscillation has no angular momentum. It is thus possible to distinguish traveling states by transferring momentum along a particular direction. In this context, the breather is more closely related to phonons than to collective tunneling.

the chain Hamiltonian were determined from the INS transitions observed at highest and lowest energies, namely (539 ± 4) and (472 ± 4) μeV, corresponding to the extremes of the tunneling band: $V_0 = 3.66$ and $V_c = 5.46$, in meV units. Then, the INS band at (517 ± 4) μeV corresponds to the $|0\rangle \rightarrow |1\rangle$ transition of the breather mode, Eq. (72), with $E_{0\mathrm{B}} = 16.25$ and $\hbar\omega_c = 4.13$, in meV units. As the full width at half maximum of the breather waveform is $\approx 5L$, the pinning potential is negligible. The transitions observed at 539, 514, ≈ 500 and 472 μeV correspond to fourth order transitions ($n_{\mathrm{NK}} = 4$) of N-ksolitons in Eq. (73) with N ranging from 22 to 24. As $E_{0\mathrm{K}} = 11.5$ meV, these pseudoparticles have internal energies ranging from 253 to 287.5 meV.

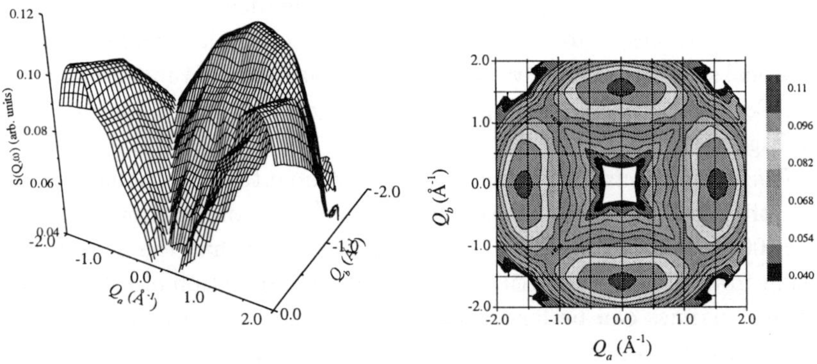

Fig. 20. Landscape view (left) and isointensity contour map (right) of $S(Q_a, Q_b, \omega)$ measured in the (a, b) plane of a single crystal of 4-methylpyridine at 1.7 K. Neutron energy-loss integrated over (500 ± 60) μeV. After Ref. 101.

The pronounced anisotropy of the scattering function measured for momentum transfer parallel to the (a, b) rotational plane with a properly oriented single crystal is distinctive of the rotational dynamics in one-dimension (see Fig. 20). Numerical analysis with Gaussian profiles of cuts along a or b reveals two components centered at (1.55 ± 0.2) and (1.0 ± 0.2), in Å^{-1} units. The intensity ratio is $\approx 75 : 25$. The most intense component corresponds to the breather state anticipated at $|Q_a|$ or $|Q_b| = 2\pi/L \approx 1.57$ Å^{-1}, according to Eq. (71). The weaker peak corresponds to N-ksolitons anticipated at $Q_{\parallel} = 0$ and $|Q_{\perp}| \approx 1$ Å^{-1}. The full width at half height of 0.74 Å^{-1} for the components is directly related to the zero-point fluctuation of the L parameter. On the other hand, the kinetic energy of the pseudoparticles is independent of L, see Eqs. (72) and

(73). In real space, these excitations must be represented as superposition of dispersionless wavepackets localized at each site and whose group velocity is zero. The full-width at half-height (FWHH) of 2.65 Å is much smaller than the width of the breather waveform, and even much smaller than the lattice parameter. Therefore, the hope that nonlinearity could give rise to energy localization and transport is not realized in the context of quantum methyl rotation.

4.5.2. *Strong vibrational coupling: Hydrogen bonding*

The concept of "hydrogen bond" appeared at the beginning of the twentieth century to account for chemical, spectroscopic, structural, thermodynamical and electrical properties. It was recognized that under certain conditions an atom of hydrogen is attracted by rather strong forces to two atoms, instead of only one, so that it may be considered to be acting as a bond between them. However, the location of the H atom and the physical origin of the binding energy long remained matters of controversies, until the development of the quantum-mechanical theory of valence. L. Pauling wrote in his renowned book: "It is now recognized that the hydrogen atom, with only one stable orbital (the 1s orbital), can form only one covalent bond, that the hydrogen bond is largely ionic in character, and that it is formed only between the most electronegative atoms".[113] Consequently, hydrogen bonds can be described as involving resonance among the three structures $X{-}H\cdots Y$; $X^-{-}H^+\cdots Y$ and $X^-\cdots H^+{-}Y$. Electronic structures, proton transfer and vibrational dynamics are intimately correlated.

In a premonitory view, L. Pauling emphasized the main motivations for physicists, chemists and biologists to study hydrogen bonds in a great variety of systems in different states. "Because of its small bond energy and the small activation energy involved in its formation and rupture, the hydrogen bond is especially suited to play a part in reactions occurring at normal temperatures. It has been recognized that hydrogen bonds restrain protein molecules to their native configurations, and I believe that as the methods of structural chemistry are further applied to physiological problems it will be found that the significance of the hydrogen bond for physiology is greater than that of any other single structural feature."[113]

Despite the spectacular amount of knowledge accumulated during almost a century, a comprehensive view of hydrogen bonding phenomena is still far from being achieved.[114,115] It is extremely difficult to obtain an unambiguous estimate of the specific contribution of hydrogen bonds in complex systems because the bond energy is in the same range as other weak interactions, such as van der Waals or dispersion. Thermal energy at room temperature is also similar. In many solvents, including the very important case of water, there is a manifold of various hydrogen bonds that cannot be unraveled easily. Consequently, although most experimental and theoretical works give bond energies in the range 2-5 kcal/mol,[116] these values should be treated with caution.

Hydrogen bonds are still, nowadays, difficult to model because there is no unique way to partition the binding energy resulting from simultaneous changes of the electronic and vibrational wavefunctions upon hydrogen bond formation. The basis of the interaction in hydrogen bonds is still regarded as essentially electrostatic in nature. The correlation between the strength and the length of hydrogen bonds is a distinctive source of nonlinear effects. In this section, we consider hydrogen-bonded crystals containing O-H$\cdots$O or N-H$\cdots$O entities for which the magnitude of the coupling can be estimated experimentally from the empirical correlation between spectroscopic (OH or NH stretching frequencies) and crystallographic data (O$\cdots$O or N$\cdots$O distances, Figs. 21*a* and 22*a*)[117,118]

Because the hydrogen atom is much lighter than oxygen or nitrogen atoms, and because the OH or NH stretching frequencies are greater than those of the O$\cdots$O or N$\cdots$O stretching (vibrons) by about one order of magnitude, dynamics can be treated within the adiabatic approximation. The adiabatic potentials for the slow coordinate (ΔR defined with respect to the equilibrium position) in the ground and first excited states of the fast stretching coordinate (x) are largely determined by the dissociation threshold along ΔR and the coupling between the two coordinates.

As the binding energy of a hydrogen bond is typically on the order of 1000 cm^{-1}, the adiabatic potential in the ground state can be represented with a Morse function for an ideally isolated complex. In crystals, the stacking does not permit dissociation. In the ground state, only the $|00\rangle \to |01\rangle$ transition is observed (the two quantum numbers refer to the fast and slow coordinates, respectively) and the adiabatic potential is defined only around the equilibrium position, as a quasiharmonic potential.

In the excited state the potential is shifted towards short distances ($\Delta R < 0$) by the strong coupling. Both the minimum and the dissociation threshold are shifted and the stronger the coupling is, the larger is the displacement. Because the positions of the surrounding atoms are not changed, the plateau corresponding to the dissociation threshold may become visible. Since the original idea of Stepanov[119] (1946) different models have been proposed for these nonlinear dynamics.[120–128]

With infrared or Raman, the profiles of intensity are determined by the Taylor series expansion of the relevant operator ($\mathcal{M}$) with respect to the fast and slow coordinates:

$$\sum_{n,m} \frac{\partial^{n+m}\mathcal{M}}{\partial x^n \partial \Delta R^m}. \tag{74}$$

This introduction to hydrogen bond dynamics is meant to emphasize that these systems have great potential to demonstrate nonlinear effects. Moreover, as already stressed in the introduction, infrared and INS spectroscopy techniques are very sensitive to proton dynamics in hydrogen bonds. On the other hand, it is important to realize that the dynamics is never one dimensional in nature, because strong coupling also exists for the proton bending modes, but in the opposite way: the bending frequencies increase with decreasing lengths. Therefore, at thermal equilibrium, the binding energy in the H-stretching excited state is largely counterbalanced by the "*anti-binding*" energy in the bending excited states. The upper adiabatic potentials shown in Figs. 21*b* and 22*b* would be misleading if they were regarded as a proof of existence of stable "self-trapping" states with long lifetimes. Further examination of the adiabatic potential hypersurface around the upper minimum would reveal the intrinsic instability of the excited stretching state.

Strong coupling: OH$\cdots$O

In Fig. 21*a*, one can distinguish roughly three domains corresponding to strong ($R_{\mathrm{O}\cdots\mathrm{O}} < 2.6$ Å), moderate ($2.6 < R_{\mathrm{O}\cdots\mathrm{O}} < 2.7$ Å), and weak ($R_{\mathrm{O}\cdots\mathrm{O}} > 2.7$ Å) hydrogen bonds. For each domain, the average slope ($\Delta\nu/\Delta R \approx 12000, 5000$ and $1500\ \mathrm{cm}^{-1}$/Å, respectively) characterizes the strength of the coupling.

For weak hydrogen bonds the upper minimum is only slightly shifted, the electrical anharmonicity is weak and only the $|00\rangle \to |10\rangle$ transition is observed. For moderately strong hydrogen bonds the upper minimum is significantly shifted and discrete stationary energy levels are observed as combination bands $|00\rangle \to |1N\rangle$. The spectral profile of the OH stretching mode can be partially resolved into individual components whose relative intensities depend on the magnitude of both the anharmonic coupling and the electrical anharmonicity.[120,127] For stronger hydrogen bonds ($R_{\mathrm{O}\cdots\mathrm{O}} \approx 2.6$Å), the coupling increases dramatically, dynamics in the upper

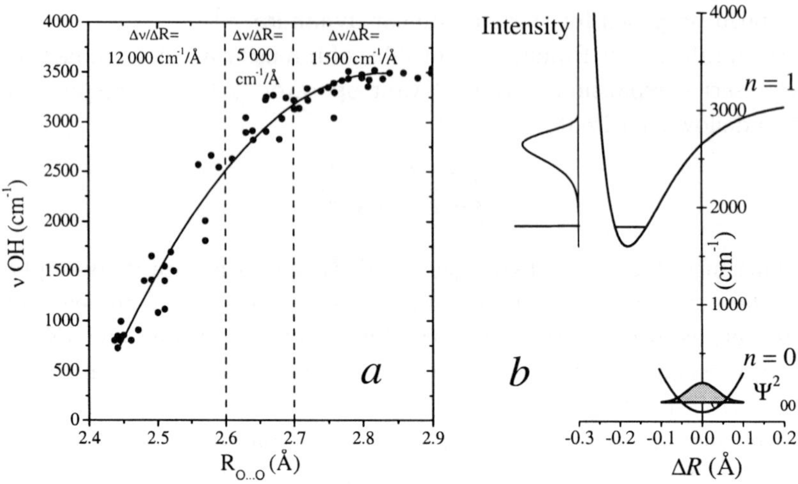

Fig. 21. a: relation between OH stretching frequencies and $R_{\mathrm{O}\cdots\mathrm{O}}$ distances, after Ref. 117. b: schematic view of the adiabatic potential functions for the slow ν O···O mode (ΔR coordinate) in the states $n = 0$ and $n = 1$ of the fast ν OH mode and the intensity profile in the first order approximation for the dipole moment.

state become unstable and give a broad continuum of intensity extending itself over several hundreds of wavenumbers. However, in the vicinity of the upper minimum, $\Delta E_1/\Delta R \approx 0$, the $|10\rangle$ state can be stationary and hence gives rise to a sharp $|00\rangle \rightarrow |10\rangle$ transition, referred to as "*zero-phonon*" (see Fig. 21*b*).[121] This model applies to various systems consisting of dimers like $KHCO_3$[123,124] or benzoic acid,[128] or infinite chains of hydrogen bonds, like cesiumdihydrogenphosphate (CsH_2PO_4).[125] For even shorter hydrogen bonds, the upper minimum is further shifted away and the sharp zero-phonon component is no longer observed. Higher order terms of the transition moment operator become prevalent and a great variety of spectral profiles can be observed.[117,123]

The spectral profile schematically represented in Fig. 21*b* corresponds to the first order term of the transition moment ($n = 1$ and $m = 0$) in Eq. (74). The integrated intensity is proportional to the magnitude of the transition moment, $\partial\mathcal{M}/\partial x$, whereas the intensity profile is determined by the overlap integral of the wavefunctions for the slow coordinate ΔR in the fundamental and excited states (Franck-Condon profile).

The zero-phonon band is a clear signature of the strong vibrational coupling that is distinctive of the hydrogen bond itself. The relative intensity of the zero-phonon transition with respect to the total intensity is $\ln(I_{zph}/I_{tot}) \approx -\Delta^2/(2u^2)$, where Δ is the distance between the minima in the lower and upper states and u^2 is the mean square amplitude, as defined in Eq. (14). For a rough estimation, the oscillator effective mass for the O···O stretching mode is ≈ 8 amu and $u^2 \approx 0.01$ Å^2 for a frequency at 200 cm^{-1}. For the systems under consideration, the relative intensity of the zero-phonon band is $\sim 10^{-2}$ and $|\Delta| \approx 0.3$ Å.

Zero-phonon bands are observed only at low temperature and vanish rapidly above 50 K.[128] At low temperature, the FWHH ≈ 10 cm^{-1} means that the lifetime of the zero-phonon state is much longer than 3 ps. This rather long lifetime arises from the very small integral overlap with the ground state. Thermally induced disorder, especially the population of higher O···O states, opens new relaxation channels which shorten dramatically the lifetime.

The whole spectral profile in Fig. 21*b* arises from internal coupling of the OH···O system, exclusively. In the hydrogen bonded crystals under consideration, further dynamical correlations between equivalent entities in the unit cells are negligible and there is no visible Davydov splitting. This is confirmed by examination of the band profiles of H entities surrounded by deuterated entities (isotope dilution).[117,125] Therefore, bandwidths are not related to localization and transport.

It would be an error to conclude, from a superficial examination of the upper adiabatic potential, that a dramatic shortening of the O···O distance (self-trapping) takes place in the zero-phonon state. Because infrared and Raman profiles correspond to coherent excitation of a virtually infinite number of unit cells at the center of the Brillouin-zone (see Table 1), the simultaneous shortening of an infinite number of H-bonds is not allowed by the crystal environment. Furthermore, according to the Franck-Condon principle, ν OH transitions are extremely fast on the time scale of O···O vibrations. They occur without any significant rearrangement of the heavy atoms. It is exactly because there is no relaxation of the ΔR coordinates that the zero-phonon band is so sharp. Otherwise, relaxation on the time scale of the O···O vibrations would be ten times faster and the zero-phonon band would vanish, as it does at high temperature.

Weak coupling: NH···O

NH···O hydrogen bonds are much weaker than OH···O bonds: the N···O distances are longer and the coupling ($\Delta\nu/\Delta R$) is about three times

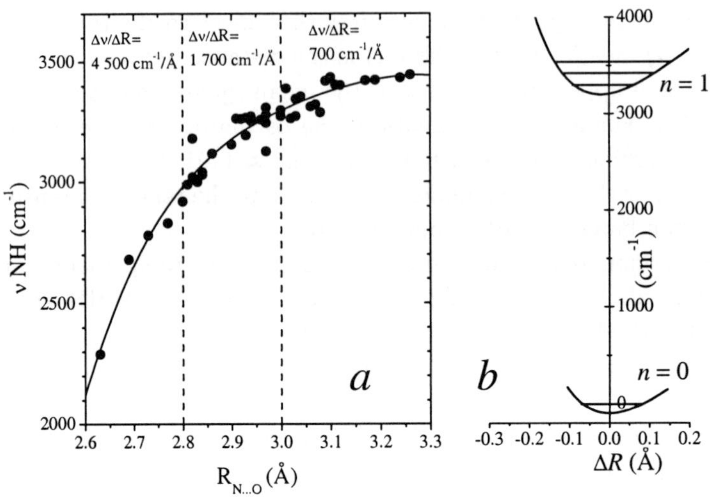

Fig. 22. Relation between NH stretching frequencies and $R_{N\cdots O}$ distances of crystals containing NH···O hydrogen bonds. The experimental data (dots) were reported from Ref. 118.

For amides and peptides linked by intermolecular NH···O bonds, the N···O distances are in the range of 2.8–2.9 Å where the coupling is similar to that for weak OH···O systems. Consequently, the upper minimum is much less shifted with respect to the equilibrium position, the wavefunctions for the slow mode are almost identical in the lower and upper states and the $|00\rangle \rightarrow |10\rangle$ (zero-phonon) transition is the most intense of the NH stretching profile observed in the infrared or with Raman. Transitions $|00\rangle \rightarrow |1N\rangle$ $(N > 1)$ are invisible because the wavefunctions are virtually orthogonal. The spectra are commonly interpreted in terms of Amide A band (≈ 3250 cm^{-1}) corresponding to the $|00\rangle \rightarrow |10\rangle$ transition and Amide B band (≈ 3100 cm^{-1}) arising from Fermi resonance with the overtone of the Amide II band at ≈ 1550 cm^{-1}.[129,130] These bands can be further split by dynamical correlation (Davydov splitting) or internal nonlinear coupling with other degrees of freedom.[131]

Careri and co-workers[132,133] have studied the acetanilide crystal containing infinite chains of hydrogen-bonded amide groups, with N···O distances of ≈ 2.9 Å,[134] similar to those stabilizing the α-helix structure. The interpretation of the Amide-I band is discussed below (Sec. 4.5.3). The NH stretching profile is composed of an intense band at 3295 cm^{-1} and

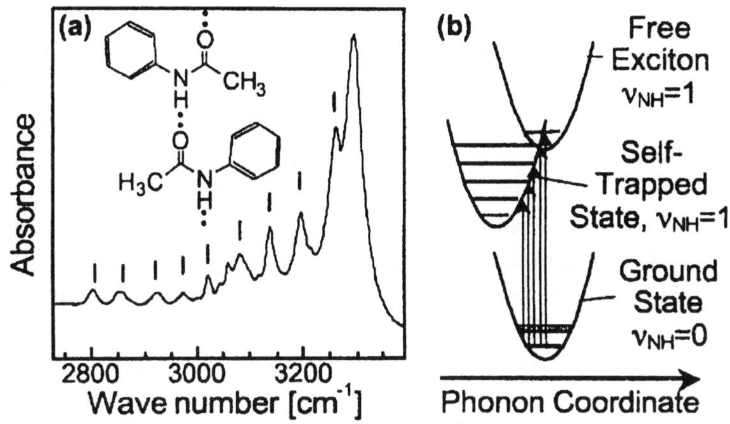

Fig. 23. Reproduced from Ref. 135. (a) absorption spectrum in the NH stretching region of crystalline acetanilide. (b) The proposed scheme of potential functions with the allowed transitions.

terms of Franck-Condon progression, proposed recently on the basis of time-resolved vibrational spectroscopy is quite puzzling.[135] First, the location of the zero-phonon transition at ≈ 2800 cm^{-1} for a N$\cdots$O distance of ≈ 2.9 Å is incompatible with the correlation curve in Fig. 22*a*. Such a large frequency shift, anticipated for N$\cdots$O distances of ≈ 2.7 Å, is quite unlikely for the acetanilide crystal. Second, the very weak intensity of the zero-phonon transition would indicate a displacement of the upper minimum of several tenths of an Å due to a strong coupling like that for OH$\cdots$O systems with O$\ldots$O ≈ 2.6 Å. Then, the sharp lines in acetanilide would contrast markedly to the broad continuum observed for O$\cdots$O systems. Third, the frequency spacing in the upper state, as seen on the spectrum, should be greater than the frequency in the ground state. Time-resolved spectroscopy suggests strong coupling with modes at 48 and 76 cm^{-1}.[135] However, the nature of these modes is not elicited. Such low frequencies are very unlikely for the N$\cdots$O stretching mode. It is then necessary to suppose alternative strong coupling mechanisms with other modes at low frequency that could override the dominant coupling between the ν NH and the $R_{\mathrm{N}\cdots\mathrm{O}}$ distances. Then dynamics of the multidimensional nonlinearly coupled crystal would be far beyond the simple model in one dimension. Fourth, at room temperature, excited states of the modes at 48 and 76 cm^{-1} should be largely populated and transitions $|0N_0\rangle \rightarrow |1N_1\rangle$ should contribute to the ν NH band profile. The satellite bands should show pronounced temperature effects. By analogy with the zero-phonon bands for OH$\cdots$O systems, they should be barely visible at room temperature. Unfortunately, there is no information on the temperature of the sample in Ref. 135.

Furthermore, the adiabatic scheme represented in Fig. 23b is confusing. The authors distinguish two adiabatic potentials crossing each other although they correspond to the same NH stretching quantum number, $n = 1$. This is dramatically erroneous. If the ν NH splitting were due to intra or intermolecular coupling, it should apply to the whole adiabatic potential and give rise to a splitting of all satellites.[128] Then, crossing should be avoided. Alternatively, if the splitting were arising from different NH stretching states (this is very unlikely regarding the anharmonicity of this mode) they should not be labelled with the same quantum number.

The concept of "exciton" in the context of vibrational spectroscopy is confusing as it refers to localized excitations analogous to electron-hole pairs. However, a "free" exciton is merely a plane wave, but, in contrast to phonons, it should have an effective mass m^*. According to Eq. (4) and with $\mathbf{k_i} \approx 10^{-4}$ Å from Table 1, the free recoiling effective mass that can be probed in the infrared at $\approx 3000\ \text{cm}^{-1}$ is $m^* \sim 10^{-10}$ uma. This value is physically meaningless because the recoil of free massive particles (Compton effect) cannot be measured with photons in the infrared or visible range. Consequently, it is impossible to prove that the observed ν NH bands are not phonons. To the best of our knowledge, there is no evidence from other sources that they could correspond to free massive pseudo particles.

The bad news is that there is no really convincing explanation for the detailed structure of the ν NH band profile of acetanilide. A conventional, but quite vague, interpretation of the NH stretching profile would be to consider that in such a complex molecular system, with 8 molecular entities in the unit cell,[134] there is a manifold of overtones and combinations which can account for the weak bands observed on the low frequency side of the NH stretching band at $\approx 3300\ \text{cm}^{-1}$ ($|00\rangle \rightarrow |10\rangle$). It can be easily suspected that as frequencies are closer to the NH band, anharmonic couplings increase intensities of overtones or combinations at the expense of the fundamental transition. Indeed, this may also account for different anharmonicity and lifetimes for fundamental and satellite bands.[135] However, a more detailed analysis of the whole spectra, including those of partially deuterated analogues, would be necessary to complete the assignment scheme. This is a very complicated problem with many degrees of freedom and parameters. Convincing models are very rare in this field. Nevertheless, as already emphasized in the previous discussion of OH···O systems, the weak bands on the low frequency side of the NH stretching mode are not distinctive of energy localization.

4.5.3. *Davydov's model*

One of the problems of bioenergetics, as formulated by some physicists, is the mechanisms by which energy, from light or chemical process, is transferred to active sites in large protein molecules. Solitons has been proposed as possible vehicles between the environment regarded as a thermal bath and the active site in enzymes and other proteins.[5,132,133,136–144] Because the excitation energy of an Amide-I bond is slightly less than one-half of the energy released by hydrolysis of one molecule of adenosine triphosphate (ATP), Davydov has proposed a model in which two quanta of Amide-I excitation energy are stabilized by phonons in a combined excitation which propagates as a solitary wave along a α-helix, thanks to the nonlinearity of the infinite chains of hydrogen bonds linking peptide units.[5,136,140] This speculated mechanism has stimulated many research works, although the even more fundamental question has received very little attention: *How is it possible to transfer the bonding energy of an ATP molecule to Amide-I bands?* Apparently, there is no straightforward answer.

The Hamiltonian proposed by Davydov can be written as:

$$H = H_{iv} + H_{ph} + H_{int} \tag{75}$$

The first term, H_{iv}, represents an infinite chain of coupled oscillators, namely the Amide-I modes coupled along the chain by dipolar interaction, H_{ph} represents the low frequency phonon mode and H_{int} couples the two subsystems. In addition to plane-wave solutions for an infinite chain of coupled oscillators, there are localized solutions, namely Davydov's solitons. The bad news is that the Davydov soliton cannot be photoinduced since the time taken for a cooperative distortion of the lattice in the formation of the soliton is much longer than the absorption time of the photon (Franck-Condon principle).[5,136,140] As an alternative model, it has been proposed that in the strong coupling regime, the excitation become localized to the region around a single site (self-trapping).[138,139] However, it is difficult to understand why the rearrangement of the lattice could be allowed by the Franck-Condon principle for self-trapping and not for Davydov's solitons.

Careri and co-workers[132,133] have been seeking nonlinear excitations in the acetanilide crystal. At room temperature, the Amide-I band is observed in the infrared at $\approx$ 1665 cm^{-1}. Upon cooling, a new band appears at $\approx$ 1550 cm^{-1} that is attributed also to the Amide-I vibration on the basis of isotope substitution. It was concluded that this new band could be due to a soliton-like excitation anticipated from Davydov's model. Alternative interpretations have been proposed, such as vibron solitons,[145] or vibronic analog of a polaron due to a coupling with phonons at low frequency.[138,139] However, neutron scattering experiments did not confirm the existence of Davydov-like soliton, or topological solitons, or proton tunneling.[143]

Time-resolved experiments have revealed that the band at 1665 cm^{-1} is quite harmonic, whilst the band at 1650 cm^{-1} shows large anharmonicity.[146] As strange as it may seam, it has been stated that anharmonicity is equivalent to localization (namely, self-trapping arising from nonlinear coupling of the Amid-I band with phonons) whereas harmonicity is equivalent to delocalization. This is clearly a fallacy.

The Amide-I band splitting cannot arise from nonlinear coupling of the Amide-I band with N$\cdots$O distances, analogous to the strong coupling of the ν NH mode. First, the coupling of $\approx$ 312 cm^{-1}Å^{-1},[133] is too weak to give significant effects on the Amide-I band. Second, splitting of the Amide-I band should be representative of the ν N$\cdots$O frequency. However, this mode is certainly at much higher frequency than 15 cm^{-1}. Alternatively, strong coupling to a phonon at $\approx$ 15 cm^{-1} is quite unlikely and, to the best of our knowledge, unprecedented. In addition, the band at 1650 cm^{-1} disappears for the methyl deuterated analogue, whereas there is no visible coupling between methyl rotation and Amide-I.[143] This effect is totally unexpected within Davydov's model.

Most likely, several mechanisms must be considered. Many amides and peptides show complicated temperature sensitive structures for the Amide-I band. In polypeptides and proteins, this band is quite sensitive to secondary structures. Dynamical correlation for equivalent molecules in the crystal unit cell, changes of the N$\cdots$O distances with temperature, interaction with the Amide-II mode or overtones and combination bands, *etc.* All these possible mechanisms can account for the different anharmonicities of the two components of the Amide-I band.[135] Here again, there is no satisfactory explanation.

Recent time-resolved spectroscopic measurements have probe the lifetime of vibrational excitations in the Amide-I region of proteins.[147,148] Significantly different lifetimes of $\approx$ 30 and 5 ps have been measured for different components. The long lifetime is regarded as due to the self-trapping states. However, it is certainly difficult to reach any positive conclusion. In addition, the sound velocity along the chains of hydrogen bonds in polypeptides is $\sim 10^3$ ms^{-1}. The propagation of a nonlinear excitation during 30 ps at a much smaller velocity is $\ll 30$ Å . It is doubtful that such a small distance could ever account for efficient energy transport along protein chains.

In conclusion, the relevance of Davydov's model to acetanilide and proteins is unproven.[6] It is quite dubious that solitons or self-trapping states, would they ever exist, could survive at room temperature in biological environment.

5. Conclusion

5.1. *Vibrational spectroscopy and nonlinear dynamics*

Vibrational spectroscopy techniques can probe nonlinear dynamics in a great variety of systems. For isolated molecules nonlinearity appears primarily as deviations from the harmonic approximation. Normal and local modes are complementary representations of molecular vibrations. In principle, but there are exceptions due to chemical complexity, the formers are relevant at low energy (small quantum numbers), for small displacements around the equilibrium position. The original idea that local modes account for highly excited vibrational states, close to the dissociation threshold of single bonds, is lessened for complex molecules by the high probability of coupling with the large density-of-states. In any case, the whatever preferred representation must account for the molecular symmetry arising from indistinguishable nuclei.

In crystals, nonlinearity gives rise to distinctive effects: bound phonon-states for phonon-phonon interaction, resonance Raman for phonon-electron coupling, soliton dynamics for coupled quantum-rotors and zero-phonon excited states for hydrogen bonding.

5.2. *Optical vibrational spectroscopy and energy localization*

The idea that vibrational spectroscopy with photons could probe energy localization arising from nonlinearity, in systems free of defects or impurities, is a source of ambiguities and errors. This is in conflict with the uncertainty principle. The properties of the incident radiation probing the sample (energy, wavelength, coherence length, time-scale, *etc*) are such that vibrational spectroscopy probes extended eigenstates, except in chaotic regimes, a usual manifestation of strong coupling. In quantum mechanics, the classical concept of "particle-trajectory" supposing simultaneous knowledge of position and momentum does not hold. Equivalent atoms in molecules (for example carbon atoms and protons in benzene or protons in stannane) and equivalent sites in crystals (for example CO bonds in acetanilide) are indistinguishable. They cannot be excited or probed individually.

5.2.1. *Molecules*

(1) The main source of localization is the chemical complexity.
(2) Energy localization is not a consequence of the local mode representation. This is graphically confirmed by calculation of the density probabilities in highly excited states.
(3) The symmetry elements of the eigenstates are imposed by the molecular symmetry, irrespective of the preferred normal or local mode representation. Eigenstate localization may occur only upon a change of the molecular symmetry.
(4) Energy localized, in a single or a small number of highly excited degrees of freedom, can be represented with a superposition of eigenstates. This is probably the best approximation to energy localization. Such superpositions evolve with time. Quantum beats arising from nearly degenerate states merge into chaotic regime due to coupling with the molecular density-of-states.

5.2.2. *Crystals*

(1) There is no evidence of vibrational energy localization in crystals free of defects, or impurities, or isotope mixtures, at least in the systems reported in these lecture-notes.
(2) Optical techniques probe states at the center of the Brillouin-zone. They correspond to an "infinity" of unit cells oscillating coherently.
(3) The creation of solitons or self-trapping localized states with photons in the infrared-visible range is forbidden by the Franck-Condon principle.
(4) In the 4-methylpyridine crystal, dynamics of the pre-existing quantum sine-Gordon solitons can be probed with photons.

5.3. ***Inelastic neutron scattering spectroscopy of solitons***

(1) The INS technique probes soliton dynamics in the quantum regime, simultaneously in energy and momentum.

(2) Localized dispersionless waveforms in the classical regime turn into plane waves.

(3) Free translational motion along the chain turns into stationary states due to diffraction. The hope that solitons could be efficient vehicles to transport energy is not realized in the quantum regime.

(4) The continuum of classical momentum for free solitons turns into singularities of the quantum momentum. Localization in momentum space is the most distinctive feature of soliton dynamics in the quantum regime.

5.4. *Vibrational spectroscopy and dynamical models*

(1) Dynamical models are ideal systems with a restricted number of degrees of freedom, usually single molecules or one-dimensional lattices. Real systems are more complex and it is rarely possible to isolate a particular sub-system for the sake of confronting theory and experiments. This is quite clear for hydrogen-bonded systems. A counter example is the remarkable adequacy of the sine-Gordon equation to infinite chains of coupled rotors. In this case, collective rotation is markedly isolated from the lattice dynamics. The contact between the model and the sample is so close that the crystal itself "performs" the resolution of the Hamiltonian. This crystal-based simulation reveals new quantum dynamics, largely unforeseen in previous theoretical and numerical works. Conversely, theoretical evidences that nonlinear waves and solitons can exist in multidimensional systems are rather tiny and there is virtually no guideline for experimental searches.

(2) Most of the theoretical works and numerical simulations deal with classical mechanics. However, the classical concepts of energy localization and particle trajectories cannot be transposed to the quantal world without cautions, owing to the particle-wave duality. For vibrational spectroscopy studies, there is no definite proof that intrinsic energy localization may correspond to eigenstates.

References

1. A. J. Sievers and S. Takeno, Phys. Rev. Letters **61**, 970 (1988).
2. Y. S. Kivshar and B. A. Malomed, Rev. Modern Phys. **61**, 763 (1989).
3. R. S. MacKay and S. Aubry, Nonlinearity **7**, 1623 (1994).
4. A.Scott, *Nonlinear science; Emergence and dynamics of coherent structures* (Oxford University Press, 1999).
5. A. S. Davydov, Annalen der Physik **43**, 93 (1986).

6. P.-A. Lindgå rd and A. M. Stoneham, J. Phys.:Condens. Matter **15**, V5 (2003).
7. G. P. Tsironis, Chaos **13**, 657 (2003).
8. E. Fermi, J. Pasta, and S. Ulam, Technical Report No. LA-1940, Los Alamos (unpublished).
9. R. Rajaraman, *Solitons and instantons. An introduction to solitons and instantons in quantum field theory* (North-Holland, Amsterdam, 1989).
10. S. Flach and C. R. Willis, Phys. Rep. **195**, 181 (1998).
11. S. Aubry, Physica D **103**, 201 (1997).
12. R. Dashen, B. Hasslacher, and A. Neveu, Phys. Rev. D **11**, 3424 (1975).
13. R. Dashen, B. Hasslacher, and A. Neveu, Phys. Rev. D **10**, 4130 (1974).
14. W. Z. Wang, J. Tinka Gammel, A. R. Bishop, and M. I. Salkola, Phys. Rev. Letters **76**, 3598 (1996).
15. P. W. Atkins, *Quanta: A handbook of concepts*, 2nd ed. (Oxford University Press, Oxford, 1991).
16. R. T. Birge and H. Sponer, Phys. Rev. **28**, 259 (1926).
17. G. Herzberg, *Molecular Spectra and Molecular Strucutre II. Infrared and Raman Spectra of Polyatomic Molecules* (D. Van Nostrand Company, Inc., Princeton, N.J., 1964).
18. E. B. Wilson, J. C. Decius, and P. C. Cross, *Molecular Vibrations The Theory of Infrared and Raman Spectra* (McGraw-Hill Book Company, INC., New-York, 1955).
19. K. Nakamoto, *Infrared spectra of inorganic and coordination compounds*, 2nd ed. (John Wiley & Sons, New York, 1970).
20. D. A. Long, *The Raman effect. A unified treatment of the theory of Raman scattering by molecules.* (John Wiley & Sons, Ltd., New York, 2002).
21. J. R. Ferraro, K. Nakamoto, and C. W. Brown, *Introductory Raman Spectroscopy*, 2nd ed. (Academic Press, Amsterdam, 2003).
22. C. V. Raman and K. S. Krishnan, Nature **121**, 501 (1928).
23. S. W. Lovesey, *Nuclear scattering*, Vol. I of *Theory of neutrons scattered from condensed matter* (Clarendon Press, Oxford, 1984).
24. L. Landau and L. Lifchitz, *Mécanique quantique*, Vol. III of *Physique Théorique* (Editions MIR, Moscou, 1966).
25. C. Cohen-Tannoudji, B. Diu, and F. Laloë, *Mécanique Quantique* (Hermann, Paris, France, 1977).
26. L. Golic and J. C. Speakman, J. Chem. Soc. 1745 (1965).
27. F. Fillaux and J. Tomkinson, Chem. Phys. **158**, 113 (1991).
28. J. L. Dunham, Phys. Rev. **41**, 721 (1932).
29. P. M. Morse, Phys. Rev. **34**, 57 (1929).
30. J. W. Ellis, Phys. Rev. **33**, 27 (1929).
31. R. Mecke, Z. Phys. **101**, 405 (1936).
32. B. R. Henry and W. Siebrand, J. Chem. Phys. **49**, 5369 (1968).
33. T. E. Martin and A. H. Kalantar, J. Chem. Phys. **49**, 235 (1968).
34. R. Wallace, Chem. Phys. **11**, 189 (1975).
35. R. L. Swofford, M. E. Long, and A. C. Albrecht, J. Chem. Phys. **65**, 179 (1976).

36. R. J. Hayward and B. R. Henry, Chem. Phys. **12**, 387 (1976).
37. B. R. Henry, Acc. Chem. Res. **10**, 207 (1977).
38. W. R. A. Greenlay and B. R. Henry, J. Chem. Phys. **69**, 82 (1978).
39. D. F. Heller, Chem. Phys. Letters **61**, 583 (1979).
40. R. G. Bray and M. J. Berry, J. Chem. Phys. **71**, 4909 (1979).
41. M. S. Burberry, J. A. Morrell, A. C. Albrecht, and R. L. Swofford, J. Chem. Phys. **70**, 5522 (1979).
42. H. S. Mø ller and O. S. Mortensen, Chem. Phys. Letters **66**, 539 (1979).
43. P. R. Stannard, M. L. Elert, and W. M. Gelbart, J. Chem. Phys. **74**, 6050 (1981).
44. I. A. Watson, B. R. Henry, and I. G. Ross, Spectrochimica Acta **37A**, 857 (1981).
45. D. F. Heller and S. Mukamel, J. Chem. Phys. **70**, 463 (1979).
46. C. K. N. Patel, A. C. Tam, and R. J. Kerl, J. Chem. Phys. **71**, 1470 (1979).
47. B. R. Henry, M. A. Mohammadi, and J. A. Thomson, J. Chem. Phys. **75**, 3165 (1981).
48. M. S. Child and R. T. Lawton, Faraday Discuss. Chem. Soc. **71**, 273 (1981).
49. M. L. Sage and J. Jortner, Adv. Chem. Phys. **47, Pt. I**, 293 (1981).
50. M. S. Child and L. Halonen, Adv. Chem. Phys. **57**, 1 (1984).
51. L. A. Findsen, H. L. Fang, R. L. Swofford, and R. R. Birge, J. Chem. Phys. **84**, 16 (1986).
52. R. H. Page, Y. R. Shen, and Y. T. Lee, J. Chem. Phys. **88**, 4621 (1988).
53. R. H. Page, Y. R. Shen, and Y. T. Lee, J. Chem. Phys. **88**, 5362 (1988).
54. H. G. Kjaergaard and B. R. Henry, J. Chem. Phys. **96**, 4841 (1992).
55. H. G. Kjaergaard, H. Yu, B. J. Schattka, B. R. Henry, and A. W. Tarr, J. Chem. Phys. **93**, 6239 (1990).
56. H. G. Kjaergaard, B. R. Henry, and A. W. Tarr, J. Chem. Phys. **94**, 5944 (1991).
57. Y. Zhang, S. J. Klippenstein, and R. A. Marcus, J. Chem. Phys. **94**, 7319 (1991).
58. H. G. Kjaergaard, J. D. Goddard, and B. R. Henry, J. Chem. Phys. **95**, 5556 (1991).
59. T. Sako, K. Yamanouchi, and F. Iachello, Chem. Phys. Letters **299**, 35 (1999).
60. T. Sako, K. Yamanouchi, and F. Iachello, J. Chem. Phys. **113**, 6063 (2000).
61. T. Sako, K. Yamanouchi, and F. Iachello, J. Chem. Phys. **113**, 7292 (2000).
62. T. Sako, K. Yamanouchi, and F. Iachello, J. Chem. Phys. **114**, 9441 (2001).
63. F. Iachello, Chem. Phys. Letters **78**, 581 (1981).
64. F. Iachello and R. D. Levine, J. Chem. Phys. **77**, 3076 (1982).
65. O. S. van Roosmalen, F. Iachello, R. D. Levine, and A. E. L. Dieperink, J. Chem. Phys. **79**, 2515 (1983).
66. F. Iachello, A. Levitan, and A. Mengoni, J. Chem. Phys. **95**, 1449 (1991).
67. E. L. Sibert III, W. P. Reinhart, and J. T. Hynes, J. Chem. Phys. **77**, 3583 (1982).
68. E. L. Sibert III, W. P. Reinhart, and J. T. Hynes, J. Chem. Phys. **77**, 3595 (1982).

69. T. Sako and K. Yamanouchi, Chem. Phys. Letters **264**, 403 (1997).
70. R. Wallace, Chem. Phys. **11**, 189 (1975).
71. M. Halonen, L. Halonen, H. Bürger, and S. Sommer, J. Chem. Phys. **93**, 1607 (1990).
72. A. Tabyaoui, B. Lavorel, G. Pierre, and H. Bürger, J. Mol. Spectrosc. **148**, 100 (1991).
73. M. Halonen, L. Halonen, H. Bürger, and W. Jerzembeck, J. Chem. Phys. **108**, 9285 (1998).
74. G. Kopidakis and S. Aubry, Physica B **296**, 237 (2001).
75. A. A. Maradudin, E. W. Montroll, and G. H. Weiss, *Theory of lattice dynamics in the harmonic approximation, Solid state physics advances in research and applications* (Accademic press Inc., New York and London, 1963).
76. O. Madelung, *Introduction to solid-state theory, Springer series in solid-state sciences* (Springer-Verlag, Berlin, 1978).
77. P. Brüesch, *Phonons: Theory and experiments I, Springer series in solid-state sciences* (Springer-Verlag, Berlin, 1982).
78. R. S. MacKay, in *in Nonlinear dynamics and chaos: where do we go from here? Eds J. Hogan, A. Champneys, B. Krauskopf, M. di Bernardo, M. Homer, E. Wilson and H. Osinga* (IOOP, 2002), Chap. Many-body quantum mechanics, pp. 100–134.
79. M. P. Gush, W. F. Hare, E. J. Allin, and J. C. Welsh, PHys. Rev. **106**, 1101 (1957).
80. J. H. Eggert, H. Mao, and R. J. Hemley, Phys. Rev. Letters **70**, 2301 (1993).
81. A. Ron and D. F. Hornig, J. Chem. Phys. **39**, 1129 (1963).
82. M. H. Cohen and J. Ruvalds, Phys. Rev. Letters **23**, 1378 (1968).
83. P. Jakob, Physica D **119**, 109 (1998).
84. P. Jackob and B. N. Persson, J. Chem. Phys. **109**, 8641 (1998).
85. J. C. Kimball, C. Y. Fong, and Y. R. Shen, Phys. Rev. B **23**, 4946 (1981).
86. S. C. Huckett, B. Scott, S. P. Love, R. J. Donohoe, C. J. Burns, E. Garcia, T. Franckcom, and B. I. Swanson, Inorg. Chem. **32**, 2137 (1993).
87. S. P. Love, L. A. Worl, R. J. Donohoe, S. C. Huckett, and B. I. Swanson, Phys. Rev. B **46**, 813 (1992).
88. B. I. Swanson, J. A. Brozik, S. P. Love, G. F. Strouse, A. P. Schreve, A. R. Bishop, W.-Z. Wang, and M. I. Salkola, Phys. Rev. Letters **82**, 3288 (1999).
89. S. P. Love, S. C. Huckett, L. A. Worl, T. M. Franckcom, S. A. Ekberg, and B. I. Swanson, Phys. Rev. B **47**, 11107 (1993).
90. N. K. Voulgarakis, G. Kalosakas, A. R. Bishop, and G. P. Tsironis, Phys. Rev. B **64**, 020301(R) (2001).
91. G. Kalosakas, A. R. Bishop, and A. P. Schreve, Phys. Rev. B **66**, 094303 (2002).
92. F. Fillaux, A. Cousson, and D. Keen, Phys. Rev. B **67**, 054301 (2003).
93. F. Fillaux, J. Tomkinson, and J. Penfold, Chem. Phys. **124**, 425 (1988).
94. A. Novak, P. Saumagne, and L. D. C. Bock, J. Chim. Phys.-Chim. Biol. **60**, 1385 (1963).
95. K. Nakamoto, Y. A. Sarma, and K. Ogoshi, J. Chem. Phys. **43**, 1177 (1965).
96. G. Lucazeau and A. Novak, J. Raman Spectrosc. **1**, 573 (1973).

97. S. Ikeda and F. Fillaux, Phys. Rev. B **59**, 4134 (1999).
98. S. Ikeda, S. Kashida, H. Sugimoto, Y. Yamada, S. M. Bennington, and F. Fillaux, Phys. Rev. B **66**, 184302 (2002).
99. F. Fillaux, J. P. Fontaine, M.-H. Baron, G. J. Kearley, and J. Tomkinson, Chem. Phys. **176**, 249 (1993).
100. E. K. Morris, Ph.D. thesis, Université d'Orsay, 1997.
101. F. Fillaux, B. Nicolaï, W. Paulus, E. Kaiser-Morris, and A. Cousson, submitted to Phys. Rev. B (2003).
102. F. Fillaux and C. J. Carlile, Phys. Rev. B **42**, 5990 (1990).
103. F. Fillaux, C. J. Carlile, and G. J. Kearley, Phys. Rev. B **44**, 12280 (1991).
104. F. Fillaux, C. J. Carlile, J. Cook, A. Heidemann, G. J. Kearley, S. Ikeda, and A. Inaba, Physica B **213&214**, 646 (1995).
105. F. Fillaux, C. J. Carlile, and G. J. Kearley, Phys. Rev. B **58**, 11416 (1998).
106. U. Ohms, H. Guth, W. Treutmann, H. Dannohl, A. Schweig, and G. Heger, J. Chem. Phys. **83**, 273 (1985).
107. C. J. Carlile, B. T. M. Willis, R. M. Ibberson, and F. Fillaux, Z. Kristallogr. **193**, 243 (1990).
108. B. Nicolaï, G. J. Kearley, A. Cousson, W. Paulus, F. Fillaux, F. Gentner, L. Schröder, and D. Watkin, Acta Cryst. Section B Structural Science **57**, 36 (2001).
109. B. Nicolaï, A. Cousson, and F. Fillaux, Chem. Phys. **290**, 101 (2003).
110. P. J. Caudrey, J. D. Gibbon, J. C. Eilbeck, and R. K. Bullough, Phys. Rev. Letters **30**, 237 (1973).
111. M. J. Ablowitz, D. J. Kaup, A. C. Newell, and H. Segur, Phys. Rev. Letters **30**, 1262 (1973).
112. N. LeCalvé, B. Pasquier, G. Braathen, L. Soulard, and F. Fillaux, J. Phys. C: Solid State Phys. **19**, 6695 (1986).
113. L. Pauling, *The nature of the chemical bond* (Cornell University Press, Ithaca, New York, 1960).
114. P.Schuster, G. Zundel, and C. Sandorfy, *The hydrogen bond. Recent developments in theory and experiments* (North-Holland, Amsterdam, 1976), Vol. I, II and III.
115. S. Scheiner, *Hydrogen Bonding: A Theoretical Perspective* (Oxford University Press, Oxford, 1997).
116. C. L. Perrin and J. B. Nielson, Annu. Rev. Phys. Chem. **48**, 511 (1997).
117. A. Novak, Structure and bonding (Berlin) **18**, 177 (1974).
118. A. Lautié, F. Froment, and A. Novak, Spectroscopy Letters **9**, 289 (1976).
119. B. I. Stepanov, Nature 157 (1946).
120. Y. Maréchal and A. Witkowski, J. Chem. Phys. **48**, 3697 (1968).
121. S. F. Fischer, G. L. Hofacker, and M. A. Ratner, J. Chem. Phys. **52**, 1934 (1970).
122. G. L. Hofacker, Y. Maréchal, and M. A. Ratner, in *Dynamical properties of hydrogen bonded systems*, Vol. 1 of *The hydrogen bond Recent developments in theory and experiments* (North-Holland, Amsterdam, 1976), Chap. 6, pp. 295–357.
123. F. Fillaux, Chem. Phys. **74**, 395 (1983).

124. F. Fillaux, Chem. Phys. **74**, 405 (1983).
125. F. Fillaux, B. Marchon, and A. Novak, Chem. Phys. **86**, 127 (1984).
126. F. Fillaux, B. Marchon, A. Novak, and J. Tomkinson, Chem. Phys. **130**, 257 (1989).
127. O. Henri-Rousseau and P. Blaise, *S. Rice, I. Prigogine (Eds.), Advances in Chem. Phys.* (Wiley, New York, 1998), Vol. 103, pp. 1–186.
128. F. Fillaux, M. H. Limage, and F. Romain, Chem. Phys. **276**, 181 (2002).
129. T. Miyazawa, S. Shimanoushi, and S. Mizushima, J. Chem. Phys. **29**, 611 (1958).
130. T. Miyazawa, in *Poly-α-aminoacids, G. Fasman (Ed.)* (Dekker, New York, 1967), p. 69.
131. F. Fillaux, Chem. Phys. **62**, 287 (1981).
132. G. Careri, U. Buontempo, F. Carta, E. Gratton, and A. C. Scott, Phys. Rev. Letters **51**, 304 (1983).
133. G. Careri, U. Buontempo, F. Galluzzi, A. C. Scott, E. Gratton, and E. Shyamsunder, Phys. Rev. B **30**, 4689 (1984).
134. S. W. Johnson, J. Eckert, R. K. McMullan, and M. Muller, J. Phys. Chem. **99**, 16253 (1995).
135. J. Edler, P. Hamm, and A. C. Scott, Phys. Rev. Letters **88**, 067403 (2002).
136. A. S. Davydov, J. Theor. Biol. **38**, 559 (1973).
137. A. C. Scott, Phys. Rev. A **26**, 578 (1982).
138. D. M. Alexander, Phys. Rev. Letters **54**, 138 (1985).
139. D. M. Alexander and J. A. Krumhansl, Phys. Rev. B **33**, 7172 (1986).
140. A. S. Davydov, Phys. Stat. Sol. (B) **138**, 559 (1986).
141. S. Ichinose, Phys. Stat. Sol. (B) **138**, 73 (1986).
142. A. F. Lawrence, J. C. McDaniel, D. B. Chang, B. M. Pierce, and R. R. Birge, Phys. Rev. A **33**, 1188 (1986).
143. M. Barthes, R. Almairac, J.-L. Sauvajol, J. Moret, R. Curat, and J. Dianoux, Phys. Rev. B **43**, 5223 (1991).
144. Y. Zolotaryuk and J. C. Eilbeck, Phys. Rev. B **63**, 054302 (2001).
145. S. Takeno, Prog. Theor. Phys. **75**, 1 (1986).
146. J. Edler and P. Hamm, J. Chem. Phys. **117**, 2415 (2002).
147. W. Fann, L. Rothberg, M. Roberson, S. Benson, J. Madey, S. Etemad, and R. Austin, Phys. Rev. Letters **64**, 607 (1990).
148. R. H. Austin, A. Xie, L. vander Meer, M. Shinn, and G. Neil, J. Phys.: Condens. Matter **15**, S1693 (2003).

CHAPTER 3

SLOW MANIFOLDS

R.S.MacKay
Mathematics Institute, University of Warwick,
Coventry CV4 7AL, U.K.
E-mail: mackay@maths.warwick.ac.uk

A key mathematical concept for the understanding of the dynamics of coherent structures in typical (as opposed to integrable) spatially extended systems is that of a slow manifold. The theory of slow manifolds is reviewed and extended here, with special emphasis on the Hamiltonian context. For a broad range of dynamical systems with a small separation of timescales parameter ε, given any integer $r \geq 1$, existence of a slow manifold and slow vector field εX on it, containing all nearby equilibria and with error field of order $(\varepsilon|X|)^r$, is proved. In the Hamiltonian case, Hamiltonian slow dynamics is constructed and the theory is extended to slow manifolds with an internal oscillation. Applications are given to a variety of problems, including the interactions of localised excitations in several types of spatially extended system. In particular, the theory helps one to understand the interaction and propagation of discrete breathers.

1. Introduction

Definition 1: A *slow manifold* for a smooth (i.e. C^k for some $k \geq 1$) vector field V on a manifold P is a smooth submanifold M together with a tangential vector field U on it (called the *slow vector field*), such that

(1) U is close to V on M,
(2) in local coordinates $(x, y) \in M_j \times E$, where $\bigcup_j M_j$ is an open cover of M and E is a Banach space (complete normed linear space) representing displacements from M, and writing $V = (u, v)$ in these coordinates, then $\| (\partial v/\partial y)^{-1} \|$ is small compared to significant timescales for the system $\dot{x} = U(x)$ on M (say the time for $V - U$ to change by an

order-one factor along the orbits of U).

If $U = V$ on M then M is an *invariant* slow manifold (some authors add this to the definition of slow manifold but it will be seen to be too restrictive).

Sometimes I am sloppy in the use of the term "slow manifold" and mean just M rather than the pair (M, U). The ambient space P may be infinite-dimensional, in which case M may be of infinite dimension or codimension.

For example, systems of the form

$$\begin{aligned} \dot{x} &= \varepsilon X(x, y) \\ \dot{y} &= Y(x, y) \end{aligned} \tag{1}$$

for ε small have a slow manifold $M = \{(x, y) : Y(x, y) = 0, \|(\partial Y/\partial y)^{-1}\| \ll 1/\varepsilon\}$, which is locally a graph $y = \eta(x)$ (by the implicit function theorem), carrying slow dynamics $\dot{x} = \varepsilon X(x, \eta(x))$.

The main advantage of finding a slow manifold is *dimension reduction*: all the fast variables are eliminated by "slaving" to the slow ones, giving a differential-algebraic system of lower order. Another advantage is improved initialisation from imperfect data: one can push the initial conditions onto the slow manifold to reduce unwanted rapid oscillations in numerical integrations. The utility of a slow manifold, however, depends on its accuracy, so it can be important to find accurate ones.

Much has been written about slow manifolds. They are fundamental to relaxation oscillators, slaving effects, the realisation of holonomic constraints, and interactions of coherent structures. The main mathematical issues are:

(1) given a candidate slow manifold M, to determine a suitable slow vector field U on it (some projection of V),
(2) given one slow manifold M, to find a better one $\tilde{M}$, in the sense that condition (1) of the definition is satisfied more strongly,
(3) to decide whether or not an invariant slow manifold exists, i.e. one for which V is everywhere on M tangent to M, and so U can be taken to be simply the restriction of V to M.
(4) to decide for how long true orbits remain close to those of the approximate dynamics on the slow manifold.

I will work with systems like Eq. (1) where the fast motion happens on timescales of order 1 and the slow timescale is long ($1/\varepsilon$). Many systems of different forms, however, can be treated similarly, or put into this form by change of variables, e.g. the van der Pol oscillator in the relaxation regime is usually written as

$$\begin{aligned} y_1' &= -x_1 \\ x_1' &= y_1 - \lambda(x_1^3/3 - x_1), \qquad \lambda \text{ large,} \end{aligned} \tag{2}$$

but can be converted to

$$\begin{aligned} \dot{x} &= -\varepsilon y \\ \dot{y} &= x - y^3/3 + y \end{aligned} \tag{3}$$

by $\varepsilon = 1/\lambda^2$, $x = y_1/\lambda$, $y = x_1$ and the time change $\dot{z} = z'/\lambda$ for general functions z.

Although much of what follows is well known to experts, I have not found a comprehensive account in the literature which goes as far as I have; in fact, I believe some of the results here are new. Some precursors are Refs 1-5 (also Ref. 6 is a significant precursor for the sections on Hamiltonian slow manifolds with internal oscillation). In particular, it turns out that I have followed similar lines to van Kampen, but with significant extensions. For some relatively recent lists of references to slow manifolds, see Refs 7 and 8. Additional topics that I would like to cover one day are slow manifold approaches to the Born-Oppenheimer approximation in quantum dynamics of molecules (for excellent recent work on this, see Ref. 9) and to the derivation of stochastic equations of motion for slow degrees of freedom as a result of their coupling to chaotic fast ones (on which there is a large literature, e.g. Ref. 10, but for which I'm not aware of a good treatment in the Hamiltonian case).

Here is an outline of the contents. In Sections 2 and 3, I'll highlight the special features of the normally hyperbolic and Hamiltonian cases, respectively. In Section 4 I'll introduce an iterative scheme to produce r^{th} order slow manifolds from a 0th order one. I'll adapt the procedure in Section 5 to produce Hamiltonian ones for Hamiltonian systems. In Sections 6 and 7 I compare my method with some precursors. Section 8 is a short aside on an extension from Hamiltonian to Poisson systems. Section 9 introduces the concept of slow manifold with internal oscillation, which is of crucial importance for a variety of applications. Theory and examples are given for $U(1)$-symmetric Hamiltonian systems in Section 10 and for general Hamiltonian systems in Section 11. Section 12 gives methods to obtain bounds on the time evolution of initial conditions on or near a slow manifold. These lecture-notes close in Section 13 with some comments on the effects of weak damping on Hamiltonian cases.

2. Normally Hyperbolic versus General Case

The theory of slow manifolds is in very good shape if there is a *normally hyperbolic* slow manifold, i.e. the spectrum of $(\partial v/\partial y)$ at all its points avoids a neighbourhood of the imaginary axis. This is the case in typical slaving problems. A textbook example is Michaelis-Menten enzyme kinetics, e.g. Ref. 11:

$$\begin{aligned}\dot{c} &= s - (s+K)c \\ \dot{s} &= \varepsilon(-s + (s+K-U)c),\end{aligned} \tag{4}$$

for which c collapses rapidly towards the slow manifold $c = s/(s+K)$, which then gives the (non-mass action) rate law $\dot{s} \approx -\varepsilon Us/(s+K)$. Another is the Fitzhugh-Nagumo model of a neuron, e.g. Ref. 11:

$$\begin{aligned}\dot{v} &= f(v) - w - w_0 \\ \dot{w} &= \varepsilon(v - \gamma w - v_0),\end{aligned} \tag{5}$$

with $f(v) = Av(v-\alpha)(1-v)$, for which v relaxes rapidly onto the attracting parts of the curve $w = f(v) - w_0$, followed by slow evolution along this curve until near a turning point.

For such problems one can do much better than just approximations. The theory of normally hyperbolic invariant manifolds[12] applies, which proves that there is a locally invariant slow manifold nearby, together with many other useful results (smoothness, persistence, forwards and backwards contracting foliations, local linearisability in the normal direction, and local maximality and uniqueness in case of full invariance; furthermore there are strong results on the approach and departure from normally hyperbolic invariant slow manifolds[13]). So exact dimension reduction is possible.

In principle, this exact dimension reduction could be used in numerics. It would be extremely useful, for example, in biochemical reaction networks, neurophysiological circuits and chemically reacting flows, where there are typically hundreds of variables but a large fraction are fast and could be eliminated. Indeed Ref. 14 discusses how to use the existence of normally hyperbolic invariant slow manifolds in numerical computations, but I think more could be done. Ref. 15 computes "intrinsic low-dimensional manifolds" (ILDM), which are "second order" slow manifolds in the terminology I will introduce, but in general not invariant[c]. There is good computational work for special cases of normally hyperbolic manifolds, e.g. local

[c]Although Deuflhardt[16] appreciates this fact, Maas's paper does not mention it, so here is an example: applied to a system of the form $\dot{x} = \varepsilon x, \dot{y} = -y + \delta(x)$, his method

unstable[17] and centre manifolds[18] of equilibria and periodic orbits, and inertial manifolds.[19,20,21] I am aware of only one paper on a computational approach to general normally hyperbolic manifolds,[22] though even this treats only discrete time systems and the case of trivialisable backwards and forwards contracting normal bundles.

In contrast, the situation is much more delicate for slow manifolds which are not normally hyperbolic. This is the case in many classical mechanical problems. For example, a pendulum on a stiff spring behaves roughly like a rigid pendulum but with possible fast oscillations in the length superposed. Similarly, the classical motion of a hydrocarbon can be regarded as slow motion of the carbon skeleton with fast oscillation of CH stretches and bends superposed. Other examples are: compressible fluid dynamics separates at low Mach number into incompressible fluid dynamics and acoustic waves (e.g. Ref. 23), water waves separate into gravity waves with surface tension ripples (e.g. Ref. 24), and many nonlinear classical field theories possess propagating coherent structures (such as kinks and monopoles) surrounded by small radiation (e.g. Refs. 25-28). Mid-latitude atmospheric dynamics provides an example of a system with three ranges of timescale: geostrophic motion on a 6-hour scale, inertia-gravity waves on a 5-minute scale, and acoustic waves on scales of a few seconds or less (e.g. Ref. 29).

For a general system with a slow manifold which is not normally hyperbolic, it is unlikely that there is an invariant slow manifold nearby, because typical smooth perturbations of a system with an invariant manifold which is not normally hyperbolic are believed to destroy it[d].

Example 2: A fragile invariant slow manifold

A simple illustration of the fragility of invariant slow manifolds is furnished by the system

$$\begin{aligned}\dot{\theta} &= \varepsilon\omega(I) \\ \dot{I} &= 0 \\ \dot{\xi} &= \eta + \delta g(\theta) \\ \dot{\eta} &= -\xi,\end{aligned} \tag{6}$$

yields ILDM $y(x) = \delta(x) - \varepsilon x\delta'(x)/(1+\varepsilon)$, on which the deviation of $\dot{y}$ from $y'(x)\dot{x}$ is $\varepsilon x^2\delta''(x)/(1+\varepsilon)$.

[d] The only precise result of this form that I know, is somewhat weaker: if for a C^1 compact invariant submanifold M for a C^1 diffeomorphism f there is a neighbourhood U of M and a neighbourhood V of f such that for all $g \in V$ the maximal invariant set for g in U is a C^1 submanifold M_g and M_g is C^1-close to M, then M is normally hyperbolic.[30]

where $\theta \in S^1 = \mathbb{R}/2\pi\mathbb{Z}$, $I \in [1,2]$, $\omega(I) = I$ (variants are possible, hence the more general notation), $\delta \in \mathbb{R}$, and $g(\theta) = \sum_{m\in\mathbb{Z}} g_m e^{im\theta}$ is a function analytic in a strip $|\Im\theta| \le \kappa$ with all Fourier coefficients g_m nonzero, e.g.

$$g(\theta) = \frac{1}{2-\cos\theta}. \tag{7}$$

The analyticity implies that $|g_m| \le Ce^{-\kappa|m|}$ with $C = \sup\{|g(\theta)| : |\Im\theta| \le \kappa\}$. The case $\delta = 0$ has an invariant slow manifold $M = \{(\theta, I, 0, 0) : \theta \in S^1, I \in [1,2]\}$. Let us seek an invariant slow manifold for $\delta \neq 0$ as a Fourier series $\xi(\theta, I) = \sum_{m\in\mathbb{Z}} \xi_m e^{im\theta}$, $\eta(\theta, I) = \sum_{m\in\mathbb{Z}} \eta_m e^{im\theta}$. The equations for each order m of Fourier coefficients separate and there is a unique solution

$$\begin{aligned}\eta_m &= \frac{\delta g_m}{m^2\varepsilon^2\omega(I)^2 - 1}, \\ \xi_m &= -im\varepsilon\omega(I)\eta_m,\end{aligned} \tag{8}$$

if the denominator is nonzero. If the denominator has a zero, however, then there is no solution. This happens where $\omega(I) = \pm 1/m\varepsilon$. Now $\omega(I)$ goes from 1 to 2, and m ranges through the whole of $\mathbb{Z}$. Thus for all $\varepsilon \le 1$, the invariant slow manifold goes to infinity at a set of $N \sim 1/\varepsilon$ values of I and there is no solution valid uniformly on $I \in [1,2]$.

Nevertheless, for this example there exists a very nearly invariant slow manifold uniform in I, e.g. for any $\alpha \in (0, 1/\omega_{\max})$, where $\omega_{\max} = 2$, define $M(\alpha)$ by

$$\eta_m = \frac{\delta g_m}{m^2\varepsilon^2\omega(I)^2 - 1} \text{ for } |m| < \frac{\alpha}{\varepsilon}, 0 \text{ otherwise}, \tag{9}$$

$$\xi_m = -im\varepsilon\omega(I)\eta_m. \tag{10}$$

The normal component of the vector field on $M(\alpha)$ is of order $\delta e^{-\kappa\alpha/\varepsilon}$, i.e. exponentially small in ε, and $M(\alpha)$ is within order δ of the unperturbed one (in any C^k).

Thus sensible goals in the general case are, given one slow manifold, (i) to try to construct a better one, i.e. one for which condition (1) of the definition holds more strongly, and (ii) to determine how small any deviations from invariance can be made. These issues will be addressed in Section 4. There are also related goals for numerical computation, which I will not address here, but some references are Refs 31-33, 8.

3. Hamiltonian versus General Case

Many potential applications of slow manifolds are Hamiltonian systems, especially most of those which are not normally hyperbolic. Recall that a *Hamiltonian system* is a vector field V on a manifold P defined by $\Omega(V,\xi) = dH(\xi)\ \forall\xi \in TP$, for some function $H : P \to \mathbb{R}$ (called the "Hamiltonian") and "symplectic form" Ω on P. A *symplectic form* is a closed ($\int_{S^2}\Omega = 0$ for all contractible 2-spheres in P), non-degenerate ($\Omega(\xi,\eta) = 0\ \forall\xi \implies \eta = 0$), antisymmetric ($\Omega(\eta,\xi) = -\Omega(\xi,\eta)$) bilinear (linear in each argument) form on the tangent space TP (Ω takes two tangent vectors at a point of P and returns a real number). The "canonical" example is $\Omega = \sum_{j=1}^{d} dq_j \wedge dp_j$ on $\mathbb{R}^d \times \mathbb{R}^d$, which gives Hamilton's equations $\dot{q}_j = \frac{\partial H}{\partial p_j}, \dot{p}_j = -\frac{\partial H}{\partial q_j}$. Ref. 34 is a good text on the differential topology formulation of Hamiltonian dynamics.

Hamiltonian systems have special properties, mostly associated with preservation of the symplectic form. So in the Hamiltonian case, one would hope to construct slow manifolds on which the slow dynamics is itself Hamiltonian, and we will see in Section 5 that this is indeed often possible.

It should be remarked that if one restricts to the Hamiltonian context, there are invariant manifolds which survive all small Hamiltonian perturbations but are not normally hyperbolic. The obvious examples are the regular energy levels, but even within an energy level there can be persistent invariant manifolds, e.g. KAM tori (*KAM tori* are invariant tori on which the motion is smoothly conjugate to a constant vector field with incommensurate direction, e.g. Refs 34, 35) and the stable and unstable manifolds of KAM tori of dimensions strictly between 1 and the number d of degrees of freedom. All the examples I know, however, are non-symplectic whereas I believe the interesting case for slow manifolds (in Hamiltonian systems) is *symplectic submanifolds*, i.e. on which the symplectic form is non-degenerate, because that condition is necessary to define a natural Hamiltonian dynamics on them. It would be a good project to try to prove a generic non-persistence result for symplectic submanifolds of Hamiltonian systems (but without requiring the local maximality of Ref. 30). A result in this direction was obtained by Fermi (see Ref. 36 for an appraisal). In the context of infinite-dimensional systems exhibiting coherent structures and radiation, Ref. 29 suggests that such a result should follow from the fact that typical motions of coherent structures generate radiation; an initial context in which to address this might be kinks moving in a lattice, e.g. Ref. 37. Perhaps the variational principle of Ref. 38 could be useful to make a general proof.

As a start, it would be interesting to study the following Hamiltonian version of Example 2, for which I give the beginnings of an analysis.

Example 3: A fragile Hamiltonian invariant slow manifold

Let $H(I,\theta,\xi,\eta) = \frac{\varepsilon}{2}I^2 + \frac{1}{2}(\xi^2+\eta^2) + \delta g(\theta)\eta$ and $\Omega = d\theta \wedge dI + d\xi \wedge d\eta$, where $\theta \in S^1$, $I \in [1,2]$, $\xi,\eta \in \mathbb{R}$, $\delta \in \mathbb{R}$ and g is an analytic function in a strip $|\Im\theta| \le \kappa$. For $\delta = 0$ there is an invariant slow manifold $\xi = \eta = 0$. It consists of periodic orbits given by $I =$ constant, of period $2\pi/\varepsilon I$. Their Floquet multipliers are $\exp \pm 2\pi \mathrm{i}/\varepsilon I$. Given $\beta > 0$ there is a $\delta_0(\varepsilon,\beta) > 0$ such that the periodic orbits with Floquet multipliers at least β from $+1$ persist for $|\delta| \le \delta_0$ (moving by at most order $|\delta|/\beta\varepsilon$), so the parts of the invariant slow manifold where $|I - \frac{1}{m\varepsilon}| \le \frac{\beta}{m^2\varepsilon}$ survive such perturbation. These form $N \sim 1/\varepsilon$ invariant annuli (for $\beta < 1/2$). Generic Hamiltonian perturbation, however, destroys periodic orbits with a Floquet multiplier $+1$, so we can expect the family of periodic orbits to be punctured around $I = 1/m\varepsilon$ for each $m \in \mathbb{Z}$. When $\delta = 0$, the set of periodic orbits near $I = 1/m\varepsilon, \xi = \eta = 0$, of period near $2\pi/\varepsilon I$, is (modulo rotation in θ) the union of the line $\xi = \eta = 0$ and the plane $I = 1/m\varepsilon$. The perturbation δ can be expected to turn this into a broken pitchfork, as shown in Figure 1(a-b). Taking all integers m between $1/2\varepsilon$ and $1/\varepsilon$ yields a "ladder" of broken pitchforks as in Figure 1(c). The same phenomenon occurs in classical models of stretches of water molecules (as was observed by Mark Child and explained by me in similar terms in 1993, though never published; see Ref. 39, however, for related numerics).

One might ask whether there could still be an invariant slow manifold C^1-close to the unperturbed one, without the gaps, but the answer is no, because it would be symplectic, two-dimensional and without equilibria, so the motion on it would be Hamiltonian of one degree of freedom and consist entirely of periodic orbits of nearby period. One can connect the invariant annuli, however, into an almost invariant slow manifold: any surface C^1-close to $\xi = \eta = 0$ and connecting the invariant annuli will do, such as the dotted curve in Figure 1(c).

To examine whether the specific perturbation by δ above really punctures the family of periodic orbits, and how close to invariant one can make a slow manifold, one would have to do a more detailed calculation. Here is a sketch.

Near $I = 1/m\varepsilon, \xi = \eta = 0$, use coordinates (x,y), defined by $x + iy = (\xi + i\eta)e^{im\theta}$, and $K = I - 1/m\varepsilon + \frac{m}{2}r^2$, where $r^2 = x^2 + y^2$. Then $H = \frac{1}{2m^2\varepsilon} + 2K/m + \frac{\varepsilon}{2}(K - \frac{m}{2}r^2)^2 + \delta g(\theta)\Im(e^{-im\theta}(x+iy))$ and $\Omega = d\theta \wedge dK + dx \wedge dy$. This is in a suitable form to apply the method of averaging (e.g. Ref. 40), i.e. a near-identity coordinate change to reduce the θ-dependence of H while keeping the same expression for Ω. Using the

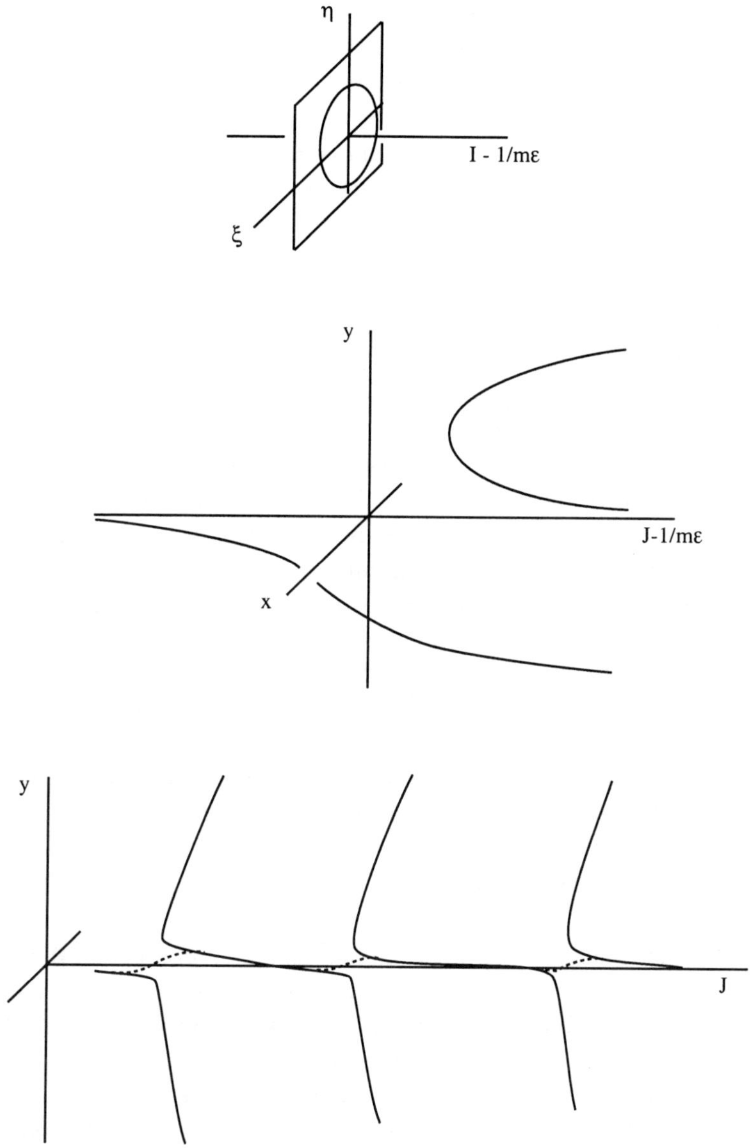

Fig. 1. Expected effect of δ on the set of periodic orbits near $I = 1/m\varepsilon, \xi = \eta = 0$ with period near $2\pi/\varepsilon I$: (a) $\delta = 0$, (b) $\delta \neq 0$, where (x, y) are coordinates rotating in the (ξ, η)-plane at rate m with respect to θ, and $J = I + \frac{x^2+y^2}{2\varepsilon I}$, (c) part of the resulting ladder of broken pitchforks and an example of a way to make an almost invariant slow manifold (dotted curve).

same symbols for the new coordinates, but rotating in the (x, y)-plane if necessary, the Hamiltonian takes the form $H = \bar{H} + \tilde{H}$, with $\bar{H}$ independent of θ, $\bar{H} \approx \frac{1}{2m^2\varepsilon} + 2K/m + \frac{\varepsilon}{2}(K - \frac{m}{2}r^2)^2 + f(\delta)y$ for some f which goes to 0 with δ and is of size at most order $e^{-\kappa m}$, and the gradient of $\tilde{H}$ is of smaller order. Note that the above form for $\bar{H}$ is already obtained in one iteration of the averaging method (with $f(\delta) = \delta|g_m|$), but further iterations are required to reduce $\tilde{H}$ to smaller order, and in general the iterations introduce additional terms of the same order into f.

Let us first analyse the dynamics of $\bar{H}$. Since $\bar{H}$ is independent of θ, its dynamics conserve K, so we can study the dynamics in (x, y) as a function of K. Its equilibria are given by $x = 0, \frac{\varepsilon}{2}m^2y^3 - \varepsilon mKy + f = 0$, whose solutions look like Figure 1(b) for $f \neq 0$ $(K \sim J - \frac{1}{m\varepsilon})$. The rest of the dynamics consists of periodic orbits around the elliptic critical points and separatrices of the saddle, as shown in Figure 2.

No curve connecting the two branches with r small is invariant; the normal component of the vector field reaches order f somewhere in between. Nevertheless, connecting curves can be chosen so that the normal component does not exceed this size (indeed, any connecting curve with $|r| \leq (2f/m^2\varepsilon)^{1/3}$ and small slope will do), thus giving an exponentially accurate slow manifold. The effects of $\tilde{H}$ do not change this conclusion, though they fuzz out the dynamics of $\bar{H}$, because K is not exactly conserved. H is still conserved, however, and thus Figure 2 changes to a one-parameter family of surface of section plots (at say $\theta = 0$, parametrised by H), with the principal changes being that the separatrices of the saddle become stochastic layers, and invariant circles with near-rational winding ratio become zones of instability.

Ref. 41 treats numerically a similar example, proposed in Refs 42 and 43, and claims existence of an invariant "fractal slow manifold" by appealing to KAM theory, but in my opinion the existence problem has nothing to do with KAM theory, just persistence of periodic orbits, and the authors are incorrect to say that one has to avoid all rational resonances not just integer ones, so in fact their system has an invariant slow manifold consisting of a collection of annuli, as above; they are correct, however, that KAM theory can be used in their example to bound the distance orbits can go from the slow manifold and to establish nonlinear stability of large parts of it. For some analysis of this model, see Ref. 44. Another related example was studied in Ref. 45, where results closely analogous to mine were obtained independently, and also illustrated by some nice numerics. Note that most

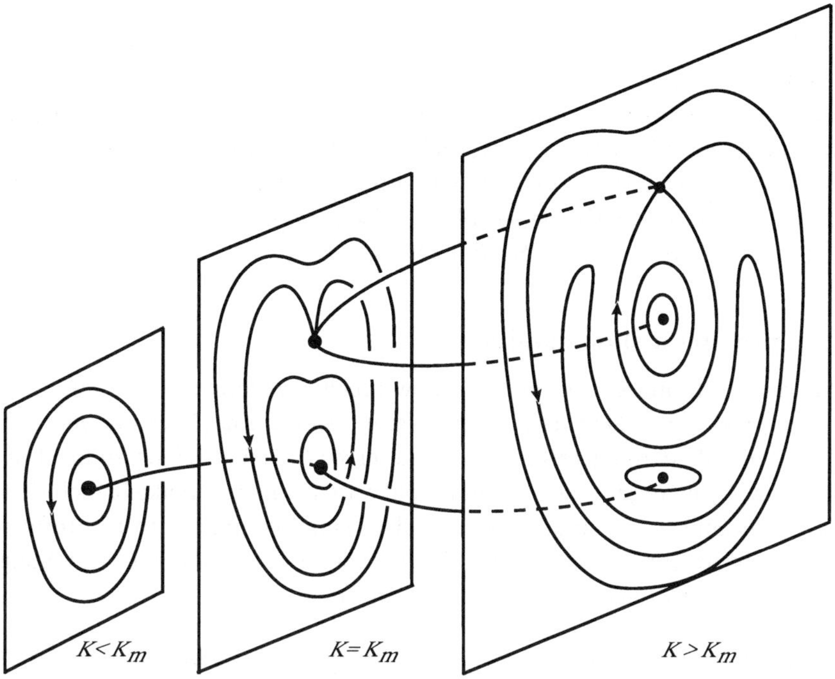

Fig. 2. The dynamics of $\bar{H}$ in the (x,y)-plane for Example 3 for three values of K. The bifurcation value $K_m \sim \frac{3m}{2}(\frac{2f}{m^2\varepsilon})^{2/3}$, and the equilibria occur at $y \sim \frac{f}{m\varepsilon K}$ for $K \ll -K_m$, $y \sim (\frac{2f}{m^2\varepsilon})^{1/3}$ and $-2(\frac{2f}{m^2\varepsilon})^{1/3}$ for $K = K_m$, and $y \sim \frac{f}{m\varepsilon K}$ and $\pm\sqrt{\frac{2K}{m}}$ for $K \gg K_m$.

orbits of Hamiltonian slow dynamics on any 2D symplectic submanifold are periodic, because it is a 1 degree of freedom Hamiltonian system (so every regular compact connected component of a level set is a periodic orbit), which confirms Lorenz's intuition.

Another special context worth mentioning for slow manifolds is families of Hamiltonian systems perturbed weakly, but not necessarily Hamiltonianly, by slow evolution of the parameters of the family, studied in Ref. 46.

Returning to the Hamiltonian context, with normally elliptic behaviour it is natural to try to go beyond just finding good slow manifolds, by seeking to describe all the dynamics in a neighbourhood of a slow manifold. If there is just one fast degree of freedom, this is provided by the standard theory of adiabatic invariants (e.g. Ref. 35), and indeed the dynamics can be reduced by an excellent approximation to a 1-parameter family of slow Hamiltonian systems, parametrised by the adiabatic invariant for the fast degree of freedom. Extension to more than one fast degree of freedom introduces problems with resonances between them, e.g. Refs. 2, 40, 35. Nevertheless, following initial ideas of Ref. 6, I have derived a good theory for slow manifolds with one internal oscillation, irrespective of the number of other fast degrees of freedom, subject just to non-resonance of the internal oscillation with the set of linearised frequencies for the other fast modes. This is described in Sections 9–11.

4. Improving a slow manifold

This section treats the question of how much a general slow manifold can be improved in a general dynamical system. Condition (1) of my definition of slow manifold requires only that the error field $V - U$ be small, but I show that by changing the slow manifold a little one can make it include all nearby equilibria and make the error field small compared to U. Iterating the procedure makes the error field of order U^r for any $r \in \mathbb{N}$ (though perhaps with a prefactor that grows with r).

Theorem 4: Every sufficiently good slow manifold can be improved to one which contains all nearby equilibria and has small angle to the vector field.

Definition 5: I call a slow manifold with these properties *first order.*

Proof: Treat the local situation first. So we can take local coordinates $(x, y) \in M_j \times E$ on P, for some open subset $M_j \subset M$ and Banach space E, and start from

$$\begin{aligned} \dot{x} &= \varepsilon X(x, y) \\ \dot{y} &= A(x)y + \delta(x) + R(x, y), \end{aligned} \tag{11}$$

with $||A(x)^{-1}|| \leq K$ of order 1, $||\delta||_{C^1} \leq \delta_1$ small, and $R(x, y) = O(y^2)$. The equation

$$A(x)\eta + \delta(x) + R(x, \eta) = 0 \tag{12}$$

has a locally unique solution $\eta(x) \approx -A(x)^{-1}\delta(x)$. In particular, all nearby equilibria are contained in the graph $y = \eta(x)$. Then put $y = Y + \eta(x)$ to obtain

$$\dot{Y} + \varepsilon D\eta(x)X(x, Y + \eta) = A(x)Y + R(x, Y + \eta) - R(x, \eta). \tag{13}$$

Define $\tilde{X}(x,Y) = X(x, Y + \eta(x))$, $\tilde{A}(x) = A(x) + \frac{\partial R}{\partial y}(x,\eta) - \varepsilon D\eta \frac{\partial X}{\partial y}(x,\eta)$ and $\tilde{\delta}(x) = -\varepsilon D\eta(x) X(x,\eta(x))$ to obtain

$$\begin{aligned} \dot{x} &= \varepsilon \tilde{X}(x,Y) \\ \dot{Y} &= \tilde{A}(x)Y + \tilde{\delta}(x) + \tilde{R}(x,Y), \end{aligned} \tag{14}$$

where $\tilde{R} = O(Y^2)$. The term $\tilde{\delta}$ has size of order $\varepsilon|X(x,\eta(x))|\delta_1$, so the angle between the new slow manifold and the vector field is of order $\varepsilon\delta_1$.

To make the result global, we need to face the fact that there might be no diffeomorphism from $M \times E$ to a neighbourhood of M, e.g. P a Möbius band and M a circle which goes once round P, or $P = \mathbb{C}P^N$ (*complex projective space*, the set of ratios $z_0 : z_1 : \ldots : z_N$ of non-zero $(N+1)$-tuples of complex numbers) and M the subset where $z_2 = \ldots = z_N = 0$ (which will occur in Example 14). Here is one way to proceed, though one might want to improve on it for explicit calculations.

Choose a transverse foliation to M in a neighbourhood, i.e. a decomposition $\bigcup_{x\in M} F_x$ of a neighbourhood of M into smoothly varying smooth submanifolds F_x such that the tangent space T_xP at each point $x \in M$ decomposes as $TF_x \oplus T_xM$. This can be done by choosing any normal bundle and a "spray" (e.g. the geodesic spray of a Riemannian metric) and extending the normal bundle to a neighbourhood using the spray (the tubular neighbourhood theorem, e.g. Ref. 47). Choose local trivialisations of the foliation (i.e. coordinates $(x,y) \in M_j \times E$ for an open cover $\bigcup_j M_j$ of M and Banach space E, so that F_x is the subset with first coordinate x and M is $y = 0$), so we can say what $\dot{y}$ means in each of these, defining vector fields V_x^j on the leaf F_x for each j such that $x \in M_j$. Choose a partition of unity for M, i.e. a set of functions $\phi_j : M \to \mathbb{R}$ such that $\phi_j = 0$ outside M_j and $\sum_j \phi_j(x) = 1\ \forall x \in M$, e.g. Ref. 47. Then study the vector field $V_x = \sum_j \phi_j(x)V_x^j$ on each leaf F_x. It is still of the form $A(x)y + \delta(x) + R(x,y)$ in any of the local trivialisations, so has a locally unique zero $y = \eta(x)$. Since it is unique it gives the same point (even though with different coordinates) in all the coordinate systems j for which $x \in M_j$. Thereafter one continues as in the local case. □

Definition 6: A slow manifold is r^{th} *order* if it contains all nearby equilibria and the normal component of the vector field is of order the rth power of the tangential component on it.

Theorem 7: If the vector field is C^r for some $r > 1$, the above procedure can be iterated, reducing the size of the normal component of the vector field to order $(\varepsilon|X(x,\eta(x))|)^r\delta_r$, where δ_r is a C^r norm of δ, giving an rth order slow manifold.

For an rth order slow manifold with $r > 1$, the approximation is especially good near equilibria. In particular, the linearisation about an equilibrium is an exact subsystem of the full linearised dynamics as soon as $r \geq 2$.

The possibility of reducing the error to $O(\varepsilon^r)$ is standard folklore (among those who know it!), but I feel the improvements to $O((\varepsilon|X|)^r)$ and including all nearby equilibria are significant (unless $|X|$ is bounded away from zero, of course, when the two are equivalent). Firstly they give extra accuracy near equilibria, and secondly in problems like the interaction of widely separated localised coherent structures (where there may be no explicit ε or δ parameters but they could be introduced by scaling) the first improvement gives extra accuracy for large separation. One could argue that the difference between the two results is trivial because by restricting attention to a neighbourhood N of an equilibrium the effective value of ε is reduced to $\varepsilon \sup_{x\in N}|X(x)|$, but one would first have to show that one could make the slow manifold contain all nearby equilibria, and the point that one can obtain position dependent accuracy which improves near equilibria does not seem to have been made before, as far as I am aware.

The above process of improving a slow manifold involves bounding a derivative each time, however, so an r-dependent prefactor arises in the error bound, which means that in general the error can not be reduced arbitrarily for given ε by increasing r. Nevertheless, if the system is analytic, one should be able to obtain reduction of the error to something of the form $\exp -C/(\varepsilon|X(x,\eta(x)|)$, for example by applying the Neishtadt-Nekhoroshev method[48,49] of estimation of the prefactor to determine an optimal number of iterations as a function of $\varepsilon|X|$. Reduction of the error to $O(\exp -C/\varepsilon)$ has been achieved for analytic systems in the case of graphs,[50] but without the crucial factor $|X|$.

5. Symplectic slow manifolds

For Hamiltonian systems, there is extra structure, namely the symplectic form Ω. Thus it is natural to ask for slow manifolds on which the slow dynamics is also Hamiltonian. This more or less requires the slow manifold to be symplectic, i.e. the symplectic form restricted to its tangent space is non-degenerate, else one would have to invent a relevant symplectic form on it from some other considerations, or perhaps make do with a Poisson structure (I'll discuss this option in Section 8). So in this section we'll study symplectic slow manifolds.

Suppose we have a state space P with a symplectic form Ω and a Hamiltonian H, and a symplectic submanifold M of P. The Hamiltonian vector field V is defined by $\Omega(V,\xi) = dH(\xi)$ for all tangent vectors ξ. The submanifold M comes with a natural Hamiltonian vector field U tangent to it, given by $\Omega_M(U,\xi) = dH_M(\xi)$ for all $\xi \in TM$, where Ω_M and H_M are the restrictions of Ω and H to M. In nineteenth century language, U gives the equations of motion when "workless constraint forces" are applied to keep the state on M. So we will choose U for the dynamics on M. We assume that the pair (M, U) satisfies the condition for a slow manifold. We want to improve M as in the previous section to a slow manifold $\tilde{M}$ which contains all nearby equilibria and has small angle to the vector field, but keeping Hamiltonian dynamics on it.

Choose a *symplectically orthogonal foliation* F to M, i.e. a decomposition of a neighbourhood of M into smoothly varying smooth submanifolds F_z labelled by points $z \in M$, such that $z \in F_z$ and the tangent space to P at each point $z \in M$ decomposes as $TP_z = TF_z \oplus T_zM$ and

$$\Omega(\xi,\eta) = 0 \;\forall \xi \in TF_z, \eta \in T_zM. \tag{15}$$

One concrete way to make such a foliation is to construct the symplectically orthogonal bundle to M, uniquely defined at each of its points by (15), and extend it to a neighbourhood by affine subspaces in local coordinates (one would need to ensure that the overlap maps in the intersection of different charts are affine too); this gives a foliation of any neighbourhood smaller than the smallest radius of curvature of M. A more sophisticated but perhaps less implementable way is to choose a spray on P, as in the general case, and extend the symplectically orthogonal bundle to a foliation of a neighbourhood of M by the spray. The use of symplectically orthogonal projection to describe slow dynamics in Hamiltonian systems was proposed independently about the same time as me by the author of Ref. 28 (in a PDE context; in addition he addressed the many technical issues that arise there such as choice of suitable function spaces and existence of solutions).

The normal motion being fast corresponds to $||(D^2H|_{F_z})^{-1}|| \le K$ for some K of order 1, and V close to TM corresponds to $|DH_z| \le \delta$ for some δ small. Thus by the implicit function theorem there exists a locally unique critical point $\tilde{z}(z)$ of $H|_{F_z}$, and it depends smoothly on $z \in M$. Hence we obtain a C^1-close submanifold $\tilde{M}$ of constrained critical points of H; in particular it contains all nearby true critical points of H. Note that in contrast to the general case, we did not need trivialisations and a partition of unity, because we are looking only for a critical point of a function on each leaf, not a zero of a projection of a vector field. Since $\tilde{M}$ is C^1-close to M, $|\Omega(\xi,\eta)| \leq K\delta_1|\xi||\eta|$ for $\xi \in T\tilde{M}, \eta \in TF_z$, where δ_1 is the C^1 norm of $DH|_M$. Furthermore, the restriction $\Omega_{\tilde{M}}$ of Ω to $\tilde{M}$ remains non-degenerate. Let $H_{\tilde{M}}$ be H restricted to $\tilde{M}$ and define U by $\Omega_{\tilde{M}}(U,\xi) = dH_{\tilde{M}}(\xi)$ for all $\xi \in T\tilde{M}$.

We now prove that $V - U$ is small compared to U. Firstly, $\Omega(V,\xi) = dH(\xi)$ for all ξ. In particular, for $\xi \in T\tilde{M}, \Omega(V,\xi) = dH_{\tilde{M}}(\xi) = \Omega(U,\xi)$; and for $\xi \in TF_z, \Omega(V,\xi) = dH(\xi) = 0$ and $\Omega(U,\xi) = O(\delta_1|U||\xi|)$. So $\Omega(V-U,\xi) = 0$ for all $\xi \in T\tilde{M}$ and is $O(|U|\delta_1|\xi|)$ for all $\xi \in TF_z$. It follows that $V - U = O(|U|\delta_1)$, as claimed.

For a C^r system, this procedure can be iterated to reduce the error to $O(\delta_r|U|^r)$, where δ_r is the C^r norm of $DH|_M$. In the analytic case I expect it is possible to reduce the error to $O(\exp -C/|U|)$, for example by truncation of the iteration at optimum order. This was achieved for the case of just one fast degree of freedom in Ref. 7, but without ensuring that the slow manifold contains all nearby equilibria and with error estimate only $O(\exp -C/\varepsilon)$ with $\varepsilon = \sup_{x\in M} |U(x)|$, rather than pointwise.

Note that it is not necessary to choose the foliation to be exactly symplectically orthogonal; deviations of the same order as the change from M to $\tilde{M}$ can be allowed and will still give the same order of improvement.

Now I give some examples, worked out to various stages of completeness.

Example 8: The Spring Pendulum

This consists of a light spring with one end fixed but free to rotate about it in a vertical plane, and a mass m on the other end, moving in the plane under the influence of the spring and gravity g. It has Hamiltonian

$$H(x,y,p_x,p_y) = \frac{1}{2m}(p_x^2 + p_y^2) + mgy + \frac{k}{2}(r-a)^2, \tag{16}$$

where (x,y) denote horizontal and vertical components of position of the mass relative to the fixed end of the spring, $r = \sqrt{x^2+y^2}$, and (p_x,p_y) the components of its momentum. The symplectic form Ω is the canonical one

$dx \wedge dp_x + dy \wedge dp_y$. The equations of motion are:

$$\begin{aligned}\dot{x} &= \frac{p_x}{m} \\ \dot{y} &= \frac{p_y}{m} \\ \dot{p}_x &= -k(r-a)\frac{x}{r} \\ \dot{p}_y &= -mg - k(r-a)\frac{y}{r}.\end{aligned} \tag{17}$$

The parameter regime of interest is g small. Then there is an obvious symplectic slow manifold M_0 corresponding to the spring length being fixed to a. Parametrised by the angle θ of the pendulum to the downward vertical and the angular momentum $p_\theta = xp_y - yp_x$, M_0 can be expressed as:

$$\begin{aligned}x &= a\sin\theta \\ y &= -a\cos\theta \\ p_x &= \frac{1}{a}\cos\theta\ p_\theta \\ p_y &= \frac{1}{a}\sin\theta\ p_\theta.\end{aligned} \tag{18}$$

One computes that $\Omega_{M_0} = d\theta \wedge dp_\theta$ and $H_{M_0} = \frac{1}{2ma^2}p_\theta^2 - mga\cos\theta$, so the zeroth order slow dynamics is simply that of a rigid pendulum. It is slow (a typical inverse time is $\sqrt{g/a}$) compared to the frequency (of order $\sqrt{k/m}$) for the linearised normal motion, provided $p_\theta = O(\sqrt{g})$, so it is sensible to write $p = \sqrt{g}P$ and work with P instead of p. One can check that in (x, y, P_x, P_y) coordinates the error vector field is small (order $\sqrt{g}$), but this is of the same size as the vector field of H_{M_0}, so the approximation is not reliable enough. Furthermore, M_0 does not contain the equilibria of the spring pendulum, which are at $x = p_x = p_y = 0$, $y = \pm a - \frac{mg}{k}$. So we will compute a first order symplectic slow manifold.

For symplectically orthogonal foliation, choose the leaves F_{θ,p_θ} defined by

$$\begin{aligned}x\cos\theta &= y\sin\theta \\ xp_y - yp_x &= p_\theta.\end{aligned} \tag{19}$$

They can be parametrised by (r, p_r), the radius and radial momentum ($p_r = (xp_x + yp_y)/r$). Then $H = \frac{1}{2m}(\frac{p_\theta^2}{r^2} + p_r^2) - mgr\cos\theta + \frac{k}{2}(r-a)^2$. So making H critical on F_{θ,p_θ} yields the first order symplectic slow manifold M_1 defined by $p_r = 0$ and $r = \rho(\theta, p_\theta)$, the root of $\frac{p_\theta^2}{mr^3} + mg\cos\theta = k(r-a)$ near a.

Then

$$H_{M_1} = \frac{p_\theta^2}{2m\rho(\theta,p_\theta)^2} - mg\rho(\theta,p_\theta)\cos\theta + \frac{k}{2}(\rho(\theta,p_\theta)-a)^2 \tag{20}$$

and $\Omega_{M_1} = d\theta \wedge dp_\theta$ still (because $p_r = 0$). The resulting slow vector field U is $\dot\theta = \frac{p_\theta}{m\rho^2}$, $\dot p_\theta = mg\rho\sin\theta$. Computing the error vector field $V-U$ on M_1, one finds its components to be $\dot\theta = 0$, $\dot p_\theta = 0$, $\dot p_r = 0$, $\dot r = -\frac{gp_\theta \sin\theta}{k\rho^2+3p_\theta^2/m\rho^2}$, and so bounded by the quadratic expression $\dot\theta \dot p_\theta / k\rho$ in the components of the slow vector field, confirming that M_1 is a first order slow manifold, and showing that it is actually second order. This fluke arises because M_1 happens to still be symplectically orthogonal to the foliation (as $p_r = 0$ on M_1).

Since M_1 is a second order slow manifold, one can in particular obtain the exact slow part of the linearised dynamics about the equilibria by taking the quadratic part of $H|_{M_1}$ there (note that the third term of (20) contributes to this).

Example 9: Interaction of kinks

For a nonlinear wave equation $u_{tt} = u_{xx} - V'(u)$ with V possessing (at least) two minima at the same height, a *kink* is a solution asymptotic to one minimum as $x \to -\infty$ and to a minimum with a larger value of u as $x \to +\infty$. An *antikink* is the opposite. Here I give the beginnings of an analysis of the interaction of a widely separated kink and antikink, using my approach (it has been studied before by many authors using other methods, e.g. Ref. 51).

The equations can be expressed in Hamiltonian form by defining $\pi = u_t$, Hamiltonian $H(u,\pi) = \int \frac{1}{2}\pi^2 + \frac{1}{2}u_x^2 + V(u)\, dx$ and symplectic form $\Omega((\dot u,\dot\pi),(u',\pi')) = \int \dot u \pi' - u'\dot\pi \, dx$ for a general pair of variations. For concreteness, take $V(u) = \frac{1}{4}(u^2-1)^2$. This has kink and antikink solutions $u(x,t) = \pm\tanh\frac{\gamma}{\sqrt2}(x-vt-x_0)$, with associated momentum field $\pi(x,t) = \mp\frac{\gamma v}{\sqrt2}\operatorname{sech}^2\frac{\gamma}{\sqrt2}(x-vt-x_0)$, for any $x_0 \in \mathbb{R}, v \in (-1,1)$, where $\gamma = \frac{1}{\sqrt{1-v^2}}$. Linearising about the equilibria $u = \pm1, \pi = 0$ gives waves with frequencies at least $\sqrt{V''(\pm1)} = \sqrt2$.

It suffices to study the interaction in the centre of mass frame, so let us restrict attention to slow manifolds consisting of pairs of even functions

(u, π). A natural initial guess M_0 at a slow manifold is the set of pairs

$$u(x) = 1 + \tanh \frac{\gamma}{\sqrt{2}}(x - x_0) - \tanh \frac{\gamma}{\sqrt{2}}(x + x_0) \tag{21}$$
$$\pi(x) = -\frac{\gamma v}{\sqrt{2}} \left(\operatorname{sech}^2 \frac{\gamma}{\sqrt{2}}(x - x_0) + \operatorname{sech}^2 \frac{\gamma}{\sqrt{2}}(x + x_0) \right),$$

for $|x_0| \gg 1, v$ small. This looks like the superposition of a kink at x_0 with velocity v and an antikink at $-x_0$ with velocity $-v$. To obtain the slow equations of motion on M_0 we compute the restrictions of H and Ω to M_0.

The restriction of H to M_0 has already been computed:[51]

$$H|_{M_0}(x_0, v) = H_0(x_0, v) = \frac{1}{2} m_1(x_0, \gamma) v^2 + V_1(x_0, \gamma) + V_2(x_0, \gamma), \tag{22}$$

where the functions m_1, V_1, V_2 are given in equations (B.8a), (B.5a), (B.5b) respectively, of Ref. 51. The symplectic form, however, does not appear to have been computed before (Ref. 51 obtained equations of motion by assuming $\dot{x}_0 = v$ and using conservation of H_{M_0}). I obtain $\Omega|_{M_0} = S(x_0, v)\, dx_0 \wedge dv$, where $S = \gamma^2 m_1 + \gamma^5 v^2 S_1(x_0, \gamma)$ with S_1 a certain integral decaying exponentially like $e^{-2\sqrt{2}|x_0|}$ as $x_0 \to \pm\infty$. Now $m_1(x_0, \gamma) = \gamma(M + K(x_0, \gamma))$, with[51] $M = \frac{4\sqrt{2}}{3}$ (which is just twice the rest energy of a kink) and K exponentially decaying in x_0. So $S = \gamma^3 M + v^2 S_2(x_0, \gamma)$ with S_2 exponentially decaying in x_0. The leading term suggests that instead of the coordinate v on M_0 it might be better to use $p = \gamma M v$, because then $\Omega|_{M_0} = (1 + p^2 S_3)\, dx_0 \wedge dp$, with S_3 exponentially decaying in x_0, but I'll continue with v. The resulting slow equations are:

$$\dot{x}_0 = \frac{1}{S} \frac{\partial H_0}{\partial v} \tag{23}$$
$$\dot{v} = -\frac{1}{S} \frac{\partial H_0}{\partial x_0}.$$

To leading order in v and $e^{-2\sqrt{2}|x_0|}$, it is sufficient to keep only the leading terms of m_1 and S and to ignore the v dependence of V_1 and V_2 (via γ). Then we obtain the approximate slow dynamics

$$\dot{x}_0 = v \tag{24}$$
$$\dot{v} = -12\sqrt{2} \operatorname{sgn}(x_0) e^{-2\sqrt{2}|x_0|}$$

(which says that the kink and antikink attract exponentially). For v small and x_0 large, this is indeed slow compared to the minimum frequency of linearised vibrations about the tails ($\sqrt{2}$). It is also slow compared to the linearised vibrations about a kink (the only addition to the spectrum is a shape mode at frequency $\sqrt{3/2}$), and hence presumably slow compared to the linearised frequencies normal to the slow manifold.

Next one should estimate the error vector field, to see how reliable is Eq. (24). Diogo Pinheiro computed that it is of the same order in $(v, e^{-2\sqrt{2}|x_0|})$ as the slow vector field, so M_0 is almost a first order slow manifold except that there is not a small prefactor. It is possible, however, that the slow dynamics is more accurate than the slow manifold. To investigate this, one could apply one iteration of my method, which would produce a second order slow manifold, and see if it makes a significant difference to the slow vector field, and this is under investigation.

Example 10: Motion of a kink in a lattice

For a spatial discretisation of a nonlinear wave equation with kinks, there are still stationary kinks (generally one centred on a site and one on a bond, with energies differing by the so-called Peierls-Nabarro barrier), but it is believed to be unlikely that there are travelling ones, except perhaps for some isolated speeds (for a recent article on this which includes references to the basic but not yet rigorous intuition, see Ref. 52)[e]. Numerics (e.g. Ref. 54), however, show existence of approximate travelling kinks, which travel as if in a periodic potential, of period equal to the lattice spacing. There should be a slow manifold describing this, so it would be interesting to apply my method. Some discretisations have no Peierls-Nabarro barrier, but still appear not to have true travelling kinks,[55] though there is also a completely integrable discretisation of the sine-Gordon PDE.[56] An interesting change of coordinates to study the motion of kinks in a lattice and their radiation was obtained in Ref. 57. Also the interaction of travelling kinks with impurities in the lattice has been studied.[58] For more sophisticated work on the motion of kinks, in particular the radiation damping effects which place an ultimate limit on the accuracy of a slow manifold, see Ref. 37.

Note that motion near a normally elliptic symplectic slow manifold of a Hamiltonian system can differ substantially from that on the manifold. There can be adiabatic invariants for the normal oscillations which produce substantial net forces on the slow degrees of freedom. For example, for a charged particle in a strong magnetic field, there is a slow manifold

[e]Note that travelling pulses do exist in chains of FPU type for all speeds above the sound speed,[53] but the problem with discretisations of nonlinear wave equations is that for every speed except zero there is a phonon with the same phase speed.

corresponding at lowest order to motion along the field lines at constant speed, but nearby trajectories execute helices around the field lines with adiabatic invariant $\mu = v_\perp^2/2B$, whose instantaneous centre s experiences a force $-\mu\, \partial B/\partial s$ along the field line, leading to mirroring at places where the field is too strong (compared to the ratio of the kinetic energy to μ).

6. The Methods of Collective Coordinates

The method of the previous section bears a resemblance to two traditional methods for constructing slow dynamics, both often called the method of collective coordinates. To clarify the situation, I propose to call them the methods of "coordinate change" and "constrained Lagrangians".

The method of coordinate change consists of constructing a canonical coordinate change for a Hamiltonian system such that some of the new coordinates represent degrees of freedom of coherent structures and the rest represent "radiation". If the coordinates are well chosen, the submanifold corresponding to zero radiation is found to be almost invariant, at least the part of it where the motion is slow. It was proposed by Sakita and collaborators, but formulated most clearly in Ref. 59, became a standard technique in quantum field theory, and has been used occasionally in classical mechanics (e.g. Ref. 60). I am not aware, however, of a systematic procedure for choosing the coordinate change, nor for improving a choice. Also the restriction to canonical coordinate changes makes more work in my opinion than allowing the symplectic form to become nonstandard.

The method of constrained Lagrangians has been widely used in classical mechanics, e.g. Refs 55, 58, 61-65. Here I assess its validity or otherwise, which I have not seen examined adequately before. It is best formulated in the Lagrangian rather than Hamiltonian context.

Lagrangian systems are those that arise from a variational principle of the form $\delta \int L(q,\dot{q})\, dt = 0$ for some function $L : TQ \to \mathbb{R}$ on a tangent bundle TQ (the points of TQ consist of a point q of Q together with a tangent vector v there) satisfying the *Legendre condition* that $v \mapsto p = \frac{\partial L}{\partial v}$ is invertible at each $q \in Q$, giving rise to the *Euler-Lagrange equations* of motion $\frac{\dot{\partial L}}{\partial v_i} = \frac{\partial L}{\partial q_i}, \dot{q}_i = v_i$ in local coordinates q_i on Q. Any Lagrangian system can be converted to an equivalent Hamiltonian system on the cotangent bundle T^*Q (a *cotangent vector* p at $q \in Q$ is a linear map from $T_qQ \to \mathbb{R}$), by defining $H(q,p) = p \cdot v - L(q,v)$, where $v(q,p)$ is defined by the Legendre condition and $p \cdot v$ is the natural pairing of cotangent and tangent vectors, and using the canonical symplectic form $\Omega = -d(p \cdot dq)$ on T^*Q (note that this is my choice of where to put the inescapable minus sign of Hamiltonian mechanics). Similarly any Hamiltonian system on a cotangent bundle with $p \mapsto v = \frac{\partial H}{\partial p}$ invertible at each $q \in Q$ defines an equivalent Lagrangian system with $L(q, v) = p \cdot v - H(q, p)$.

In the method of constrained Lagrangians, one proposes as slow manifold the tangent bundle TQ_0 of a submanifold Q_0 of the configuration space Q and computes equations of motion on it by restricting the variational principle to TQ_0. These are generally assumed to be good approximate slow dynamics for the original system, but I have not seen this assumption justified in general[f]. Indeed there are plenty of choices for Q_0 for which the resulting dynamics has nothing to do with the original dynamics. Even for plausible choices of Q_0, the ansatz may need significant correction, e.g. Ref. 66. In general, to assess the accuracy of the slow manifold, one has to check the difference between the two vector fields. Equivalently, one has to check the size of the constraining force, though this is something that is not directly accessible in Lagrangian dynamics (which is why Lagrange was so proud of his formulation of mechanics: it allows one to compute constrained dynamics with no need to compute the constraining forces). The difference between the two vector fields is determined by dH on tangent vectors in TTQ transverse to TTQ_0, since in the notation of Ref. 68 the vector field Z at $v \in T_qQ$ is determined by $\Omega_L(v)(Z, w) = dH(v) \cdot w \ \forall w \in T_vTQ$ and the vector field Z_0 at $v \in T_qQ_0$ is determined by $\Omega_L(v)(Z_0, w) = dH(v) \cdot w \ \forall w \in T_vTQ_0$.

Any slow manifold TQ_0 for a Lagrangian system can be improved to a first order one TQ_1, as follows. Construct a foliation $\bigcup_{q \in Q_0} F_q$ transverse to Q_0. The normal dynamics being fast compared to the tangential dynamics implies that for each $q \in Q_0$ there is a locally unique critical point $\tilde{q}(q)$ of the restriction of $L(\tilde{q}, 0)$ to F_q. Then $Q_1 = \{\tilde{q}(q) : q \in Q_0\}$ is a smooth submanifold containing all equilibria in a neighbourhood of Q_0 and a straightforward calculation shows the error vector field for the resulting constrained Lagrangian on TQ_1 is relatively small.

A severe disadvantage of the method of constrained Lagrangians is that although TQ_0 can be improved to a first order slow manifold TQ_1, there is no way in general to improve it to a second order slow manifold. The reason

---

[f] Some cases close to completely integrable can be justified by perturbation theory for completely integrable systems, e.g. Ref. 67, and there are proofs for some other specific systems, e.g. Ref. 26; there are also methods based on "elimination of secular perturbations or unbounded fluctuations",[51] but I do not find the arguments totally convincing.

is that second order slow manifolds of the Hamiltonian context are in general not the cotangent bundle of a submanifold of the configuration space, e.g. q_0 an equilibrium does not imply there is a second order slow manifold containing all small tangent vectors at q_0, as can be seen in the spring pendulum. Indeed there are some problems, e.g. in atmospheric dynamics, where even the natural zeroth order slow manifolds are not cotangent bundles (see next section).

7. Velocity Splitting

Since Ref. 65, there has been quite an industry, e.g. Ref. 29 and references therein, producing symplectic slow manifolds of the special form $p = P(q)$ for Hamiltonian systems of the standard mechanical form $H(q,p) = \frac{1}{2}p^T M^{-1}p + W(q)$ on a cotangent bundle T^*Q, with the canonical symplectic form, by a technique called "velocity splitting" (they take M to be the identity matrix, but I will allow any constant invertible symmetric matrix in my exposition, and I expect one could extend to non-constant M). Here I compare the velocity splitting method with mine.

Their procedure is firstly to imagine that workless constraint forces are introduced to force $p = P(q)$, and to compute the resulting evolution $\dot{q} = U(q)$ of q by the standard theory of constrained Hamiltonian systems (as in the second paragraph of section 5). The result is that

$$U = -\Omega_0^{-1}(\nabla W + DP^T M^{-1} P), \tag{25}$$

where $\Omega_0 = DP - DP^T$ (the matrix representing the restriction of the symplectic form to the submanifold) is assumed invertible (note that this requires the dimension of Q to be even). This is just the condition for $p = P(q)$ to be symplectic, and the resulting dynamics is precisely that I would use on the slow manifold in Section 5 (though my U would also have a p-component $\dot{p} = DP\ U$).

Secondly, they compare U with $M^{-1}P$, proving that they are identical vector fields iff the constraint forces are zero. The difference $U^S = U - M^{-1}P$ is called the *velocity split* (this is just the configuration space component of my $U - V$, but the p-components take care of themselves in this simple context). So a slow manifold with identically zero velocity split is invariant.

Thirdly, some have proposed an iterative scheme to improve the slow manifold to one of a higher order of accuracy, namely to replace the field P by $\tilde{P} = MU$, but I am not aware of any argument for why this should be any better. In fact, it is not any better unless DP happens to satisfy some special conditions.

To see this, consider the step that their iteration takes along the symplectically orthogonal foliation given by affinely extending the symplectically orthogonal subspaces. The symplectically orthogonal subspace at $(q, P(q))$ is $\{(\xi, DP^T\xi) : \xi \in T_qQ\}$. It is convenient to assume DP invertible (as holds at generic points) and write this as $\{(DP^{-T}\eta, \eta) : \eta \in T_q^*Q\}$. In the iterative step the displacement of the slow manifold is vertically by MU^S. To first order, this is equivalent to a displacement along the symplectically orthogonal foliation by

$$\eta = (I - DP\ DP^{-T})M\ U^S = -DP^{-T}\Omega_0 MU^S. \qquad (26)$$

Now DH acting on a general symplectically orthogonal vector $(DP^{-T}\eta', \eta')$ is $\nabla W^T DP^{-T}\eta' + P^T M^{-1}\eta'$. So $DH|_F$, when regarded as a vector to be dotted with vertical vectors η', is $DP^{-1}\nabla W + M^{-1}P$. Comparing with (25) shows that $DH|_F = -DP^{-1}\Omega_0 U^S$. My method takes the step along F which makes $DH|_F$ zero. To first order this is the Newton step $-D^2H|_F^{-1}DH|_F(q, P(q))$, and only steps which agree with this to first order give first order accuracy. Now $D^2H|_F = DP^{-1}D^2W\ DP^{-T} + M^{-1}$ with respect to the vertical coordinate η'. So the Newton step is

$$(DP^{-1}D^2W\ DP^{-T} + M^{-1})^{-1}DP^{-1}\Omega_0 U^S. \qquad (27)$$

We see that the two steps (26) and (27) are the same for general U^S iff

$$(DP^{-1}D^2W DP^{-T} + M^{-1})DP^{-T}\Omega_0 M = -DP^{-1}\Omega_0. \qquad (28)$$

If the submanifold is slow then D^2W should be negligible here, so neglecting it and writing $B = DP^{-1}DP^T$, the condition for equality becomes $B^{-1}M = MB$.

To appreciate what sort of restriction this condition imposes on DP, consider the case where M is a multiple of the identity. Then it says $B^2 = I$, so $B = C^{-1}EC$, where E is any diagonal matrix of ± 1 and C any invertible matrix. It follows that $C\ DP^T = E\ C\ DP$. Partition the coordinates into blocks labelled + and − according to the sign of E on them, and write out the equations for the blocks of C and DP. A simple calculation shows that Ω_0 acting on $[C_{++}^T, C_{+-}^T]$ gives zero, so Ω_0 is degenerate unless the dimension of the + space is zero. But $E = -I$ implies DP is antisymmetric. So if the mass matrix M is a multiple of the identity then the only case when the step of Ref. 29 pushes the error one order higher is when DP happens to be antisymmetric. It seems that it is close to antisymmetric for most of their applications, so their iteration does lead to improvement, but for a fortuitous reason. It would be interesting to examine the consequences of $B^{-1}M = MB$ on DP for general M.

My method has the advantages that it applies to all forms of Hamiltonian system and to all forms of symplectic submanifold and the iteration in one step includes all nearby equilibria in the slow manifold and in r steps obtains r^{th} order relative accuracy for the slow dynamics. It would be interesting to apply my method to some of the systems where velocity splitting has been used.

8. Poisson slow manifolds

In some cases, like 3D atmospheric dynamics, the requirement that the candidate slow manifolds be symplectic is too strong: as pointed out in Ref. 69, for these systems the symplectic form always degenerates somewhere on the slow manifold or even everywhere. Theiss proposed a fix via Dirac's theory of constrained Hamiltonian systems, which selects a unique vector field U satisfying $\Omega(U,\xi) = dH(\xi)$ for all ξ tangent to the manifold, by some additional requirements. This can be usefully reformulated in terms of Poisson dynamics.

A *Poisson system* is a vector field U on a manifold M such that for every F in the algebra $A = C^\infty(M,\mathbb{R})$, $dF\ U = \{H,F\}$, where H is a particular function in A and $\{\cdot,\cdot\}$ is a Poisson bracket on M, i.e. an antisymmetric bilinear map $A \times A \to A$ satisfying the Jacobi identity $\{J,\{K,L\}\} + \{K,\{L,J\}\} + \{L,\{J,K\}\} = 0$ and Leibniz rule $\{J,KL\} = \{J,K\}L + K\{J,L\}$ for all $J,K,L \in A$. Every Hamiltonian system can be put in Poisson form because to every symplectic form Ω is associated a Poisson bracket $\{F,G\} = \Omega(X_F,X_G)$, where for general function F, X_F denotes the Hamiltonian vector field of (F,Ω), and then $dF X_H = \{H,F\}$.

Given any submanifold M (not necessarily symplectic) one can restrict the Hamiltonian and the Poisson bracket to it (by choosing any transverse foliation and extending functions on M to be constant along the leaves) and obtain a Poisson system on M. If the Poisson bracket is degenerate on M (meaning there exists a nonconstant function C on it such that $\{C,K\} = 0$ for all functions K) then M decomposes into symplectic leaves which are invariant under all Poisson flows on it and on which the Poisson bracket is non-degenerate and so the flow on the leaves is Hamiltonian.

9. Slow manifolds with Internal Oscillation

In Hamiltonian systems, there are often submanifolds consisting of approximate periodic orbits, and one wishes to know to what extent there is an invariant manifold nearby on which the motion is approximately periodic on order-one times but with a slow drift to be determined. I call such a manifold a *slow manifold with internal oscillation.*

A couple of examples where this situation arises are the interaction of discrete breathers with frequencies close to rational ratio, and the interaction of Q-balls.

Example 11: Interaction of discrete breathers

Discrete breathers (DB) are time-periodic spatially localised motions of dynamical networks of units. Ref. 6 addressed the question of what happens if one takes as initial condition something close to a superposition of two (or more) widely separated discrete breathers of close frequencies (or more generally, frequencies close to a low order rational ratio). How do their relative phases evolve? Do they exchange energy? For Hamiltonian systems of the form $H(q,p) = \sum_j H_j(q_j,p_j) + \varepsilon H'(q,p)$ with ε small, the H_j nonisochronous and suitable coupling H', we showed that there is an almost invariant symplectic slow manifold consisting of states which look like superpositions of such discrete breathers, and an effective Hamiltonian on it, such that the true evolution is close to that of the effective Hamiltonian. More results on these questions will be presented in Examples 13 and 14 for the special case of "discrete self trapping systems" and in Examples 15 and 16 for general Hamiltonian networks.

Example 12: Interaction of Q-balls

Q-balls are solutions of the form

$$u(x,t) = e^{\mathrm{i}\omega t} f_\omega(r) \tag{29}$$

with ω, f real, $f'(0) = 0$, and $f(r) \to 0$ as $r \to \infty$, where $r = |x|$ relative to some origin, for complex nonlinear field equations

$$u_{tt} = \Delta u - u\beta(|u|^2), u(x,t) \in \mathbb{C}, x \in \mathbb{R}^N, \tag{30}$$

where Δ is the Laplacian and β real. By Poincaré invariance of the field equation, the above stationary Q-balls can be Lorenz boosted to obtain moving Q-balls. To demonstrate the Hamiltonian structure of the field equation, introduce the momentum field $\pi = u_t$, the Hamiltonian $H(u,\pi) = \int \frac{1}{2}(|\pi|^2 + |\nabla u|^2) + V(|u|^2)\, d^N x$, where $V(z) = \int_0^z \beta(s)\, ds$, and symplectic form $\Omega((\dot{u},\dot{\pi}),(u',\pi')) = \int \Re(\bar{\dot{u}} \cdot \pi' - \bar{u}' \cdot \dot{\pi})\, d^N x$. A nice exposition is given in the introduction to Ref. 28 (including how to evaluate the functional derivatives of H). Ref. 70 investigated the question of what happens if two Q-balls of close frequency are fired at each other. The authors found numerically that they attract if the relative phase is 0, repel if the relative phase is π, and exchange "charge" if the relative phase is in between. The method of the next section can be expected to supply an effective Hamiltonian to explain these results.

The case of Hamiltonian systems for which an open subset (rather than a proper submanifold) of the state space consists of approximate periodic orbits was treated by Poincaré,[71] Chirikov[72] and Moser,[73] who obtained slow equations of motion for the parameters of the periodic orbits (in Poincaré's case this was the slow evolution of the parameters of the instantaneous Kepler ellipses for the planets). A recently studied system of this type is described in Ref. 74 and it would be interesting to apply the Poincaré-Chirikov-Moser method to it. Such situations are not the point of my lecture-notes, however; rather I am addressing how to obtain proper submanifolds with slow dynamics (modulo internal oscillation).

In Ref. 6 we treated a case when a proper submanifold consists of approximate periodic orbits and the normal motion is fast compared to the drift between periodic orbits. In the next two sections I give a better treatment, which has the advantage that it can be iterated to arbitrary order of accuracy.

Slow manifolds with internal oscillation can also arise for general (as opposed to Hamiltonian) dynamical systems. In particular, if a slow manifold with internal oscillation is normally hyperbolic then there exists an invariant slow manifold nearby, because internal oscillation implies that any tangential contraction or expansion is at a subexponential rate.

A generalisation which I will not treat in detail here, is to slow manifolds with more internal dynamics than one oscillation, for example quasiperiodic or chaotic. One approach to this is to extend the loop dynamics of Section 11 to the dynamics of other embedded manifolds (cf. Kuksin, private communication), to which similar symmetry reduction can be applied. Averaging with respect to quasiperiodic dynamics is in general much more delicate than for periodic dynamics, e.g. Refs 2, 35, 40, because the quasiperiodic frequency vector can come close to low order commensurability, thereby generating additional slow directions. Averaging with respect to chaotic dynamics can lead to non-Hamiltonian stochastic effective dynamics for the slow variables (I'll write something on this in Ref. 75).

10. Internal oscillation: $U(1)$-symmetric Hamiltonians

Slow manifolds with internal oscillation (other than normally hyperbolic ones) are most easily treated in the special case of Hamiltonian systems with a $U(1)$ symmetry, e.g. the complex field for Q-balls above and discrete self-trapping systems to be introduced below. Note that symmetry of a Hamiltonian system means both the Hamiltonian and the symplectic form are invariant under an action of the symmetry group.

Example 13: Discrete Self-Trapping Systems (DST)

Discrete self-trapping systems (e.g. Ref. 62) have equations of the form

$$\mathrm{i}\dot{\psi}_s = -\gamma_s|\psi_s|^2\psi_s - \sum_{r\in S} C_{sr}\psi_r, \tag{31}$$

where s ranges over a discrete space S, each $\psi_s \in \mathbb{C}$, $\gamma_s \in \mathbb{R}$ and C is real symmetric. Note that for C diagonal, one obtains decoupled units with solution $\psi_s(t) = Ae^{\mathrm{i}\omega_s(|A|^2)t}$, for any $A \in \mathbb{C}$, with the frequency function

$$\omega_s(n) = C_{ss} + \gamma_s n. \tag{32}$$

A special case of the DST is the discrete nonlinear Schrödinger equation (DNLS), where $S = \mathbb{Z}$ and $C_{rs} = C_{00}$ for $r = s$, C_{10} for $|r - s| = 1$ and 0 otherwise (one can also generalise the DST to allow complex Hermitian C and nonlinear terms of the form $\beta_s(|\psi_s|^2)\psi_s$ for real functions β_s, and to other forms of coupling, e.g. the Ablowitz-Ladik equation[62]).

The DST have a Hamiltonian formulation with Hamiltonian

$$H = -\frac{1}{4}\sum_s \gamma_s|\psi_s|^4 - \frac{1}{2}\sum_{r,s} C_{rs}\bar{\psi}_r\psi_s, \tag{33}$$

and symplectic form $\Omega = \frac{\mathrm{i}}{2}\sum_s d\psi_s \wedge d\bar{\psi}_s$, i.e.

$$\Omega(\psi, \psi') = \Im(\sum_s \bar{\psi}_s\psi'_s). \tag{34}$$

Note that in computing the equations of motion from this formulation, ψ_s and $\bar{\psi}_s$ are to be thought of as independent variables (equivalent to two linear combinations of the real and imaginary parts). The DST have the $U(1)$ symmetry $\psi \mapsto e^{\mathrm{i}\theta}\psi$ for all $\theta \in S^1$.

They have many discrete breathers of the special form $\psi_s(t) = e^{\mathrm{i}\omega t}\phi_s$, ω real (i.e. $U(1)$ orbits), where ϕ is a spatially localised solution of an associated ω-dependent real static problem. The (signed) quantity ω is called the *frequency* of the $U(1)$-symmetric DB.

For a Hamiltonian system with a $U(1)$ symmetry, by Noether's theorem (e.g. Ref. 68) there is an associated conserved quantity A. For (30) it is the *charge* $Q = \int \Im(\bar{u}\pi)\, d^N x$. For DST it is the *number* $N = \sum_s |\psi_s|^2$ (called "norm" by Ref. 62, but it is not of degree 1 so is not a norm though its square is; the term "number" refers to the integer number of quanta in the quantum mechanical case, but classically it can be any non-negative real). In this case, one can reduce the study of the dynamics to that on the quotient spaces P_a, with $a \in \mathbb{R}$, i.e. the $U(1)$-equivalence classes with $A = a$. The Hamiltonian and symplectic form on P_a are just those induced by the original problem. If the internal oscillation corresponds to $U(1)$-orbits, then this reduction turns the problem of a slow manifold with internal oscillation into a standard slow manifold problem, because the periodic orbits are mapped to equilibria of the reduced system. For any order r of accuracy, one obtains a one-parameter family of *effective Hamiltonians* H_a for the reduced system by the method of Section 5.

Example 14: Targeted energy transfer (TET) in a DST

The setting[76] is a DST for which the coupling (off-diagonal part of C) is weak and there is an $a > 0$ and two units i, j with frequency functions such that $\omega_i(n), \omega_j(a-n)$ are (i) close for all $n \in [0, a]$, and (ii) far from the set $F = \{C_{ss} : s \in S \setminus \{i, j\}\}$ of linearised frequencies about the other units (note that in contrast to the next section, there is no need here to require $k\omega$ to avoid F for more $k \in \mathbb{Z}$ than $k = 1$). Then let H_a be the restriction of H to the $U(1)$-equivalence classes of configurations ψ with $N(\psi) = a$ (I'll write the equivalence relation as $\psi \sim e^{i\theta}\psi$ for all $\theta \in S^1$), and Ω_a the associated symplectic form. Let M_0 be the set of $U(1)$-equivalence classes of superpositions $\psi = e^{i\theta_i}\sqrt{n}\,\delta_i + e^{i\theta_j}\sqrt{a-n}\,\delta_j$ with $\theta_i, \theta_j \in S^1, n \in [0, a]$, where for general $k \in S$, δ_k is the configuration with value 1 on unit k and 0 elsewhere. Topologically, M_0 is a 2-sphere, because it is diffeomorphic to the complex projective space $\mathbb{C}P^1$. For coordinates on the sphere I'll use height $n \in [0, a]$ and longitude $\theta = \theta_j - \theta_i$, though of course this coordinate system has a singularity at the poles $n = 0, a$. It is a symplectic submanifold, with Ω_{M_0} the standard area on the 2-sphere $(d\theta \wedge dn)$.

For zero coupling, M_0 is an invariant manifold for H_a. The motion on the sphere is a differential rotation about the polar axis at rate

$$\omega_d(n) = \omega_j(a-n) - \omega_i(n), \tag{35}$$

which by assumption (i) is slow. The poles are equilibria, representing all the excitation being on unit i or j, respectively. The normal motion is oscillation at frequencies $C_{ss} - \bar{\omega}(n), s \in S \setminus \{i, j\}$, where $\bar{\omega}(n)$ depends on choice of trivialisation of the normal bundle around the orbit but can be taken to be close to $\omega_i(n)$ for $n \geq a/2$ and to $\omega_j(a-n)$ for $n < a/2$ (the normal bundle to this $\mathbb{C}P^1$ in $\mathbb{C}P^N$ is an example of a non-trivialisable bundle, having first Chern number $c_1 = 1$, e.g. Ref. 77, so there is no natural choice). Thus by assumption (ii), the normal motion is fast. So M_0 is an invariant symplectic slow manifold for no coupling.

For small coupling, M_0 is still a symplectic slow manifold, though not in general invariant. Let us compute the constrained motion on it. We have

$$\begin{aligned} H_{M_0} &= -\frac{1}{2}(\gamma_i n^2 + \gamma_j (a-n)^2) \\ &\quad -C_{ii}n - C_{jj}(a-n) - 2C_{ij}\cos\theta\sqrt{n(a-n)}, \end{aligned} \tag{36}$$

and $\Omega_{M_0} = d\theta \wedge dn$, so

$$\begin{aligned} \dot{\theta} &= \omega_d(n) - \frac{a-2n}{\sqrt{n(a-n)}}C_{ij}\cos\theta \\ \dot{n} &= 2C_{ij}\sqrt{n(a-n)}\sin\theta. \end{aligned} \tag{37}$$

Note that if $C_{ij} \neq 0$ the equilibria move away from the poles (to see this it is best to change to coordinates $x = \sqrt{n(a-n)}\cos\theta, y = \sqrt{n(a-n)}\sin\theta$). The resulting dynamics may not be meaningful, however, as the errors can be of similar order if C_{ij} (or $\varepsilon^{d(i,j)}$ in the case of nearest neighbour coupling ε) is comparable to $\omega_d(n)$. So let's apply the procedure of Section 5 to obtain a better slow manifold M_1.

For foliation choose the leaves $F_{\theta,n}$ to be the set of $U(1)$-equivalence classes of $\psi = e^{i\theta_i}\sqrt{(1-\eta)n}\ \delta_i + e^{i\theta_j}\sqrt{(1-\eta)(a-n)}\ \delta_j + \tilde{\psi}$ such that $\theta_j - \theta_i = \theta, \tilde{\psi}_i = \tilde{\psi}_j = 0$, and $\eta = N(\tilde{\psi})/a$. Then we seek the unique critical point of H_a on $F_{\theta,n}$ with η small. Differentiating with respect to $\bar{\psi}_s, s \in S \setminus \{i,j\}$, the equations for the critical point are:

$$\begin{aligned} &\frac{G}{a}\tilde{\psi}_s - \gamma_s|\tilde{\psi}_s|^2\tilde{\psi}_s - \sum_{r\notin\{i,j\}} C_{sr}\tilde{\psi}_r \\ &\quad = C_{si}e^{i\theta_i}\sqrt{(1-\eta)n} + C_{sj}e^{i\theta_j}\sqrt{(1-\eta)(a-n)}, \end{aligned} \tag{38}$$

where

$$\begin{aligned} G = {} & \gamma_i(1-\eta)n^2 + C_{ii}n + \gamma_j(1-\eta)(a-n)^2 + C_{jj}(a-n) \\ & +2C_{ij}\sqrt{n(a-n)}\cos\theta + \textstyle\sum_{r\notin\{i,j\}} C_{ir}\Re(e^{-i\theta_i}\tilde{\psi}_r)\sqrt{\frac{n}{1-\eta}} \\ & + \textstyle\sum_{r\notin\{i,j\}} C_{jr}\Re(e^{-i\theta_j}\tilde{\psi}_r)\sqrt{\frac{a-n}{1-\eta}}. \end{aligned} \tag{39}$$

To first order in $\{C_{si}, C_{sj} : s \in S \setminus \{i,j\}\}$, $\tilde{\psi}$ is given by the solution of the affine system

$$\frac{G_0}{a}\tilde{\psi}_s - \sum_r C_{sr}\tilde{\psi}_r = C_{si}e^{i\theta_i}\sqrt{n} + C_{sj}e^{i\theta_j}\sqrt{a-n}, \tag{40}$$

where $G_0 = \gamma_i n^2 + C_{ii}n + \gamma_j(a-n)^2 + C_{jj}(a-n) + 2C_{ij}\sqrt{n(a-n)}\cos\theta$ and $\eta = 0$. To obtain a first order slow manifold one should solve (39) and $N(\tilde{\psi}) = a\eta$ exactly for $\tilde{\psi}(\theta, n; a)$, and substitute the result into (33) and (34) to obtain some $H_{M_1}(\theta, n; a)$ and $\Omega_{M_1}(\theta, n; a) = \phi(\theta, n; a)\; d\theta \wedge dn$ for some function ϕ. Then the slow dynamics on M_1 is given by

$$\begin{aligned} \dot{\theta} &= \frac{1}{\phi}\frac{\partial H_{M_1}}{\partial n} \\ \dot{n} &= -\frac{1}{\phi}\frac{\partial H_{M_1}}{\partial \theta}. \end{aligned} \tag{41}$$

Many different phase portraits can occur, depending on the frequency difference function ω_d and the coupling. See Fig. 3 for some sketches. Parts (a-e) and (f-i) of the figure are sequences as the coupling is increased from zero for systems where ω_d has one sign change and no sign changes, respectively. For comparison, parts (j-m) of the figure show the case of two coupled oscillators with interchange symmetry, familiar from modelling of vibrations of water and other symmetric triatomic molecules, progressing from small to large amplitude at fixed coupling.

Because the slow dynamics is Hamiltonian and on a 2-sphere, all orbits are periodic except equilibria and separatrices (orbits asymptotic to equilibria). In particular, as a codimension-1 phenomenon achievable in some examples by suitable choice of total number a, there can be a periodic trajectory from one pole to the other and back, in which case the energy exchanges completely and periodically between the configurations they represent. A necessary and sufficient condition for this is that the poles have the same value of H_M, not be equilibria, and there be no separatrix dividing them (Ref. 76 neglects to mention the last condition). Limiting cases of codimension 3 and 5 are that one pole is a saddle and the other lies on a separatrix from it, or both poles are saddles and are connected by a separatrix.

Extensions of the TET theme have been proposed by Aubry, e.g. a transistor and an AND gate using discrete breathers, creating the concept of "breathonics".

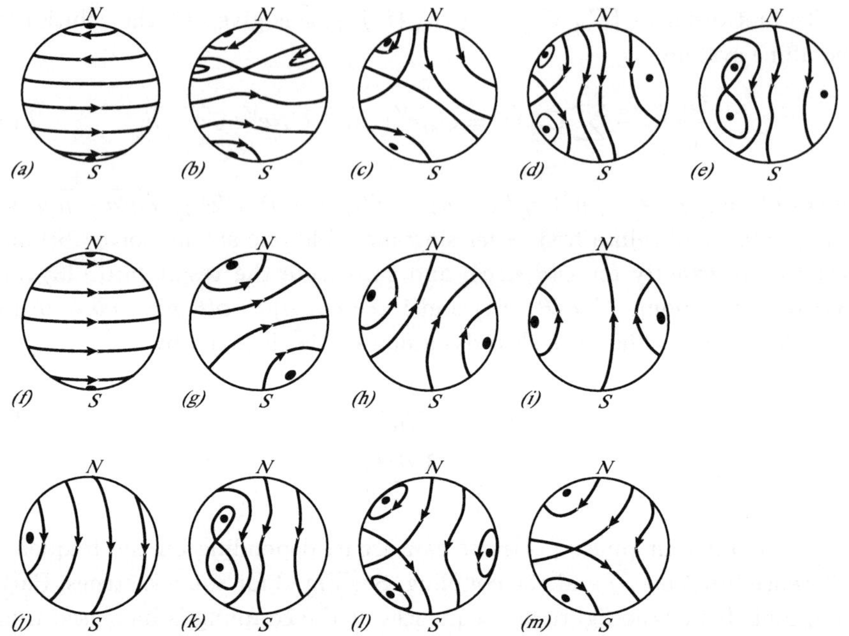

Fig. 3. Three series of examples of phase portraits for the effective Hamiltonian on a sphere: (a) uncoupled oscillators with one sign change in ω_d, (b) effect of generic weak coupling, (c) a possible effect of larger coupling, (d) yet larger coupling, (e) complete energy transfer; (f) an uncoupled case with no sign change, (g) effect of generic weak coupling, (h) a possible effect of larger coupling, (i) complete energy transfer; (j) a $\mathbb{Z}_2$-symmetric system in the linear regime (note that it exhibits complete energy transfer at a frequency proportional to the coupling, and the equilibria represent the symmetric and antisymmetric normal modes), (k) sufficient nonlinearity causes one of the normal modes to lose stability and create a pair of asymmetric equilibria, (l) greater nonlinearity makes the asymmetric equilibria take on a more local character, (m) yet greater nonlinearity stops the energy transfer.

One can also apply the method to DST systems with given coupling and two families of $U(1)$ discrete breathers $D_1(n)$, $D_2(n)$, assumed already known, parametrised by their number n, with frequencies $\omega_1(n)$ and $\omega_2(a-n)$ close for some choice of a, and non-degenerate equilibria of the reduced Hamiltonians H_n and H_{a-n} respectively; note that non-degeneracy requires conditions on both the phonon spectrum for the zero state and the frequencies of any shape modes[g] of the breathers. One makes a

[g] *Shape modes* on a spatially localised solution are oscillatory solutions of the linearised equations about it which are themselves spatially localised: for a $U(1)$ discrete breather

slow manifold by a superposition of the breathers with given total number: $M_0 = \{\psi = (e^{i\theta_1} D_1(n_1) + e^{i\theta_2} D_2(n_2))_\sim : N(\psi) = a, n_1, n_2 \geq 0, \theta_1, \theta_2 \in S^1\}$. It is again diffeomorphic to $\mathbb{C}P^1$ and the poles are equilibria, representing $D_1(a)$ and $D_2(a)$. One can then apply the method of Section 5 to construct a higher order slow manifold, but note that it will continue to include the exact discrete breathers $D_1(a)$ and $D_2(a)$ as equilibria at the poles, because they are present by assumption, so there is no energy transfer between them, except for the codimension-1 possibility that they are both saddles at the same energy and possess a separatrix joining them. There may, however, be energy transfer from near $D_1(a)$ to near $D_2(a)$.

As a second example, the method can be applied to compute the interaction of two widely separated Q-balls of close frequency: the effective dynamics is on a space of relative separation, relative velocity (or momentum), phase difference and frequency (or charge) difference, which has dimension $2N + 2$ for N space dimensions. This will be reported in a forthcoming paper.

One question for slow manifolds with $U(1)$ symmetry is to what accuracy one can recover the rate of change of phase of the internal oscillation. An immediate problem is to decide how the phase above one point of the reduced space is related to that above nearby points. One way is via the Hannay-Berry connection,[79] but note that in general it has non-trivial holonomy, i.e. following locally constant phase round a loop on the reduced space one may find one returns to a different phase from the initial one, the phase change being given by the integral of a certain 1-form around the loop. With this choice, I would expect the rate of change of the phase to be order r close to $\frac{\partial H_a}{\partial a}$, which is the instantaneous frequency. It would be interesting to prove (or disprove) this.

11. Internal oscillation: General Hamiltonians

Although $\mathbb{T}^d$ symmetry (and hence many $U(1)$ subgroups) arises to arbitrarily high order near a generic elliptic equilibrium point in a Hamiltonian system of d degrees of freedom (Birkhoff normal form), existence of an exact $U(1)$ symmetry is a special situation. Also the treatment of the previous section does not apply to internal oscillations of $U(1)$-symmetric systems if they are not $U(1)$ orbits. So it is important to generalise the method of the previous section. A first step in this direction was made in Ref. 6,

---

the question of existence of shape modes reduces to linearisation about an equilibrium, for which there are good methods, e.g. Ref. 78.

but there we could not see how to make iterative improvements beyond the first, which is rectified here.

The idea is to turn the problem of slow manifolds with internal oscillation into a standard slow manifold problem for *loop dynamics*, i.e. the induced dynamics on the loop space $\tilde{P} = \{z : \mathbb{R}/\mathbb{Z} \to P\}$, where each point of a loop evolves independently of the others, according to Hamilton's equations. This can be written[80] as a Hamiltonian system on $\tilde{P}$, by choosing $\tilde{H}(z) = \oint H(z(s))\, ds$ and $\tilde{\Omega}(\zeta, \xi) = \oint \Omega(\zeta(s), \xi(s))\, ds$. Now loop dynamics has a $U(1)$ symmetry: $(R_\theta z)(s) = z(s+\theta), \theta \in \mathbb{R}/\mathbb{Z}$. The associated conserved quantity is the *area* $A(z) = \int_{D_z} \Omega$ (often $A/2\pi$ is called "action", but this word is used for so many different things that I prefer to avoid it), where D_z is any disc spanning the loop z (the value of $A(z)$ might depend on the homotopy class of D_z but it does not vary under continuous deformation of D_z fixing the boundary z). So we can reduce loop dynamics to a Hamiltonian dynamics on $\tilde{P}_a = \{z/U(1) : A(z) = a\}$ and apply the Hamiltonian slow manifold theory of Section 5. A slow manifold in $\tilde{P}_a$ gives an approximately invariant manifold in P consisting of approximate periodic orbits of area a, evolving by an effective Hamiltonian.

To carry out this programme, one first needs to establish what it means for the normal dynamics to be fast in $\tilde{P}_a$. The answer is the *non-resonance condition* that the normal Floquet multipliers of the given periodic orbit should avoid $+1$. This is equivalent to all integer multiples of its frequency avoiding both the phonon spectrum and the frequencies of any shape modes. The non-resonance condition is encountered on trying to make $\tilde{H}_a$ critical on the leaves of a symplectically orthogonal foliation, e.g. representing loops by Fourier series, one needs the variation of $\tilde{H}$ with respect to all Fourier coefficients (except those representing tangents to the slow manifold) to be non-degenerate.

Examples to which this method can be applied include generalised multibreathers, targeted energy transfer in general Hamiltonian networks, and travelling discrete breathers, all of which were sketched in Ref. 6. Some examples of travelling discrete breather were treated from this improved point of view in Ref. 81. In particular, the phenomena of Fig. 3 can be expected in general Hamiltonian networks, not just DST.

Example 15: Existence of multibreathers

One of the by-products of the method of this section is existence for Hamiltonian networks of at least n families (parametrised by area) of true periodic orbits (and 2^{n-1} if all are non-degenerate) near superpositions of

n spatially separated near-resonant (i.e. frequencies in rational ratio) DBs with all amplitudes nonzero and satisfying a generic condition (to be given below), called *multibreathers* (Ref. 6 treated the particular case of DBs in a weakly coupled network). The idea is that by applying one iteration of my method, there is a first order symplectic slow manifold (with internal oscillation) consisting of loops that look like superpositions of the n DB with all possible choices of total area a, relative phases θ_i and their rates of change p_i. Thus the reduced slow manifold for given area is diffeomorphic to $\mathbb{T}^{n-1} \times I^{n-1}$, I being an interval. Since the DB are spatially separated, the effective Hamiltonian H_a has small first derivative in p. Next, I formulate the required generic condition on the DB (which we neglected to do in Ref. 6). Denote the energy of the ith DB as a function of its area a by $E_i(a)$. Then its frequency is $\omega_i = E_i'(a)$ (note that the frequency is signed). Let $\beta_i = \omega'(a)$. We are supposing the frequencies to be in close to rational ratio, say $\omega \propto k \in \mathbb{Z}^n$. Then the condition is that $\sum_i (k_i \prod_{j \neq i} \beta_j) \neq 0$. Under this condition, the effective Hamiltonian H_a has invertible second derivative in p. Thus for each θ there is a locally unique $p(\theta)$ such that the first derivative is zero. These points form a smooth submanifold diffeomorphic to $\mathbb{T}^{n-1}$. Let $V(\theta) = H_a(\theta, p(\theta))$. Then V is a smooth function on an $(n-1)$-torus so by Liusternik-Shnirelman theory (e.g. Ref. 77) it has at least n critical points, and by Morse theory if all its critical points are non-degenerate there are at least 2^{n-1}. Critical points of V correspond to true periodic orbits, giving the multibreathers. If the method is carried to second order then one obtains the exact slow part of the linearised dynamics about the periodic orbits (in contrast to Ref. 6 where we were able to obtain only approximate linearised dynamics).

Example 16: Travelling discrete breathers

The method allows one to identify parameter regimes for Hamiltonian systems on periodic lattices where DB should travel well,[81] and puts the concept of Peierls-Nabarro barrier for DB on a firm footing (extending the proposal of Ref. 82 from the $U(1)$-symmetric to the general case). The idea is that if there is a slow manifold with internal oscillation of at least first order, containing a DB and its translate by one lattice vector for some value(s) of area a, and the slow dynamics has an orbit which passes from a point close to the DB to its translate by the lattice vector, then that orbit represents a periodically travelling DB, so the original system has a solution which looks approximately like a travelling DB for a long time. The accuracy is better the slower the motion on the orbit (because the slow manifold is at least first order).

In the case of a 2D slow manifold, existence of such an orbit is a consequence of conservation of the effective Hamiltonian, provided the "torsion" (to be defined below) is non-zero; and the maximum speed along the orbit can be quantified roughly by the following concept of barrier to motion of the DB (which is defined for any dimension of slow manifold). Choose coordinates q on the slow manifold to represent the position of the DB. The derivative of the effective Hamiltonian H_a at constant q is small. Assume that the torsion β, i.e. the second derivative of H_a at constant q, is invertible. Then there is a locally unique critical point $z(q,a)$ with respect to variations with constant q and a. Let $V_a(q) = H_a(z(q,a))$. For curves γ connecting the DB to its translate at constant a let $b(\gamma)$ be the variation of H_a on the curve (its maximum minus its minimum). Define the *barrier* $B(a)$ to be the minimum of $b(\gamma)$ over such curves. Note that the given DB is a critical point of V_a and curves minimising b must pass through at least one other critical point, which will also be a DB, so $B(a)$ is given by the difference in energy of some pair of DBs. Then there are orbits of the desired type with maximum speed approximately $\sqrt{B(a)/\beta}$.

An interesting question is what significance the barrier has if the slow manifold is more than two-dimensional (probably not much if the torsion β is indefinite). The same question can be posed for the simpler case of travelling dislocations in more than 1D (where the analogue of the torsion is the kinetic energy, so usually positive definite).

Note that we do not expect exact travelling DB in general (article in preparation), but there can occur[83] exact "nanopterons", i.e. travelling solutions asymptotic to small amplitude phonons at large distance.

Another example to which it would be interesting to apply the method is the interaction of two gyrating charged particles in a 3D magnetic field. One might ask whether the method could be applied to study the interaction of kinks with shape modes, as is required to understand the phenomena reported in Ref. 51 for example. Unfortunately, in a spatial continuum like that of Ref. 51, the phonon spectrum is unbounded, so it is impossible to satisfy the non-resonance condition. The resonance effects are weak for small amplitude of the shape modes, however, so it might nevertheless be possible to construct useful approximate slow manifolds including them. Ref. 37 reports some work in such a direction, including the effects of weak spatial discretisation and estimating the radiative losses.

Note that for $U(1)$-symmetric systems, the methods of this and the previous section can be combined to treat 2-frequency quasiperiodic internal oscillations, provided one of the frequencies corresponds to $U(1)$ orbits. First quotient out the $U(1)$ symmetry as in the previous section, and then consider loop dynamics as in this section, on the reduced space (alternatively, one can consider loop dynamics on the full space and then quotient by the action of $\mathbb{T}^2$ generated by the $U(1)$ symmetry and the phase-shift along loops).

12. Bounds on time evolution

The concern of many authors in the theory and application of slow manifolds is to know for how long true solutions remain close to the solutions of the approximate dynamics, or at least to the slow manifold. There are four main ways of obtaining estimates on these.

The first is *Gronwall estimates*. The idea here is to choose a norm and then use the simple fact that $\dot{e} = f(e,t)$, representing the evolution of the error $e(t)$, with $|e(0)| \leq \eta$, $|f(0,t)| \leq \delta$ and $||\frac{\partial f}{\partial e}|| \leq M$ for $|e| \leq E$, some $E \geq \eta$, implies that $|\dot{e}| \leq \delta + M|e|$. Gronwall's lemma says that it follows that $|e(t)| \leq (\eta + \frac{\delta}{M})e^{Mt} - \frac{\delta}{M}$ as long as the right hand side does not exceed E (this might look a trivial result but the difficulty is that $|e(t)|$ is not necessarily a differentiable function of t). So the error remains at most E as long as

$$t \leq \frac{1}{M} \log \frac{EM + \delta}{\eta M + \delta}. \tag{42}$$

One can do much better than Gronwall estimates, however, if the motion near the given orbit is oscillatory or contracting, because the bound $||\frac{\partial f}{\partial e}|| \leq M$ does not make any distinction between these and the expanding case. Such an improvement is crucial for slow manifolds with oscillatory or contracting normal dynamics because M is of necessity large on the slow timescale and hence the Gronwall estimate (42) is short. One way to obtain improvement is, rather than just a norm, to use a suitable inner product $\langle\cdot,\cdot\rangle$. This is the second method (e.g. Ref. 84). Then $\dot{e} = f(e,t)$ implies

$$\frac{d}{dt}||e||^2 = 2\langle e, f(e,t)\rangle = 2\langle e, f(0,t) + \int_0^1 \frac{\partial f}{\partial e}(\lambda e, t) e \, d\lambda\rangle \leq 2\delta||e|| + 2\mu||e||^2, \tag{43}$$

where μ is an upper bound on the spectrum of the symmetric part (with respect to $\langle\cdot,\cdot\rangle$) of $\int_0^1 \frac{\partial f}{\partial e}(\lambda e, t)\, d\lambda$ (which in contrast to the norm can be zero or negative). It follows that

$$||e(t)|| \leq \begin{cases} (\eta + \frac{\delta}{\mu})e^{\mu t} - \frac{\delta}{\mu}, & \mu > 0 \\ \eta + \delta t, & \mu = 0 \\ \frac{\delta}{|\mu|} + (\eta - \frac{\delta}{|\mu|})e^{\mu t}, & \mu < 0 \end{cases} \tag{44}$$

In particular, if $\mu < 0$ then the solution remains within $\frac{\delta}{|\mu|}$ if it starts so. One can often improve the bounds further by choosing an "adapted" inner product (cf. Ref. 85 for the case of a DB of a dissipative network): in the oscillatory case this means one which is chosen to make oscillatory motion round rather than elliptical. Note, however, that in general one would have to allow the inner product to depend on position on the slow manifold and this introduces an additional term in the expression for $\frac{d}{dt}||e(t)||^2$. One can apply the result to the total error or just to the normal component. In the latter case, one can subsequently bound the tangential error in terms of the tangential derivative using the same method.

The above procedure applied to the normal component of the error is a special case of the third method, called the *energy method*. It applies to systems for which there is an "energy" function $E : P \to \mathbb{R}$ (the Hamiltonian is the natural one for many Hamiltonian systems) such that the set $E_<(K) = \{z \in P : E(z) \leq K\}$ is (or at least has as a connected component) a small neighbourhood of the slow manifold for suitable K, and $\frac{dE}{dt}$ is bounded above by some function of position in P. The nicest case is $\frac{dE}{dt} \leq f(E)$ with f Lipschitz. Then $E(t) \leq E_m(t)$, the solution of $\dot{E} = f(E)$ with given $E(0)$, and hence $z(t) \in E_<(E_m(t))$. Once the normal deviation is controlled by this bound, tangential errors can be controlled by Gronwall estimates on the tangential dynamics. For examples of this method, see Refs 4, 2, 7, 28.

The fourth method[49] is *Nekhoroshev estimates*. This applies only to Hamiltonian systems, locally near integrable. By the use of normal form transformations the dynamics in a neighbourhood can be put in a form where the rate of change of half the variables is of arbitrary order in the deviation from integrable in a large fraction of the neighbourhood, and hence the deviation from the orbits on the slow manifold can be bounded as a function of time. In the two degree of freedom case, the deviation can often be bounded for all time because the slow manifold generically has an invariant neighbourhood bounded by KAM tori; the size of the smallest such neighbourhood can be estimated. This method has been used to bound the rate of possible instability of DB [86] and KAM tori,[87] and certain Hamiltonian slow manifolds.[88]

13. Weak Damping

One question that some authors raise about slow-fast systems with oscillatory fast motion is why the observed motion is nearly devoid of fast oscillations even though the models are Hamiltonian (e.g. Ref. 89). I propose the answer is *damping*. A relatively weak amount of damping of the fast oscillations can cause them to die out on a timescale short or comparable to the slow timescale and to be only very weakly excited by the slow motion; bounds to prove this can be obtained by constructing a high order slow manifold and using the inner product method of the previous section, provided it is adapted carefully to the normal oscillations. An illustration of this effect is given in Ref. 33, where numerical damping is found to cause solutions of a Hamiltonian system to converge close to a slow manifold.

Thus in the presence of damping, one can expect to obtain an attracting invariant set with slow motion on it, but it is not necessarily a submanifold (this expectation is shared by the authors of Ref. 90, who propose to call it a "slow quasimanifold"), because candidate submanifolds may contain orbits with stronger tangential contraction than normal contraction, and so the result of Ref. 30 implies generic non-existence of an invariant submanifold. I expect the situation to be closely analogous to that for *Birkhoff attractors*[91] (see also Refs 92, 93). These arise on applying small damping to an area-preserving twist map of an annulus (equivalently, to a typical two degree of freedom Hamiltonian system). If the unperturbed system has a "zone of instability" (meaning a minimal invariant subannulus homotopic to the full annulus) which is mapped into itself by the perturbed map then for large damping the maximal invariant set in the zone of instability is diffeomorphic to a circle, but for weak damping it is an "indecomposable continuum" (a continuum is a compact connected set and it is indecomposable if it can not be written as a (non-disjoint) union of two proper subcontinua). In particular, it is not a circle, and is quite complicated. There is an extensive theory of such invariant sets, and I think it is ripe for exploitation in application areas such as those discussed here (for an introduction and application to a different area, see Ref. 94).

Acknowledgements

My interest in this topic was stimulated by Michael McIntyre and the Newton Institute session on "Mathematics of atmosphere and ocean dynamics",

July – Dec 1996; the ideas of Sections 2–4 were developed during that time. In parallel, I had become convinced that in Hamiltonian systems there should be some approximate dynamics associated with the "effective potential" that I'd derived to prove existence of multibreathers,[95] a conviction which initiated the work[6] with Ahn and Sepulchre, so starting off Section 9 and (in a backwards move) the simpler case of Section 5. Shortly afterwards, Richard Battye stimulated my interest in Q-balls, which led to Section 10. The idea of Section 11, which links Ref. 6 with the rest of the work and strengthens it greatly, came to me while preparing a talk for the SIAM Conference on Dynamical Systems and Applications in May 2001. The applications to travelling DB were developed in collaboration with Jacques-Alexandre Sepulchre and benefited from funding by the CNRS and the Research and Training Network LOCNET (EC contract HPRN-CT-1999-00163). I had intended to present the principal results of this paper at the conference in June 2001 in honour of Floris Takens (who wrote one of the important early papers[2] on the subject), but I was unable to attend. Nevertheless, I'd like to dedicate it to him. I used a draft version of these notes in my MSc module in Autumn 2002 and I'm grateful to Diogo Pinheiro and David Stern for many comments which led to improvements, and also to Michael McIntyre for comments on the final version. Finally I'd like to thank Elaine Greaves-Coelho for preparing Figures 2 and 3 and Gill Walton for assistance in transforming the manuscript into World Scientific style.

References

1. Rubin H, Ungar P, Motion under a strong constraining force, Commun Pure Appl Math 10 (1957) 65–87.
2. Takens F, Motion under the influence of a strong constraining force, Lecture Notes in Math 819, eds Nitecki Z, Robinson C (Springer, 1980), 425–445.
3. Kampen NG van, Elimination of fast variables, Phys Rep 124 (1985) 69–160.
4. Kozlov VV, Neishtadt AI, On the realisation of holonomic constraints, PMM USSR 54 no.5 (1990) 705–8.
5. Lynch P, The slow equations, Q J R Meteorol Soc 115 (1989) 201–219.
6. Ahn T, MacKay RS, Sepulchre J-A, Dynamics of relative phases and generalised multibreathers, Nonlinear Dynamics 25 (2001) 157-182.
7. Gelfreich V, Lerman L, Almost invariant elliptic manifolds in a singularly perturbed Hamiltonian system, Nonlinearity 15 (2002) 447–457.
8. Bornemann FA, Schütte C, Homogenization of a Hamiltonian system with a strong constraining potential, Phys D 102 (1997) 57–77.
9. Spohn H, Teufel S, Adiabatic decoupling and time-dependent Born-Oppenheimer theory, Commun Math Phys 224 (2001) 113–132.
10. Just W, Kantz H, Rödenbeck C, Helm M, Stochastic modelling: replacing

fast degrees of freedom by noise, J Phys A 34 (2001) 3199–3213.

11. Keener J, Sneyd J, Mathematical Physiology (Springer, 1998).
12. Fenichel N, Geometric singular perturbation theory for ordinary differential equations, J Diff Eq 31 (1979) 53–98.
13. Jones CKRT, Kaper TJ, Kopell N, Tracking invariant manifolds up to exponentially small errors, SIAM J Math Anal 27 (1996) 558–577.
14. Kreiss H-O, Problems with different time scales, Acta Num 1992 (1991) 109–139.
15. Maas U, Efficient calculation of intrinsic low-dimensional manifolds for the simplification of chemical kinetics, Comput Vis Sci 1 (1998) 69–81.
16. Deuflhard P, Heroth J, Dynamic dimension reduction in ODE models, in: Scientific Computing in Chemical Engineering, eds Keil F, Mackens W, Voss H, Werther J (Springer, 1996) 29–43.
17. Krauskopf B, Green K, Computing unstable manifolds of periodic orbits in delay differential equations, J Comp Phys 186 (2003) 230–249.
18. Jorba A, A methodology for the numerical computation of normal forms, centre manifolds and first integrals of Hamiltonian systems, Experim Math 8 (1999) 155–195.
19. Russell RD, Sloan DM, Trummer MR, Some numerical aspects of computing inertial manifolds, SIAM J Sci Comput 14 (1993) 19–43.
20. Jolly MS, Rosa R, Temam R, Accurate computations on inertial manifolds, SIAM J Sci Comp 22 (2001) 2216–38.
21. Robinson JC, Computing inertial manifolds, Disc Cont Dyn Sys 8 (2002) 815–833.
22. Broer HW, Osinga HW, Vegter G, Algorithms for computing normally hyperbolic invariant manifolds, ZAMP 48 (1997) 480–524.
23. Ebin DG, The motion of slightly compressible fluids viewed as motion with strong constraining force, Ann Math 105 (1977) 141–200.
24. Amick CJ, Kirchgässner K, A theory of solitary water waves in the presence of surface tension, Arch Rat Mech Anal 105 (1989) 1–49.
25. Stuart D, Perturbation theory for kinks, Commun Math Phys 149 (1992) 433–462.
26. Stuart D, Dynamics of Abelian Higgs vortices in the near Bogomolny regime, Commun Math Phys 159 (1994) 51–91.
27. Stuart D, The geodesic approximation for the Yang-Mills-Higgs equations, Commun Math Phys 166 (1994) 149–190.
28. Stuart D, Modulational approach to stability of non-topological solitons in semilinear wave equations, J Math Pures Appl 80 (2001) 51–83.
29. McIntyre ME, Roulstone I, Are there higher-accuracy analogues of semi-geostrophic theory? in: Large-scale atmosphere-ocean dynamics II: geometric methods and models, eds Norbury J, Roulstone I (Cambridge, 2002) 301–364.
30. Mañé R, Persistent manifolds are normally hyperbolic, Bull Am Math Soc 80 (1974) 90-91.
31. Kreiss H-O, Lorenz J, Manifolds of slow solutions for highly oscillatory problems, Indiana U Math J 42 (1993) 1169–91.
32. Kreiss H-O, Lorenz J, On the existence of slow manifolds for problems with

different timescales, Phil Trans R S Lond A 346 (1994) 159–171.
33. Lubich C, Integration of stiff mechanical systems by Runge-Kutta methods, ZAMP 44 (1993) 1022–53.
34. Arnol'd VI, Mathematical methods of classical mechanics (Springer, 2nd ed, 1989).
35. Arnol'd VI, Kozlov VV, Neishtadt AI, Mathematical aspects of classical and celestial mechanics, Ch5, in: Dynamical Systems III, ed Arnol'd VI (Springer, 1988).
36. Benettin G, Ferrari G, Galgani L, Giorgilli A, An extension of the Poincaré-Fermi Theorem on the nonexistence of invariant manifolds in nearly integrable Hamiltonian systems, Il Nuovo Cim 72B (1982) 137–148.
37. Kevrekidis PG, Weinstein MI, Dynamics of lattice kinks, Physica D 142 (2000) 113–152.
38. MacKay RS, A variational principle for invariant odd-dimensional submanifolds of an energy surface for Hamiltonian systems, Nonlin 4 (1991) 155–7.
39. Farantos SC, Exploring molecular vibrational motion with peridoic orbits, Int Rev Phys Chem 15 (1996) 345–374.
40. Lochak P, Meunier C, Multiphase averaging for classical systems (Springer, 1988).
41. Bokhove O, Shepherd TG, On Hamiltonian balanced dynamics and the slowest invariant manifold, J Atmos Sci 53 (1996) 276–297.
42. Lorenz EN, Attractor sets and quasi-geostrophic equilibrium, J Atmos Sci 37 (1980) 1685–99.
43. Lorenz EN, On the existence of a slow manifold, J Atmos Sci 43 (1986) 1547–57.
44. Camassa R, On the geometry of an atmospheric slow manifold, Physica D 84 (1995) 357–397.
45. Gelfreich V, Lerman L, Long periodic orbits and invariant tori in a singularly perturbed Hamiltonian system, Physica D 176 (2003) 125–146.
46. Lebovitz NR, Neishtadt AI, Slow evolution in perturbed Hamiltonian systems, Stud Appp Math 92 (1994) 127–144.
47. Lang S, Introduction to Differentiable Manifolds (Wiley, 1962).
48. Neishtadt AI, On the accuracy of conservation of adiabatic invariants, J Appl Math Mech 45 no.1 (1982) 58–63.
49. Nekhoroshev NN, An exponential estimate for the stability time of Hamiltonian systems close to integrable ones, Russ Math Surveys 32 (1977) 1–65.
50. Walton PB, Exponential asymptotics in slow-fast systems, PhD thesis (DAMTP, Cambridge, 2002).
51. Campbell DK, Schonfeld JF, Wingate CA, Resonance structure in kink-antikink interactions in ϕ^4 theory, Physica D 9 (1983) 1-32.
52. Aigner AA, Champneys AR, Rothos VM, A new barrier to the existence of moving kinks in Frenkel-Kontorova lattices, Physica D (2003), to appear.
53. Friesecke G, Wattis J, Existence theorem for solitary waves on lattices, Commun Math Phys 161 (1994) 391–418.
54. Peyrard M, Remoissenet M, Solitonlike excitations in a one-dimensional atomic chain with a nonlinearly deformable substrate potential, Phys Rev

B 26 (1982) 2886–99.
55. Speight JM, Ward RS, Kink dynamics in a novel discrete sine-Gordon system, Nonlinearity 7 (1994) 475–484.
56. Korepin VE, Bogoliubov NM, Izergin AG, Quantum inverse scattering method and correlation functions (Cambridge, 1993).
57. Willis C, El-Batanouny M, Stancioff P, Sine-Gordon kinks on a discrete lattice I: Hamiltonian formulation, Phys Rev B 33 (1986) 1904–11.
58. Braun OM, Kivshar YS, Nonlinear dynamics of the Frenkel-Kontorova model with impurities, Phys Rev B 43 (1991) 1060–73.
59. Tomboulis E, Canonical quantization of nonlinear waves, Phys Rev D 12 (1975) 1678–83.
60. Willis CR, El-Batanouny M, Burdick S, Boesch R, Sodano P, Hamiltonian dynamics of the double sine-Gordon kink, Phys Rev B 35 (1987) 3496–3505.
61. Whitham GB, Linear and nonlinear waves (Wiley, 1974).
62. Scott A, Nonlinear Science: Emergence and dynamics of coherent structures (Oxford, 1999)
63. Caputo JG, Flytzanis N, Soerensen MP, The ring laser configuration studied by collective coordinates, J Opt Soc Am B 12 (1995) 139–145.
64. Burzlaff J, Zakrzewski WJ, $\mathbb{C}P^2$ soliton scattering: the collective cordinate approximation, Nonlin 11 (1998) 1311–20.
65. Salmon R, Practical uses of Hamilton's principle, J Fluid Mech 132 (1983) 431–444.
66. Malomed BA, Potential of interaction between two-a dn three-dimensional solitons, Phys Rev E 58 (1998) 7928–33.
67. McLaughlin D, Scott AC, Perturbation analysis of fluxon dynamics, Phys Rev A 18 (1978) 1652–80.
68. Marsden JE, Ratiu TS, Introduction to Mechanics and Symmetry (Springer, 2nd ed, 1999).
69. Theiss J, Hamiltonian balanced models in the atmosphere, PhD thesis (DAMTP, Cambridge, 1999).
70. Battye RA, Sutcliffe PM, Q-ball dynamics, Nucl Phys B 590 (2000) 329–363.
71. Poincaré H, Les méthodes nouvelles de la mécanique céleste, tome 1 (Gauthier Villars, Paris, 1892).
72. Chirikov BV, A universal instability of many-dimensional oscillator systems, Phys Rep 52 (1979) 264–379.
73. Moser JK, Regularisation of Kepler's problem and the averaging method on a manifold, Commun Pure Appl Math 23 (1970) 609–636.
74. Barré J, Bouchet F, Dauxois T, Ruffo S, Birth and long-time stabilization of out-of-equilibrium coherent structures, Eur Phys J B 29 (2002) 577–591.
75. MacKay RS, Chaos in three physical systems, Equadiff 2003 proceedings (World Sci, to appear).
76. Aubry S, Kopidakis G, Morgante AM, Tsironis GP, Analytic conditions for targeted energy transfer between nonlinear oscillators or discrete breathers, Physica B 296 (2001) 222–236.
77. McDuff D, Salamon D, Introduction to Symplectic Topology (Oxford, 1995).
78. Baesens C, Kim S, MacKay RS, Localised modes on localised equilibria,

Physica D 113 (1998) 242-7.
79. Weinstein A, Connections of Berry and Hannay type for moving Lagrangian submanifolds, Adv Math 82 (1990) 133–159.
80. Weinstein A, Bifurcations and Hamilton's principle, Math Z 159 (1978) 235–248.
81. MacKay RS, Sepulchre J-A, Effective Hamiltonian for travelling discrete breathers, J Phys A 35 (2002) 3985-4002.
82. Kivshar YS, Campbell DK, Peierls-Nabarro barrier for highly localized nonlinear modes, Phys Rev E 48 (1993) 3077–81.
83. Iooss G, Kirchgässner K, Travelling waves in a chain of coupled nonlinear oscillators, Commun Math Phys 211 (2000) 439–464.
84. Hirsch M, Smale S, Differential equations, dynamical systems and linear algebra (Academic, 1974).
85. MacKay RS, Sepulchre J-A, Stability of discrete breathers, Physica D 119 (1998) 148–162.
86. Bambusi D, Exponential stability of breathers in Hamiltonian networks of weakly coupled oscillators, Nonlinearity 9 (1996) 433–457.
87. Perry AD, Wiggins S, KAM tori are very sticky, Physica D 71 (1994) 102–121.
88. Benettin G, Galgani L, Giorgilli A, Realization of holonomic constraints and freezing of high frequency degrees of freedom in the light of classical perturbation theory Pt II, Commun Math Phys 121 (1989) 557–601.
89. Wirosoetisno D, Shepherd TG, Averaging, slaving and balance dynamics in a simple atmospheric model, Phys D 141 (2000) 37–53.
90. Ford R, McIntyre ME, Norton WA, Balance and the slow quasimanifold: some explicit results, J Atmos Sci 57 (2000) 1236–54.
91. Birkhoff GD, Sur quelques courbes fermées remarquables, Bull Soc Math France 60 (1932) 1–26.
92. Casdagli M, Periodic orbits for dissipative twist maps, Ergod Th Dyn Sys 7 (1987) 167–173.
93. Le Calvez P, Propriétés des attracteurs de Birkhoff, Ergod Th Dyn Sys 8 (1987) 241–310.
94. Sanjuan M, Kennedy J, Ott E, Yorke JA, Indecomposable continua and the characterisation of strange sets in nonlinear dynamics, Phys Rev Lett 78 (1997) 1892–5.
95. MacKay RS, Dynamics of networks: features which persist from the uncoupled limit, in: Stochastic and spatial structures of dynamical systems, eds Strien SJ van, Verduyn Lunel SM (N Holland, 1996) 81–104.

CHAPTER 4

LOCALIZED EXCITATIONS IN JOSEPHSON ARRAYS. PART I: THEORY AND MODELING

Juan J. Mazo

Departamento de Física de la Materia Condensada
Universidad de Zaragoza, 50009 Zaragoza, Spain

Instituto de Ciencia de Materiales de Aragón (ICMA),
C.S.I.C. - Universidad de Zaragoza, 50009 Zaragoza, Spain

Instituto de Biocomputación y Física de Sistemas Complejos (BIFI),
Universidad de Zaragoza, 50009 Zaragoza, Spain

E-mail: juanjo@unizar.es

Josephson-junction arrays are excellent experimental systems for the study of nonlinear phenomena in general and nonlinear localised excitations in particular. This chapter is an introduction to the physics of vortices, kinks and breathers in Josephson arrays. Special emphasis is placed in the description of discrete breather solutions.

1. Introduction

The concept of coherent structures or coherent excitations has important consequences when applied to condensed matter systems. Spatially or temporally coherent structures appear in many nonlinear extended systems. Such structures usually can be characterized by marked particle-like properties. In the past few years, these notions have become fundamental for understanding many problems and their implications extend over different fields of the physics of continuous and discrete systems.[1]

Josephson-junction (JJ) arrays are a very well example of an experimental system where such type of structures appear. Examples are vortices, kinks and discrete breathers. This article is an introduction to the physics of these excitations with special attention given to the study of discrete breather (DB) solutions.

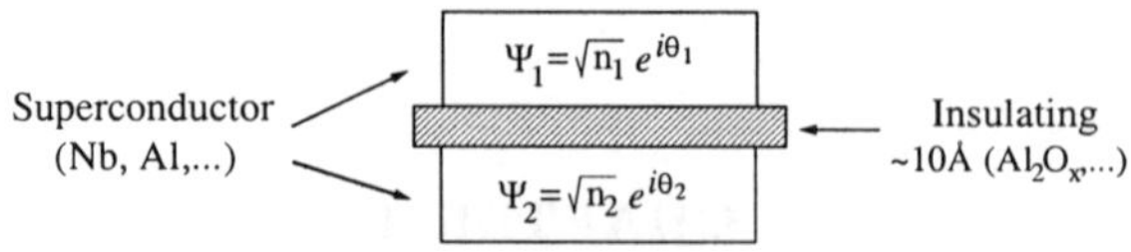

Fig. 1. Schematic of a Josephson tunnel junction.

In section 2, we will introduce the main aspects of the physics of single JJs. In section 3 we will present a method to model Josephson arrays. This section is focused in the introduction of the equations that we will use to study the different localized excitation in the arrays. In section 4 we will review some aspects of the work about vortices in two-dimensional JJ arrays and kinks in one-dimensional parallel arrays. Section 5 is the main section of this report. There, we will study discrete breathers in Josephson arrays. We focus on theoretical and numerical results since most of the experimental details are included in the chapter 5 of this book written by Alexey Ustinov.

2. The single Josephson junction

2.1. *Josephson effect*

In his work of 1962[2] Brian Josephson studied the tunnel of Cooper pairs between superconducting metals and predicted the celebrated Josephson effect. Since then, hundreds of works have been done which study the behavior of single junctions, JJ arrays, and other more complex devices with Josephson elements. Some books and references on Josephson effect in weakly coupled superconductors are Refs. 3-7

It is important to say however, that although mainly used in the context of superconductivity, the physics behind the name of "Josephson effect" applies to other weakly coupled macroscopic quantum systems.[8] Examples are the studies of the Josephson effect in weakly coupled superfluids (see Ref. 9 for instance) and the recent interest on Josephson effect in weakly coupled Bose-Einstein condensates (see Ref. 10 for instance).

2.2. *Superconducting tunnel junctions*

A Josephson tunnel junction is a solid state physics device which consist of two superconducting electrodes (usually Niobium or Aluminum) separated by a thin insulating barrier (usually an Aluminum oxide), see Fig. 1.

The physics of the junction is controlled by the value of the gauge

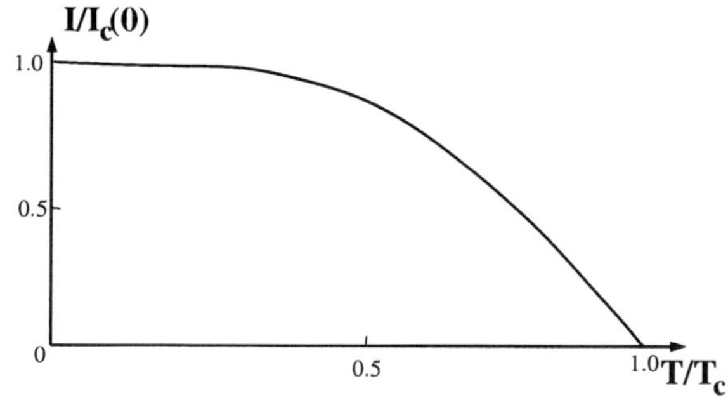

Fig. 2. Temperature dependence of the junction critical current.

invariant phase difference between the superconducting electrodes $\varphi = \theta_1 - \theta_2 - \frac{2e}{\hbar}\int_1^2 \vec{A}(\vec{r},t)d\vec{l}$, with θ_i the phase of the macroscopic wave function in electrode i ($\Psi_i = \sqrt{|\Psi_i|}e^{i\theta_i}$) and $\vec{A}$ the vector potential.

The basic equations for the Josephson effect are

$$I_s = I_c \sin\varphi \tag{1}$$

and

$$V = \frac{\hbar}{2e}\frac{d\varphi}{dt}. \tag{2}$$

They establish – *DC Josephson effect* – that at zero voltage (then φ is constant) it is possible to have a dc current across the junction. The maximum possible value of this current is I_c, the junction critical current. However – *AC Josephson effect* – in the presence of a constant voltage the junction responses with an ac current of frequency $2eV/\hbar$ (483.6 GHz/mV).

We want to mention here that the $\hbar/2e$ quotient can be also written in terms of the flux quantum unit $\Phi_0 = h/2e$. Thus $\hbar/2e = \Phi_0/2\pi$.

The potential energy associated with the supercurrent across the junction is given by

$$U_J = -E_J \cos\varphi, \tag{3}$$

with $E_J = \hbar I_c/2e$. A first requirement to observe Josephson effect is that the Josephson energy exceeds the thermal energy $E_J \gg k_B T$ ($I_c \gg 2ek_B T/\hbar$).

The junction critical current I_c has a strong temperature dependence (Fig. 2) which is usually approached by the Ambegaokar-Baratoff equation[11]

$$I_c R_n = \frac{\pi\Delta}{2e}\tanh(\Delta/2k_B T). \tag{4}$$

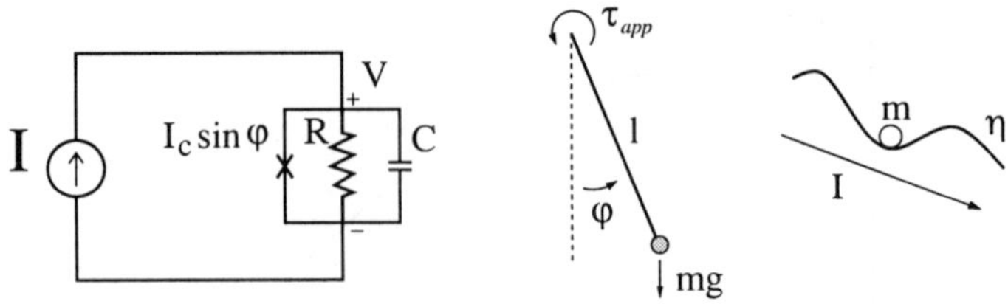

Fig. 3. RCSJ circuit model of the junction and two mechanical analogs: the forced and damped pendulum and the particle in the tilted washboard potential.

R_n is the normal state resistance of the junction and $\Delta(T)$ the superconducting gap energy. At T=0 we get $I_c R_n = \pi\Delta(0)/2e$ with $\Delta(0)=1.764 k_B T_c$ and for $T \to T_c$, $I_c R_n \simeq (2.34\pi k_B/e)(T_c - T)$.

To study the current-voltage (IV) characteristics of the junction we use the so called RCSJ (resistively and capacitively shunted junction) model[12,13] (see Fig. 3). In this model the total current through the junction is the sum of three contributions: the Josephson supercurrent, a resistive normal current (tunneling of normal carriers from one electrode to the other) and a capacitive channel (associated with the junction capacitance); $I = I_J + I_R + I_C$ with $I_J = I_c \sin\varphi$, $I_R = V/R$ and $I_C = CdV/dt$. Then

$$I = C\dot{V} + \frac{1}{R}V + I_c \sin\varphi. \tag{5}$$

If we apply the second Josephson relation $[V = (\Phi_0/2\pi)(d\varphi/dt)]$ and normalize current with respect to the junction critical current, $i = I/I_c$, time with respect to the Josephson plasma frequency $\omega_p = \sqrt{2\pi I_c/\Phi_0 C}$ and introduce the damping parameter $\Gamma = \sqrt{\Phi_0/2\pi I_c C R^2}$ [h], we obtain

$$i = \mathcal{N}(\varphi) = \ddot{\varphi} + \Gamma\dot{\varphi} + \sin\varphi. \tag{6}$$

This is the normalized equation for the dynamics of a single junction biased by an external current. This equation is identical to the equation for a forced and damped pendulum in a gravitational field or a particle in a tilted washboard potential $U(\varphi) = -E_J \cos\varphi - (\hbar I/2e)\varphi$ [mass $m = (\hbar/2e)^2 C$ and damping $\gamma = (\hbar/2e)^2(1/R)$], see Fig. 3. Both are simple mechanical analogs for the junction and illustrate that JJ devices are ideal experimental systems to probe basic nonlinear science results.

Figure 4 shows numerically computed IV curves of a junction at two different values of the damping. At large damping (figure at Γ=5) the voltage is

[h]The damping can be also defined in terms of the so-called quality factor $Q = 1/\Gamma$ or the Stewart-McCumber parameter $\beta_c = 1/\sqrt{\Gamma} = 2\pi I_c C R^2/\Phi_0$

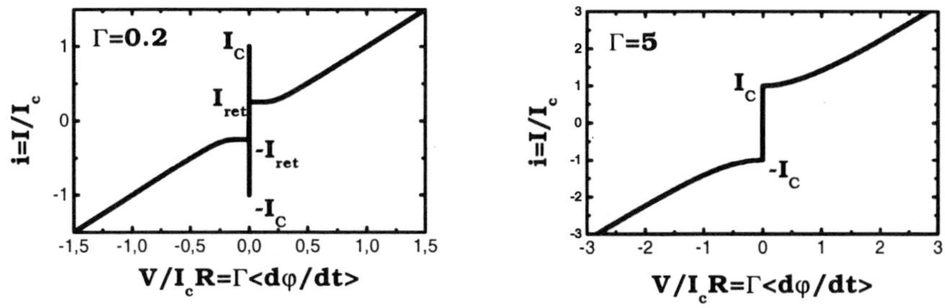

Fig. 4. IV curve for a JJ with a linear resistor biased by a dc current (Eq. 6). At $\Gamma = 0.2$ two solutions coexist for currents between the critical and the retrapping currents (underdamped case). At $\Gamma = 5.0$ (overdamped case) the voltage is a unique function of the current.

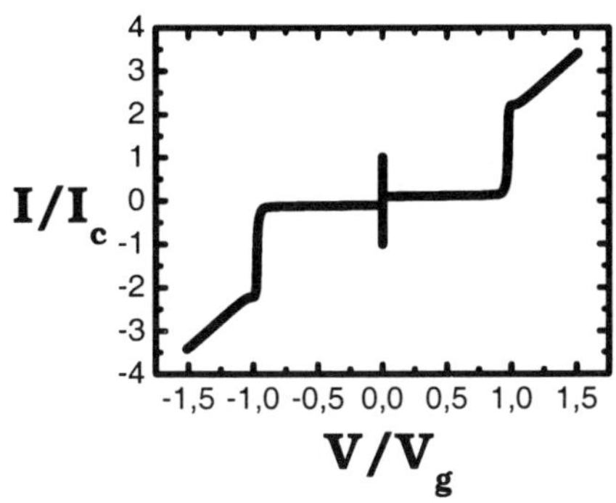

Fig. 5. Experimental IV curve of a single junction. The resistance of the junction below the gap voltage (subgap resistance) clearly differs from the resistance above the gap voltage (normal resistance).

a unique function of the current. It increases continuously from zero as soon as $I > I_c$ and approaches the ohmic relation ($I = V/R$, or $i = \Gamma\langle d\varphi/d\tau\rangle$ in normalized units) at high currents. At smaller values of the damping however the IV curve is hysteretic (see figure at Γ=0.2). Increasing the current, at $I = I_c$ the junction switches from the zero voltage state to the resistive branch $I = V/R$. If we now decrease the current, the voltage decreases continuously to zero at $I = I_{\mathrm{ret}}$. For small enough values of Γ, $I_{\mathrm{ret}}/I_c \simeq 4\Gamma/\pi$.

Figure 5 shows the experimental IV curve of a Niobium-Aluminum Oxide-Niobium tunnel junction. We observe that at I_c the voltages switches from 0 to the gap voltage $V_g = 2\Delta(T)/e$ ($V_g/I_cR_n = 4/\pi$ at small T). This voltage corresponds to the energy for breaking Cooper pairs. At larger values of the current the voltage increases and follows the ohmic dependence with a resistance given by the normal state resistance R_n. Decreasing the current the voltage decreases back to the gap voltage and then to zero at a small value of the current. This nonlinear dependence shows the existence of different dissipation regimes for voltages above and below the gap voltage. Transport above the gap voltage is governed by normal state electrons meanwhile transport below the gap is usually governed by the number of quasi-particles.

A simple approach to describe such behavior is to use the RCSJ model with a nonlinear resistance $R(V)$ such that $R{=}R_n$ if $V{>}V_g$ and $R{=}R_{sg}(T)$ if $V{<}V_g$. The sub-gap resistance usually shows a marked temperature dependence governed by $R_{sg}(T) \simeq R_n e^{\Delta/k_B T}$.

An expression for the quasi-particle tunneling current valid for $k_B T \ll \Delta$ and $V < V_g$ is given by[14]

$$I_{qp} = \frac{2}{eR_n} e^{-\Delta/k_B T} \left(\frac{2\Delta}{eV + 2\Delta} \right)^{1/2} (eV + \Delta) \sinh \left(\frac{eV}{2k_B T} \right) K_0 \left(\frac{eV}{2k_B T} \right) . \tag{7}$$

For some investigations and applications it is convenient to shunt the junction by a small resistance. In this case the equivalent resistance of the junction is small and voltage independent. It gives a large value of Γ and the overdamped limit of the RCSJ model is appropriate.

In other cases, especially when dealing with small junctions, in order to describe the behavior of the system it is essential to consider also the impedance of the external circuit.

Thermal fluctuations can be included in the model by the addition of a noise current source $\tilde{I}(t)$ with $\langle \tilde{I}(t) \rangle = 0$ and $\langle \tilde{I}(t)\tilde{I}(t') \rangle = (2k_B T/R)\delta(t - t')$. The total current, in normalized units, is

$$i = \mathcal{N}(\varphi) = \ddot{\varphi} + \Gamma\dot{\varphi} + \sin\varphi + \tilde{i} \tag{8}$$

with $\langle \tilde{i}(\tau) \rangle = 0$ and $\langle \tilde{i}(\tau)\tilde{i}(\tau') \rangle = (2\Gamma k_B T/E_J)\delta(\tau - \tau')$.

In the presence of temperature, close to I_c the junction can escape from the zero voltage state via thermal fluctuations, and close to $I_{\rm ret}$ can retrap to the zero voltage state (Fig. 6). Such process can be analyzed in terms of the escape of a single particle from a potential well. The escape rate $\Gamma_{\rm esc}$ in the classical regimen can be approached to

$$\Gamma_{\rm esc} = a_t \frac{\omega}{2\pi} \exp\left(-\frac{\Delta U}{k_B T} \right) , \tag{9}$$

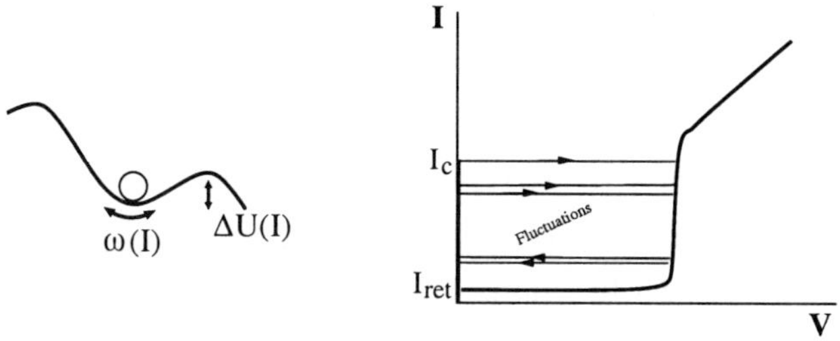

Fig. 6. The escape from the superconducting state of a junction in the presence of thermal fluctuations is a problem analogous to the escape of a particle from a well in a tilted cosine potential.

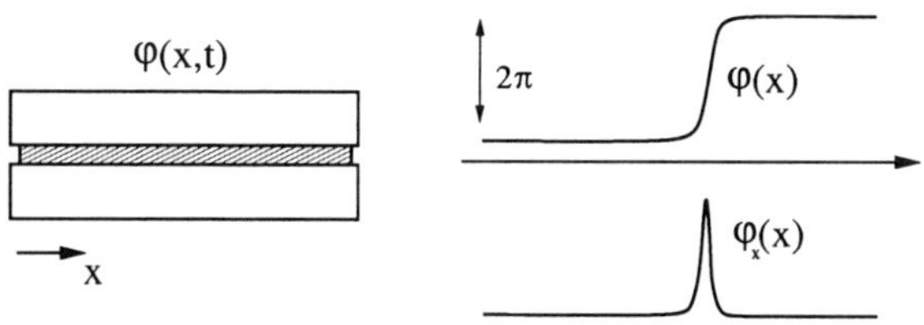

Fig. 7. Schematic of a long JJ (left) and phase and phase derivative for a soliton in the junction (right).

where a_t is a prefactor which depends on the value of the damping.[15,16]

2.3. *Long Josephson junctions*

A long Josephson junction (Fig. 7) is a junction which has one dimension (say x) long with respect to the so called Josephson penetration depth.[4] Then the phase difference is also a function of the spatial coordinate: $\varphi(x,t)$.

The electrodynamics of the junction is described by a nonlinear partial differential equation that, neglecting dissipative effects, can be written as

$$\varphi_{xx} - \varphi_{tt} = \sin\varphi. \tag{10}$$

This is the sine-Gordon equation, which has coherent localized particle-like solutions or soliton solutions. $\varphi(x)$ can be though of as the phase difference or the normalized magnetic flux. Then a soliton in the junction corresponds to a solution for which the phase difference goes from 0 to 2π; or the flux from 0 to Φ_0; that is, a fluxon of magnetic field.

When losses and bias are included the dynamics of the fluxon is described by the perturbed sine-Gordon equation

$$\varphi_{xx} - \varphi_{tt} - \sin\varphi = \alpha\varphi_t - \beta\varphi_{xxt} - \gamma. \tag{11}$$

In this lecture we are not going to deal with long JJ. For a good recent review on the subject see[17] and references within.

2.4. *Quantum effects in Josephson junctions*

In despite of its quantum mechanical origin, we have considered above that φ behaves as a classical function. This is right for large enough junctions where quantum effects can be neglected and a classical description of the junction observables is correct.

In a Josephson junction the phase difference across the junction and the charge on the junction electrode behave as quantum-mechanically conjugate variables. The Hamiltonian for the junction can be written as the addition of Josephson and charging energy

$$H = -E_J \cos\varphi + \frac{Q^2}{2C} \tag{12}$$

where $Q \sim d\varphi/dt$.

To study the dynamics of the junction we will follow canonical quantization rules and treat φ and Q as operators which satisfy the usual commutation rule $[\varphi, Q] = i2e$. Then, in phase representation, we substitute $Q/2e = N$ by $i\partial/\partial\varphi$ to get

$$H = -E_J \cos\varphi - 4E_C \frac{\partial^2}{\partial\varphi^2}. \tag{13}$$

By analogy with the problem of a particle in a periodic potential the solutions of this Hamiltonian will take the form of Bloch functions.

In the junction there are two energy scales, defined by $E_J = \hbar I_c/2e$ and $E_C = e^2/2C$, which compete. The ratio $E_J/4E_C$ measures the relative importance of the charging energy of the pairs. When E_J is dominant the quantum fluctuations in the phase are small. When $E_J \sim E_C$ the kinetic energy term induces delocalization of the phase. In the other limit, $E_J \ll E_C$ tunneling is weak and large charge transfers are energetically prohibitive (Coulomb blockade).

The classical behavior assumed in Sec. 2.1 corresponds to junctions for which $E_J \gg E_C$. Since E_J increases with the junction area and E_C decreases with the junction area, the classical description fails for small enough junctions, which have small critical currents and large capacitances.

In addition to an appropriate E_J/E_C ratio, in order to observe quantum behavior we need to have the thermal energy $k_B T \ll E_J, E_C$ and large tunneling resistances $R > h/4e^2 = 6.45\text{k}\Omega$ (to avoid quantum fluctuations).[7]

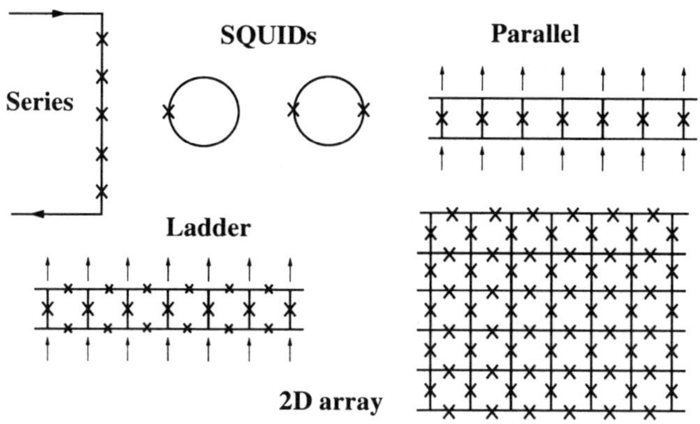

Fig. 8. Sketches of different types of Josephson arrays. Each cross represents a JJ.

Many experimental evidences of quantum mechanical behavior of small JJs were found in the late 80's and 90's. Some of the highlights were: the observation of macroscopic quantum tunneling of the phase and energy level quantization in a single JJ junction.[18] The evidence of single Copper-pair tunneling.[19] The demonstration of the Heisenberg's uncertainty principle in a superconductor.[20] More recently, it has been studied a dissipative quantum phase transition in a single junction,[21] and the quantum mechanical behavior of a single vortex in a long JJ.[22]

We want also to mention the recent works directed to operate with quantum states of single JJs and simple JJ circuits. For instance the achievement of coherent control of macroscopic quantum states in a single-Cooper-pair box.[23] The observation of the quantum superposition of distinct macroscopic states in a rf SQUID[24] and a superconducting loop with 3 junctions.[25] The manipulation of the quantum state of a superconducting tunnel junction circuit.[26] The generation and observation of coherent temporal oscillations between the macroscopic quantum states of a Josephson tunnel junction.[27] The report of quantum coherent dynamics of a superconducting flux qubit.[28] And the recent observation of quantum oscillations in two coupled charge qubits.[29]

3. Modeling Josephson arrays

Systems with superconducting wires interrupted by JJs are usually known with the name of JJ arrays. Figure 8 shows examples of such arrays. All of them are easy to fabricate and have been widely studied. The first device consist of a series array of JJs. This type of arrays have been employed for studying phase locking phenomena and build the voltage standard.[30] Superconducting loops interrupted by one or two junctions are known as SQUIDs (for superconducting quantum interference device).[4,5,31–33] SQUIDs provide a sensitive measure of magnetic flux and today are used as standard magnetic field detector in many laboratories. Parallel arrays have been designed for studying fluxon transport devices but from a fundamental point of view they are interesting because constitute an experimental realization of the Frenkel Kontorova model or the discrete sine-Gordon equation.[34–38] Ladder arrays were designed for studying the transition from one-dimensional to two-dimensional physics and allowed an experimental observation of discrete breathers,[39,40] the main topic of this lecture. Two dimensional arrays are an ideal model system to study 2D phase transitions, frustration and disorder effects, vortex dynamics, phase synchronization and other non-linear dynamics results.[41]

To derive the equations of the dynamics of the array we have to apply Kirchoff's conservation law (for the current and for the voltage) and fluxoid quantization. Fluxoid quantization condition establishes that for any loop l in the array (with at least one junction) the sum of all the phase differences around the loop is given by

$$\sum_{j\epsilon l} \varphi_j = 2\pi(n_l - f_l). \tag{14}$$

The integer n_l is the vorticity of the loop and results from the multivaluedness of the phase θ of the superconducting wave function in each superconducting island. f_l accounts for the total (external plus induced) flux of the magnetic field through the loop measured in terms of Φ_0 ($f_l = \Phi_l/\Phi_0$). In general, to compute the total induced flux in a cell one must take into account the full inductance matrix of the circuit. However, in many cases we can work with a simpler approximation and consider only the self-inductance L of each cell. The parameter $\lambda = \Phi_0/2\pi I_c L$ measures the importance of the induced fields.

We can leave out the n_l terms[i] and write

$$\sum_{j\epsilon l} \varphi_j = -2\pi(f_l^{\rm ext} + f_l^{\rm ind}) = -2\pi f_l^{\rm ext} - \frac{1}{\lambda}\frac{I^{loop}}{I_c}. \tag{15}$$

[i]In the RCSJ model the dynamical equations depend only on $\ddot{\varphi}$, $\dot{\varphi}$ and $\sin\varphi$. Then the equations of motion are independent of the n_l and we can eliminate them from our equations

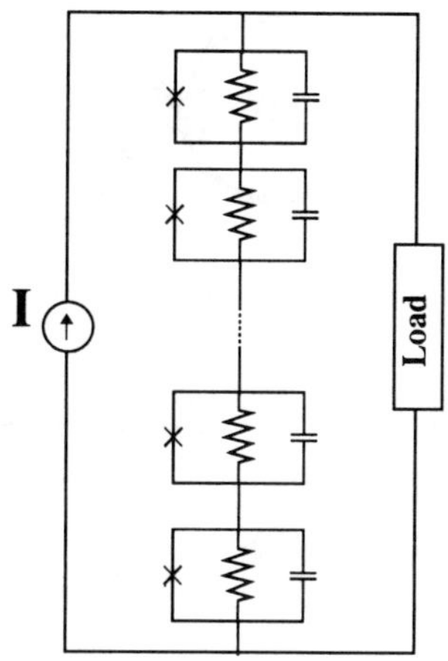

Fig. 9. Biased JJ series array shunted by an external load.

Depending on the importance of the induced fields, Josephson circuits can be divided in two general types. Circuits of the first type have $\lambda \gg 1$ so the induced flux in the loop is not important (these circuits are typically made of aluminum). Otherwise the circuits belong to the second type and the flux caused by the circulating currents is important (these circuits are typically made of niobium).

If inductive effects can be neglected, fluxoid quantization (Eq. 14) imposes a constraint to the equations and then reduces the number of independent variables of the system.

3.1. *Series arrays*

Figure 9 shows a series array of JJs biased by an external current and a load circuit. The equations of the array can be written as[42]

$$\ddot{\varphi}_k + \Gamma\dot{\varphi}_k + \sin\varphi_k + I_L(t) = I \tag{16}$$

$$V(t) = \sum_{k=1}^{N} \dot{\varphi}_k = F(I_L(t)). \tag{17}$$

Thus, the junctions behave as independent elements biased by the same current and coupled through the load circuit.

Fig. 10. rf-squid device.

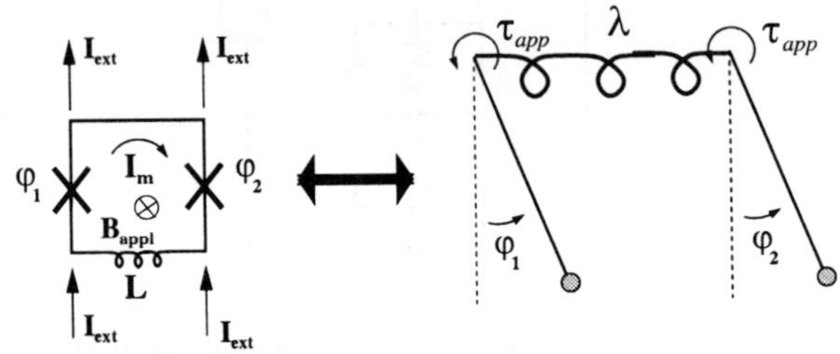

Fig. 11. dc-squid device and equivalence with the two coupled pendula system.

3.2. *rf-SQUID*

This device consist of a superconducting loop with a single JJ in it (Fig. 10). The behavior is governed by the value of total flux through the SQUID, Φ. From the fluxoid quantization condition $\varphi = -2\pi\frac{\Phi}{\Phi_0}$. Then,

$$\Phi = \Phi_{\text{ext}} - LI_c \sin 2\pi \frac{\Phi}{\Phi_0} \tag{18}$$

and the potential energy of the system is given by

$$U(\Phi) = -E_J \cos 2\pi \frac{\Phi}{\Phi_0} + \frac{(\Phi - \Phi_{\text{ext}})^2}{2L}. \tag{19}$$

Since in this circuit there is no bias current, the rf-SQUID is operated coupled to a radio-frequency circuit (resonator).

In addition to magnetic flux measurements rf-SQUIDs are important for the study of fundamental problems on quantum mechanics (see Ref. 24 and references within).

3.3. *dc-SQUID*

This device consist of a superconducting loop with two JJs (Fig. 11). Now current conservation reads:

$$\begin{aligned} i_1 &= \ddot{\varphi}_1 + \Gamma\dot{\varphi}_1 + \sin\varphi_1 = i^{\text{mesh}} + i_{\text{ext}}, \\ i_2 &= \ddot{\varphi}_2 + \Gamma\dot{\varphi}_2 + \sin\varphi_2 = -i^{\text{mesh}} + i_{\text{ext}}; \end{aligned} \tag{20}$$

and fluxoid quantization:

$$(\varphi_1 - \varphi_2) = -\frac{2\pi}{\Phi_0}\left(B_{\text{appl}}S + LI^{\text{mesh}}\right). \tag{21}$$

Normalizing, we get

$$i^{\text{mesh}} = -\lambda\left(\varphi_1 - \varphi_2 + 2\pi f_0\right) \tag{22}$$

$(f_0 = B_{\text{appl}}S/\Phi_0)$.

Then the equations for the dynamics of the array are

$$\begin{aligned}\ddot{\varphi}_1 + \Gamma\dot{\varphi}_1 + \sin\varphi_1 &= -\lambda\left(\varphi_1 - \varphi_2 + 2\pi f_0\right) + i_{\text{ext}} \\ \ddot{\varphi}_2 + \Gamma\dot{\varphi}_2 + \sin\varphi_2 &= \lambda\left(\varphi_1 - \varphi_2 + 2\pi f_0\right) + i_{\text{ext}}.\end{aligned} \tag{23}$$

Where I_{ext}=$I_{\text{total}}/2$. These equations show that the problem of two junctions connected in parallel by an inductive element is equivalent to the problem of two pendula coupled by a torsion spring (see Fig. 11).

If inductive effects can be neglected, fluxoid quantization imposes a constraint on the variables and we have that

$$\varphi_1 - \varphi_2 = -2\pi f_0. \tag{24}$$

Then, current conservation reads

$$i_{\text{ext}} = \frac{i_1 + i_2}{2} = \ddot{\varphi}_1 + \Gamma\dot{\varphi}_1 + \cos(\pi f_0)\sin\left(\varphi_1 + \pi f_0\right). \tag{25}$$

The system behaves like a single junction with the critical current modulated by the external field.

3.4. *JJ parallel array*

A JJ parallel array is formed by a set of junctions connected in parallel by superconducting wires. The mechanical analog for this system is a set of pendula connected by torsion springs (see Fig 12). An important consequence of this harmonic interaction is that all the junction must have the same dc voltage.

The equations for the array can be easily generalized from the dc SQUID. For the parallel array we get,

$$\ddot{\varphi}_j + \Gamma\dot{\varphi}_j + \sin\varphi_j = \lambda\left(\varphi_{j+1} - 2\varphi_j + \varphi_{j-1}\right) + i_{\text{ext}} \tag{26}$$

with $j = 1, ...N$. The boundary conditions are given by $\varphi_0 = \varphi_1 - 2\pi f_0$ and $\varphi_{N+1} = \varphi_N + 2\pi f_0$ for the case of open-ended arrays and $\varphi_{N+1} = \varphi_1 + 2\pi n_k$ and $\varphi_0 = \varphi_N - 2\pi n_k$ for ring arrays. In this last case the integer n_k computes the number of kinks or fluxons trapped in the array.

Equations (26) are also the equations of the dynamics of the forced and damped Frenkel-Kontorova model or the discrete sine-Gordon equation.

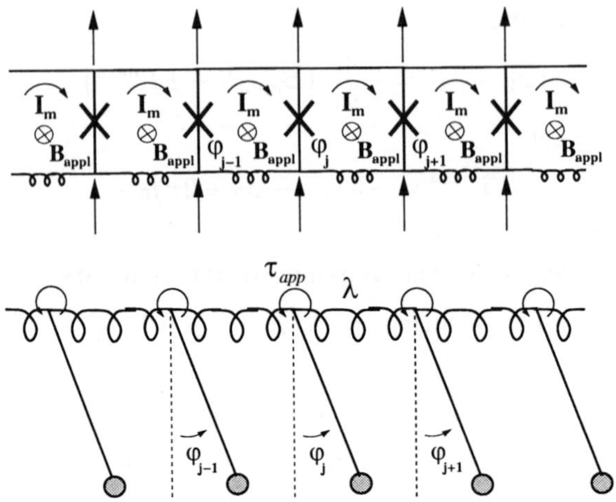

Fig. 12. JJ parallel array.

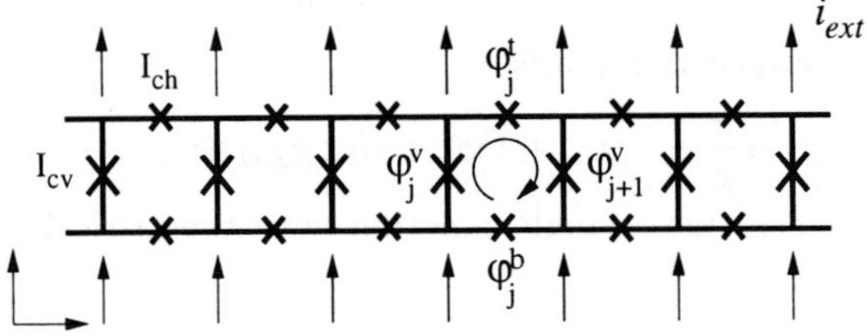

Fig. 13. JJ ladder array.

3.5. *JJ ladder array*

A Josephson ladder array (Fig. 13) is a quasi-one-dimensional array made when the superconducting horizontal wires of the parallel array are interrupted by Josephson junctions. We can think of this system as a set of pendula (the vertical junctions) connected in parallel by nonconvex springs (the horizontal junctions). Consequence of the non-convex interaction, one of the more notable differences with respect to the parallel array is that now the vertical junctions are not constrained to have the same dc voltage. Also, now fluxoid quanta can go into and escape from the array through the horizontal junctions.

Let us consider the case of an anisotropic ladder. Then the critical current of junctions in the vertical direction is different from the critical current of junctions in the horizontal direction. It can be easily made by changing

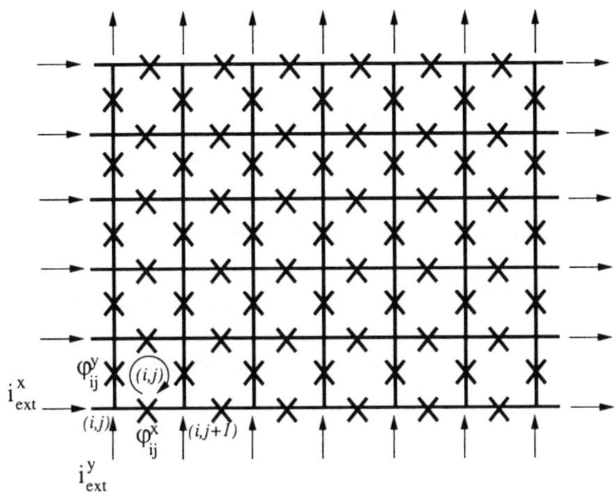

Fig. 14. 2D JJ array.

the area of the junctions. Critical current and capacitance are proportional to this area and resistance is inversely proportional to the area.

Applying current conservation and fluxoid quantization, the equations for the ladder read

$$\begin{aligned}\mathcal{N}(\varphi_j^t) &= -\frac{\lambda}{h}\xi_j,\\ \mathcal{N}(\varphi_j^v) &= \lambda(\xi_{j-1}-\xi_j)+i_{\text{ext}},\\ \mathcal{N}(\varphi_j^b) &= \frac{\lambda}{h}\xi_j.\end{aligned} \tag{27}$$

Here we have defined

$$\xi_j = -2\pi f_j^{\text{ind}} = \varphi_j^v + \varphi_j^t - \varphi_{j+1}^v - \varphi_j^b + 2\pi f_0, \tag{28}$$

where $\xi_0 = \xi_N = 0$. For a ladder with N vertical junctions, j runs from 1 to N for vertical junctions and from 1 to $N-1$ for horizontal ones. We have normalized with respect to the parameters of the vertical junctions. Thus, $h = I_{ch}/I_{cv} = C_h/C_v = R_v/R_h$ and $\lambda = \lambda_v = \Phi_0/2\pi I_{cv}L$ ($\lambda/h = \lambda_h = \Phi_0/2\pi I_{ch}L$).

3.6. *2D arrays*

Figure 14 shows a sketch of a square 2D Josephson array. Following our approach to model Josephson arrays, the junctions are coupled via the flux quantization condition with the inclusion of cell self induced magnetic fields.

In this approach, the equations for the dynamics of the array are:

$$\mathcal{N}(\varphi^x_{ij}) = \frac{\lambda}{h}(\xi_{ij} - \xi_{ij-1}) + \frac{i^{\text{ext}}_x}{h},$$
$$\mathcal{N}(\varphi^y_{ij}) = \lambda(\xi_{i-1j} - \xi_{ij}) + i^{\text{ext}}_y \qquad (29)$$

ξ_{ij} measures the intensity of the induced field

$$\xi_{ij} = -2\pi f^{\text{ind}}_{ij} = \varphi^y_{ij} + \varphi^x_{ij+1} - \varphi^y_{i+1j} - \varphi^x_{ij} + 2\pi f_0 \qquad (30)$$

and we have normalized with respect to the parameters of the y junctions. Thus $h = I_{cx}/I_{cy} = C_x/C_y = R_y/R_x$ and $\lambda = \lambda_y = \Phi_0/2\pi I_{cy}L$ ($\lambda/h = \lambda_x = \Phi_0/2\pi I_{cx}L$).

This is a model for a 2D array when only self-inductances are taken into account. Sometimes, depending on the problem to be studied, inductances are not needed at all; in some others a full inductance matrix is necessary.

In general, in a square 2D (N×N) array we have to solve equations for the dynamics of $2N^2 - 2N$ (for free boundary conditions) gauge-invariant phase differences, φ_{ij}. However, when induced fields can be neglected ($\lambda \gg 1$ limit), flux quantization condition imposes $(N-1)^2$ constrains on these variables. Then, it is more convenient to express the system equations in terms of the phase in each island, θ_i, which are N^2-1 independent variables. This is made by writing $\varphi_{ij} = \theta_i - \theta_j - A_{ij}$ [where (for a given gauge) the $A_{ij} = \frac{2\pi}{\Phi_0}\int_i^j \vec{A}(\vec{r},t)d\vec{l}$, depend only on the external magnetic field] and the dynamical equations result after applying current conservation.

In the $E_J \gg E_C$ limit of the system the total Josephson energy is the relevant energy contribution:

$$H_J = -\sum_{<ij>} E_J \cos(\theta_i - \theta_j - A_{ij}). \qquad (31)$$

In the opposite limit of ultrasmall junctions ($E_C \gg E_J$), the charging energy is the more important contribution:

$$H_Q = \frac{1}{2}\sum_{ij}(Q_i + q_i)C^{-1}_{ij}(Q_j + q_j) \qquad (32)$$

where Q_j is the charge in island j, q_j are possible offset charges or charges induced by external sources (charge frustration) and C is the capacitance matrix of the circuit.

The more complex case is that of intermediate values where both terms should be considered [j] and

$$H = \frac{1}{2}\sum_{ij}(Q_i + q_i)C_{ij}^{-1}(Q_j + q_j) - \sum_{<ij>} E_J \cos(\theta_i - \theta_j - A_{ij}). \qquad (33)$$

4. Localized excitations in Josephson arrays: Vortices and kinks

In this section we are going to summarize some aspects of the physics of vortices in 2D Josephson arrays and kinks in Josephson parallel arrays. Both are broad topics and we do not intend to review them here. A nice review with many references on the physics of 2D Josephson-junction arrays which extensively covers the role of vortices can be found in Ref. 41 For an introduction to the study of kinks in parallel arrays and similar systems see Refs. 34–38.

4.1. *Vortices in 2D arrays*

Consider a two-dimensional array of N×N superconducting islands coupled by Josephson junctions. The relevant energy of the array is the sum of the Josephson energies of the junctions, which (in the absence of magnetic fields) is given by:

$$H_J = -\sum_{<ij>} E_J \cos(\theta_i - \theta_j). \qquad (34)$$

This is the Hamiltonian of the two-dimensional XY model and thus 2D JJ arrays are a physical realization of this model. The XY model describes many systems but is particularly interesting because it shows the Kosterlitz-Thoules-Bereziinski (KTB) phase transition. Thus, many of the theoretical and experimental works with 2D arrays are focused in the observation and study of this type of phase transition.[41]

The linear excitations of Hamiltonian (34) are known as spin waves. They correspond to small amplitude and energy wave-like variations of the phase over the array. In addition to them, the system supports large energy nonlinear excitations that are called vortices (and antivortices). A vortex (see Fig. 15) is an energy localized solution defined by the value of the total phase differences along a path containing the vortex, which is equal to 2π (-2π for antivortices) [For a vortex (antivortex), $\sum(\theta_i - \theta_j) = \pm 2\pi$].

[j] The inclusion of self-induced magnetic fields would give an additional contribution to the hamiltonian of the array

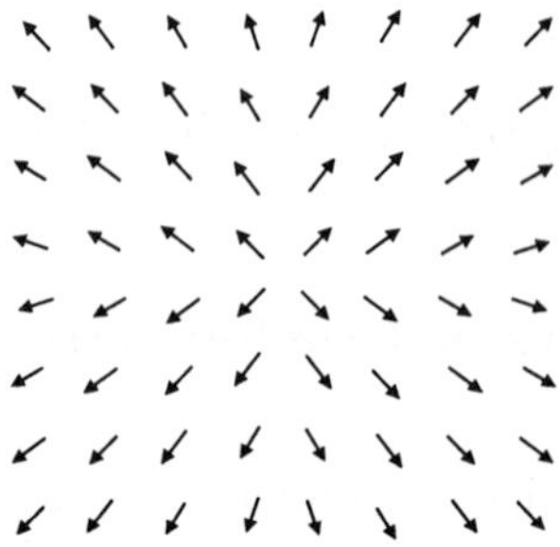

Fig. 15. Sketch of a vortex configuration in a Josephson array. Arrows represent phase of superconducting islands.

4.1.1. *Single vortex properties at zero temperature*

Vortices and antivortices are topological excitations and behave like opposite charges in a two-dimensional system. The ground state configuration for Hamiltonian (34) correspond to the configuration $\{\theta_i\}$ =const. For a large array this configuration has an energy $E = -2N^2E_J$. We can calculate also the energy of an isolated vortex in a large square array of size $L{=}Na{\to}\infty$, which is $E = \pi E_J \ln L/a - 2N^2E_J$.

In the presence of external fields the vortex has a marked single-particle behavior. When an external current is uniformly applied to the array the vortex sees an effective Lorenz-type force perpendicular to the applied current, of magnitude $F_V = (\Phi/2\pi)(I_{\rm tot}/Na)$. Due to the discreteness of the array, there exist an energy barrier that the vortex has to overcome before moving trough the array. This barrier is the Peierls-Nabarro barrier for the vortex and has a value of $E_{PN} = 0.2E_J$ for square cells and $E_J = 0.043E_J$ for triangular ones. When the vortex moves along the lattice it experiences a two-dimensional Peierls-Nabarro potential. The potential along the x direction $U_{PN} \simeq -(E_{PN}/2)\cos 2\pi x/a$. Following this picture we can compute the vortex depinning current $I_{cV} = (\pi/\Phi_0)E_{PN} = 0.1I_c$, for the square array.

In many cases we can describe the motion of a vortex in the array with an effective equation for the center of the vortex which includes an external force (caused by the external current), a viscous force (power dissipated in the resistive channels), kinetic energy (stored in the capacitors) and potential energy (Josephson energy):

$$I = I_{cV}\sin\left(\frac{2\pi x}{a}\right) + \frac{\Phi_0}{4\pi R}\frac{d(2\pi x/a)}{dt} + \frac{\Phi_0 C}{4\pi}\frac{d^2(2\pi x/a)}{dt^2}. \qquad (35)$$

4.1.2. *Array properties at non-zero temperatures*

The energy that we need to put a vortex into a large array in its ground state is $E_V = \pi E_J \ln L/a$. This energy diverges with the size of the array and thus there are not free vortices in the system at low enough temperatures. However, as soon as $T \neq 0$, the thermal generation of bound vortex-antivortex pairs is energetically favored since the energy to create one of these pairs behave as $E_p = 2\pi E_J \ln r/a$, with r the distance between the vortex and anti-vortex cores. Increasing the temperature we reach the KTB phase transition, $T_{KT} \sim \pi E_J / 2k_B$ where vortex-antivortex pairs unbind and free vortices are present in the array.

Such temperature can be estimated with a simple argument. If we add a single vortex to the array the free energy changes in

$$\Delta F_V = E_V - T\Delta S_V = \pi E_J \ln L/a - k_B T \ln (L/a)^2, \tag{36}$$

here $(L/a)^2$ is the number of places where we can put the vortex in the array. ΔF_V becomes zero at $T = \pi E_J/2k_B$.

This is the simplest description of the physics of the system. In general the results are affected by the inclusion of external magnetic fields, finite size-effects, self-induced fields, disorder,... In any of the cases the concept of vortex is essential to understand the physics of the array.

4.2. *2D arrays with small junctions*

In arrays of ultrasmall tunnel junctions the relevant energy is the charging energy,[43]

$$H_Q = \frac{1}{2} \sum_{ij} Q_i C_{ij}^{-1} Q_j \tag{37}$$

where Q_i is the charge in the island. For metallic "normal state" islands this charge is an integer multiple of the electronic charge e. In the superconducting state, at low temperatures and voltages below the gap voltages, Cooper pairs are dominant and the charge appears in multiples of $2e$.

In any of the cases (normal or superconducting arrays) at zero temperature no free charges are present in the array. The system is insulating. In arrays where mutual capacitances are dominant the charges interact logarithmically over long enough distances. Then, free charges are expected to be created by thermal activation through a KTB phase transition where pairs of opposite charges unbind, and the array becomes resistive. The estimated temperature for the transition is $T_{cn} = E_C/4\pi k_B$ for normal islands and $T_{cs} = E_C/\pi k_B$ for superconducting ones ($E_C = e^2/2C$). In real circuits this transition is strongly affected by dissipation [measured by the coefficient $\alpha_T = h/(4e^2 R_T)$], presence of offset charges, disorder,...

In small capacitance superconducting junctions we have to include both the charging and the Josephson coupling. If we ignore quasiparticles, the Hamiltonian of the system is

$$H = \frac{1}{2}\sum_{ij} Q_i C_{ij}^{-1} Q_j - \sum_{<ij>} E_J \cos(\theta_i - \theta_j), \tag{38}$$

where we have ignored also charge or phase frustration (offset or induced charges and magnetic fields).

If charging energy can be neglected (E_C=0 limit) the vortices undergo a KTB transition where vortex dipoles unbind. This transition separates a superconducting low temperature phase from a resistive high temperature one. If the Josephson coupling is weak (E_J=0 limit), the charges show a KTB transition where the dipoles, formed by a Cooper pair and a missing pair, unbind. The transition separates an insulating from a conducting phase.

At finite E_J and E_C, both charge and vortex excitations have to be considered simultaneously. The charging energy provides a kinetic energy for the vortices, and the Josephson coupling allows the tunneling of Cooper pairs and provides dynamics for the charges. If $E_C \ll E_J$ or $E_J \ll E_C$ perturbative approaches can be used.

However, to fully understand the physics of the system we have to consider also the influence of the quasiparticle tunneling. We have a quantum dissipative system.

We have also seen that vortices can be described as particles moving in a substrate potential with a mass given by the charging energy. For small junctions we can expect strong quantum mechanical effects. At finite temperatures vortex motion can be thermally excited. At low temperatures it is possible to observe quantum mechanical tunneling of vortices.

A review on quantum phase transitions and vortex dynamics in superconducting networks can be found in Ref. 44

4.3. *Kinks in parallel arrays*

We have seen in Sec. 3.4 that the equations for the dynamics of a parallel array of JJs correspond to the equations of the dynamics of the Frenkel-Kontorova model or the discrete sine-Gordon equation. Reviews of the physics of these systems are Refs. 34-36

The equations for the dynamics of the junctions in the parallel array are given by

$$\ddot{\varphi}_j + \Gamma\dot{\varphi}_j + \sin\varphi_j = \lambda\left(\varphi_{j+1} - 2\varphi_j + \varphi_{j-1}\right) + i_{\rm ext} \tag{39}$$

(with $j = 1, ...N$ and boundary conditions are $\varphi_0 = \varphi_1 - 2\pi f_0$ and $\varphi_{N+1} = \varphi_N + 2\pi f_0$ for an open array and $\varphi_{N+1} = \varphi_1 + 2\pi n_k$ and $\varphi_0 = \varphi_N - 2\pi n_k$ for a ring. n_k is the number of kinks or fluxons trapped in the ring).

The system supports different types of excitations. Linear waves are small amplitude and energy wave-like variations of the phase over the array. When the phases are small, the solution of the linearized equations (for zero damping and current) is: $\varphi_j(t) \sim \exp(i(\omega_k t - kj))$ with

$$\omega_k^2 = 1 + 4\lambda\sin^2(k/2) \quad (-\pi < k \leq \pi). \tag{40}$$

This dispersion relation is characterized by a finite band with gap $\omega_{min}{=}\omega_0{=}1$ and maximum frequency $\omega_{max} = \omega_\pi{=}(1+4\lambda)^{1/2}$.

When the phases are not small the linear approximation is not valid and the dynamics is pretty rich supporting new type of localized excitations: kinks and breathers

Kinks and anti-kinks are the only excitations which can exist in the array in the static case. Depending on the context, they are also called discrete solitons or elementary discommensurations. They correspond to solutions where the phases go from 0 to $2\pi(-2\pi)$ along the array. Since $(\varphi_{j+1} - \varphi_j) = -2\pi\Phi_j/\Phi_0$; then $(\varphi_N - \varphi_1) = 2\pi = -2\pi\Phi_{tot}/\Phi_0$ and one kink corresponds to one fluxon of magnetic field.

The existence of kinks does not depend crucially on the discreteness of the system. Moreover, in many aspects the kinks are similar to the solitons found in the continuous version of the equation (the sine-Gordon equation). However, many other properties of the kinks do depend crucially on the discreteness of the array. The existence of a discrete lattice breaks the invariance under continuous translations of the solitons of the continuous model. In the lattice, the kink is invariant only under discrete translations along the array and, due to discreteness, the kink is pinned to the lattice: there exists a minimun energy barrier the kink needs to overcome in order to move through the lattice. This energy is the so-called Peierls-Nabarro (PN) barrier (Fig. 16) and decreases rapidly with λ (Fig. 17).

We can go further and define not only a PN barrier but also a PN potential for the kink. To compute the kink potential profile we let the kink configuration to relax following the overdamped dynamics,

$$\dot{\varphi}_j = -\sin\varphi_j + \lambda\left(\varphi_{j+1} - 2\varphi_j + \varphi_{j-1}\right), \tag{41}$$

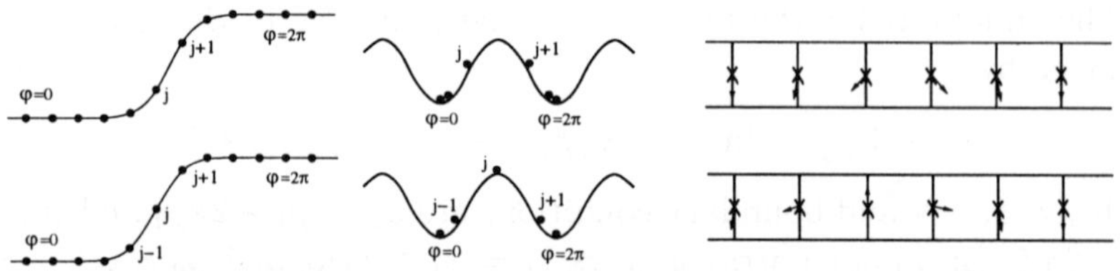

Fig. 16. Three different representations of the minimum energy configuration (top figures) and the saddle configuration, maximum of the PN potential, of a kink in the parallel array (bottom figures): phase representation (left), potential energy representation (center) and angular representation (right).

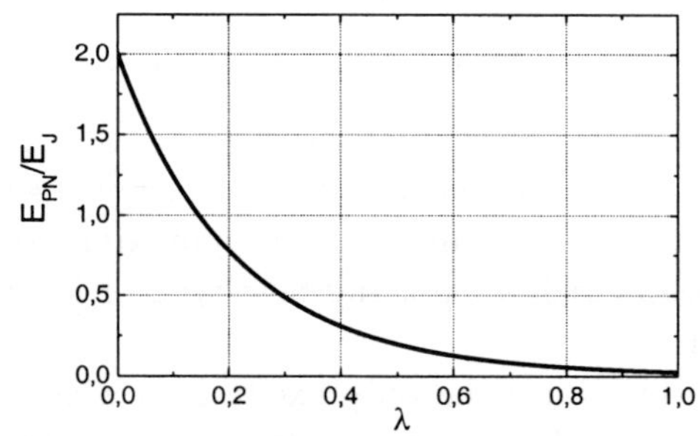

Fig. 17. E_{PN} vs the coupling parameter λ for a kink in the parallel array.

from the saddle-point configuration to the minimun energy configuration. During the relaxation we work out the energy,

$$E/E_J = \sum_j \left[(1 - \cos\varphi_j) + \frac{\lambda}{2} (\varphi_{j+1} - \varphi_j)^2 \right], \tag{42}$$

and center of mass,

$$X_{CM} = C \pm \sum_j \varphi_j, \tag{43}$$

of the configuration obtaining the potential profile $E(X_{CM})$.

In some cases is possible to identify the kink motion with the motion of a single particle over a sinusoidal periodic potential defined by:

$$V_{PN}(X) = \frac{E_{PN}}{2} (1 - \cos X). \tag{44}$$

In the presence of external bias, an effective equation of motion for the kink is given by

$$m\ddot{X} + m\Gamma\dot{X} + \frac{E_{PN}}{2} \sin X = i. \tag{45}$$

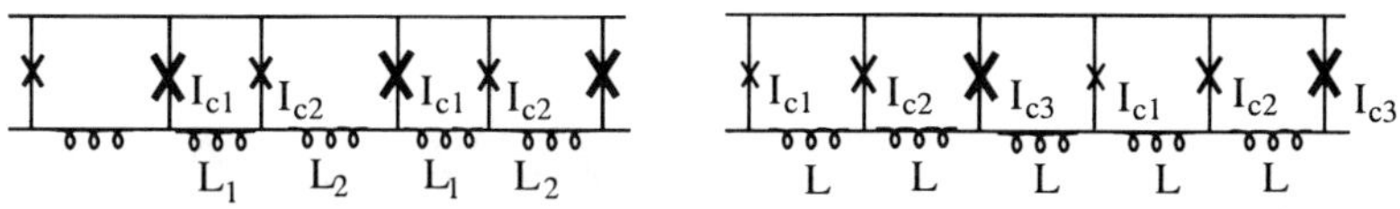

Fig. 18. Two different designs of parallel arrays where fluxons experiment a ratchet periodic potential.

The picture of the kink or fluxon as a single particle is particularly useful in arrays which are larger than the fluxon width and are driven by small currents (far from the whirling mode).

When the kink moves it radiates energy in form of small amplitude waves. This radiation is very strong for the case of underdamped arrays. There, phonons are easily excited by the kink in its wake and resonances between the kink velocity and these waves appear.[37,38]

A particularly interesting configuration is a ring of JJ connected in parallel. There, once the array is superconducting magnetic field gets trapped as an integer number of fluxons.

4.3.1. *Fluxon ratchet potentials*

We have considered arrays where all the junctions are identical and all the cells have the same size. However, it is possible to design different arrays which are adequate for studying new physical problems.

One example is the use of Josephson parallel arrays to study the dynamics of kinks subjected to substrate ratchet potentials.[47–49]

A ratchet potential is a periodic potential without inversion symmetry: $V(x) \neq V(-x)$, then it is easier to move a particle in one direction than in the other.

The equations for a non-uniform array (made of junctions with different areas and cells of different sizes) are

$$h_j\left(\ddot{\varphi}_j + \Gamma\dot{\varphi}_j + \sin\varphi_j\right) = \lambda_j(\varphi_{j+1} - \varphi_j) - \lambda_{j-1}(\varphi_j - \varphi_{j-1}) + i_{\text{ext}} \quad (46)$$

with $h_j = I_{cj}/I_c^* = C_j/C^* = R^*/R_j$, where the $*$ superscript stands for the parameters of the junction with respect to which we normalize, and $\lambda_j = \Phi_0/2\pi I_c^* L_j$.

A kink ratchet potential can be obtained with different suitable combinations of junctions and inductance. The simplest ones (see Fig. 18) are made alternating junctions of two critical currents and cells of two areas, and alternating junctions of three critical currents.

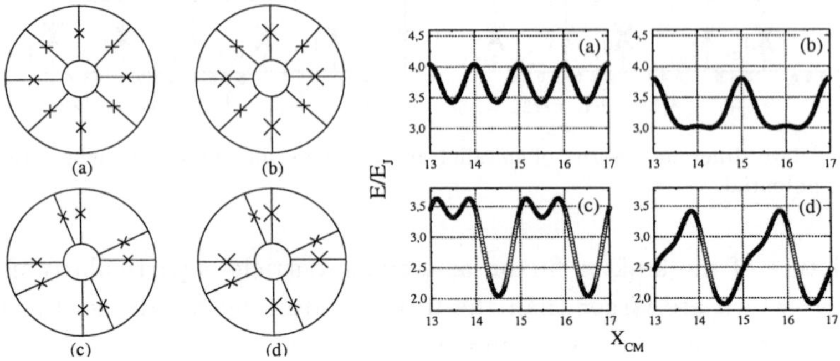

Fig. 19. Four studied arrays: (a) regular ring, (b) ring with alternating critical currents, (c) ring with with alternating cell areas, (d) ratchet ring with alternating critical currents and cell areas; and the corresponding energy profiles $E(X_{CM})$. Only configuration (d) gives a ratchet profile.

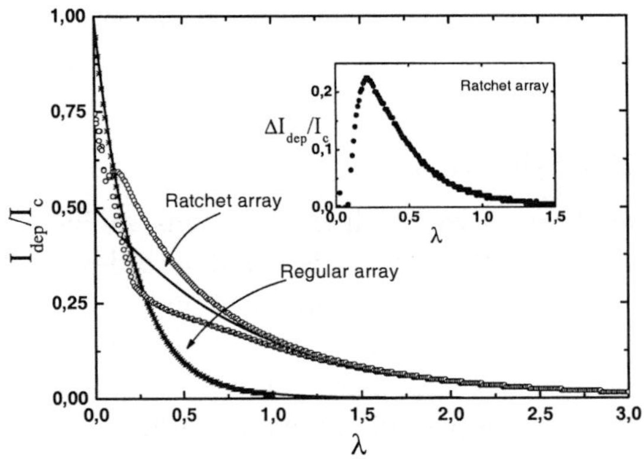

Fig. 20. Fluxon depinning currents I_{dep} as a function of λ for the regular and ratchet arrays. Solid lines stand for predictions from E_{PN} values and symbols stand for numerical computation of the depinning currents.[49] The inset shows the difference (ΔI_{dep}) between the absolute values of the two depinning currents for the ratchet array.

Figure 19 shows four different arrays and the computed PN potentials. As we should expect only array (d) shows a ratchet fluxon potential. Such arrays were built and experimentally studied.[48]

Figure 20 shows the dependence as a function of λ of the positive and negative values of the fluxon depinning current (the minimum current to move the fluxon) for the cases of the regular and ratchet arrays.[49] We see that for the ratchet array I^+_{dep} and I^-_{dep} are significantly different for values of λ between 0.1 and 0.9. The inset shows the difference ($\Delta I_{dep} = I^-_{dep} - I^+_{dep}$)

between the values of the two depinning currents for the ratchet array. As we can see there exist a moderate range of values of λ for which an important ratchet behavior is expected. The maximum of this curve is obtained for $\lambda \sim 0.2$.

4.4. *Charge solitons in 1D arrays*

We want to mention also the case of a one-dimensional array of ultra-small junctions. There the Josephson coupling is weak, $E_J \ll E_C$ and the normal resistance large $R > h/4e^2$.

Then, if a single electron (or a single pair) is added to or subtracted from an intermediate island the resulting localized state, the single charge plus the polarization cloud, is called charge soliton (or anti-soliton). It corresponds to a localize voltage profile and some of their properties have been studied by means of the sine-Gordon equation.[45,46]

5. Discrete breathers in Josephson arrays

Discrete breathers[50–54] excitations (also called intrinsic localized modes) are solutions of the dynamics of nonlinear lattices for which the energy remains exponentially localized in a few sites of the array. As we have seen, Josephson arrays are ideal experimental discrete system to study nonlinear dynamics, thus much effort in the last years has been devoted to the prediction and experimental observation[55,56] of discrete breathers in Josephson arrays. E. Trías and P. Binder wrote their thesis on this subject [57,58] and some recent review articles are Refs. [39,40,59].

Josephson circuits are externally biased dissipative systems. Thus, the breather solutions that we are going to present are attractors of the dynamics of the array. We will distinguish between two types of localized modes: oscillobreathers, where all the phases oscillate (a few of them with large amplitude and the rest with small amplitude); and rotobreathers where most phases oscillate but some other rotate. Since the main object is to experimentally study the modes, much effort has been put in the study of rotobreathers because they show a dc signal. The plasma frequency for a Josephson junction is close to 100GHz. Thus it is not possible to follow the instantaneous dynamics of the phases. A quantity easy to measure however is the dc voltage of the junction (only rotating junctions have nonzero dc voltage).

An approach to the study of DB solutions is to build the breather from some appropriate uncoupled limit (which is not going to be experimentally

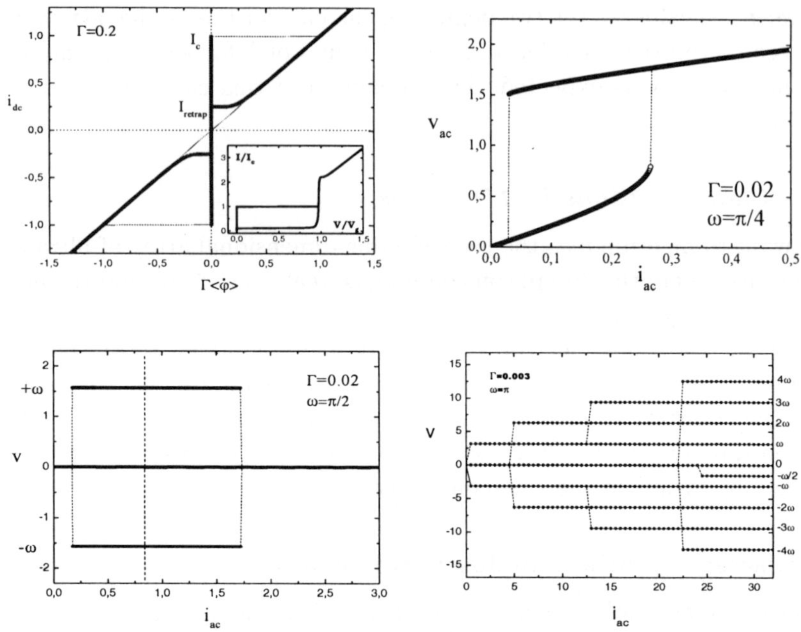

Fig. 21. Simulation of IV curves for a single junction with a linear resistor: (a) DC bias, $\Gamma = 0.2$ (inset: experimental curve showing the gap voltage branch). (b-d) Underdamped junction biased by an rf field. (b) $\Gamma = 0.02$ and $\omega = 2\pi \times 0.125$, at a certain value of the amplitude of the field, an "small" amplitude oscillating state destabilizes to a "large" amplitude one. Both states coexist for certain range of values of $i_{\rm ac}$. (c) $\Gamma = 0.02$ and $\omega = 2\pi \times 0.25$, for a certain range of values of $i_{\rm ac}$ an oscillating state coexist with two rotating states with $\langle\dot{\varphi}\rangle = \pm\omega$. (d) $\Gamma = 0.003$ and $\omega = 2\pi \times 0.5$, depending on the value of $i_{\rm ac}$ different attractors coexist with rotation velocities in general given by $\langle\dot{\varphi}\rangle = \frac{n}{m}\omega$. (The figure shows only one of the possible subharmonic ($m > 1$) states).

accessible) of the model. The idea is to set one of the junctions in a dynamical state, the other in a different one and then switch on the coupling. All junctions are identical and see the same external force. Thus we are going to start by exploring the dynamics of a single junction looking for regions where two or more states can coexist.

Figure 21(a) shows an IV curve for an underdamped junction (RCSJ model) in the presence of dc bias. We see that for currents between I_c and $I_{\rm ret}$ two different solutions are possible. One of them is a superconducting (zero voltage) state, the other a resistive one. The inset shows a measured IV curve (see also Fig. 5). In a range of external currents two states, with V=0 and V=V_g, coexist.

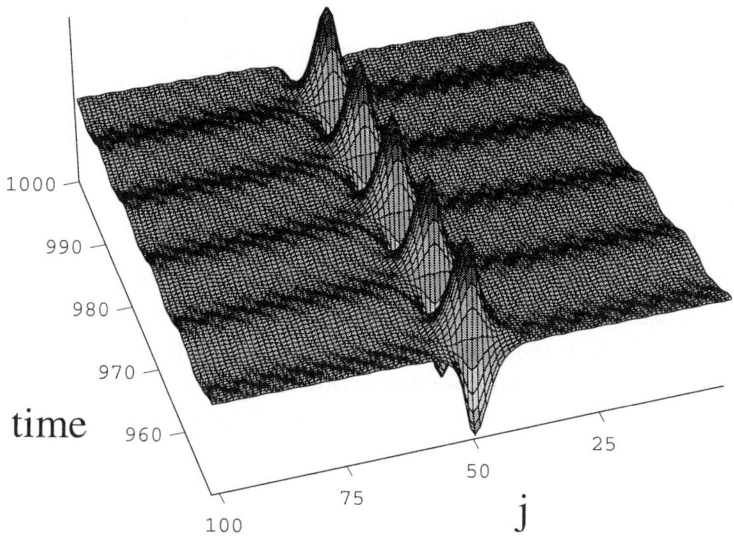

Fig. 22. Numerical simulation of an oscillobreather solution in an ac biased Josephson parallel array.

Figures 21(b-d) show IV curves for an underdamped junction (RCSJ model) in the presence of ac bias, $i(t) = i_{\rm ac} \sin(\omega t)$. (b) shows a situation where two solutions with zero dc voltage but different amplitude coexist. The dc voltage is zero so we plot $v_{\rm ac} = \sqrt{\langle \dot{\varphi}^2 \rangle - \langle \dot{\varphi} \rangle^2}$, which measures the oscillation amplitude. (c) and (d) show situations where non-zero dc voltage solutions coexist with a zero voltage one. The junction voltage is synchronized to the frequency of the external field and, in the general case, $\langle \dot{\varphi} \rangle = \frac{n}{m}\omega$.

5.1. *Oscillobreather in an ac biased parallel array*

The simplest Josephson array with breathers is the parallel array, which is described by a discrete sine-Gordon equation. This system supports only oscillobreather solutions. They are attractors of the dynamics of the array biased by ac external currents. In the breather one junction describes a large amplitude oscillation meanwhile the other follow the external force and oscillates with a small amplitude. Figure 22 shows a picture of an oscillating discrete breather solution in a Josephson parallel array. In the parallel array all the junctions have the same dc voltage (V=0 in the oscillobreather solution) since they are connected by superconducting wires.

Oscillobreather solutions can be excited in other Josephson arrays. However, due to the high frequencies involved in the Josephson effect, the oscillobreathers are very hard to experimentally detect. Thus, the experimental effort and much of the theoretical one have been focused to study rotobreathers.

5.2. *Rotobreathers in Josephson arrays*

A rotobreather corresponds to a solution where one junction rotates ($V \neq 0$) meanwhile the others oscillate (V=0). This is a voltage localized solution in the array. Such state is easy to detect by measuring the local dc voltages throughout the array. Another important advantage of the rotobreather states is that in principle they can be obtained by either biasing the array with ac currents or with a dc current. This last possibility requires a simpler experimental approach.

One may wonder if these rotating localized modes exist in JJ parallel arrays. In principle, they exist in the uncoupled limit of the model. However, the convex character of the coupling between junctions in the array shows that any localized mode will not be stable since the difference between neighboring phases can not grow without limits. Physically, in a parallel array the dc voltage is the same for all the junctions since they are connected by superconducting leads, thus preventing dc voltage localized solutions. To have rotobreathers in Josephson arrays we will need non-convex interaction terms between neighbors.

The simplest manner to overcome this difficulty is to substitute the horizontal wires connecting neighboring junctions by new Josephson elements. Such a new configuration is known as Josephson ladder.

5.3. *The ladder array*

From our perspective a Josephson ladder can be thought of as a set of parallel pendula, the vertical junctions, coupled by sinusoidal terms (the non-convex terms) provided by the horizontal junctions. The intensity of the coupling is governed by the ratio of the critical current for the horizontal junctions I_{ch} to the critical current for the vertical junctions I_{cv}. Thus, it is natural to study anisotropic ladders where the anisotropy is controlled by the parameter $h = I_{ch}/I_{cv}$. A large value of h means that coupling between vertical junctions will be strong meanwhile a small value of h means weakly coupled vertical junctions. Anisotropic arrays are fabricated by varying the area of the junctions. Critical current and capacitance are proportional

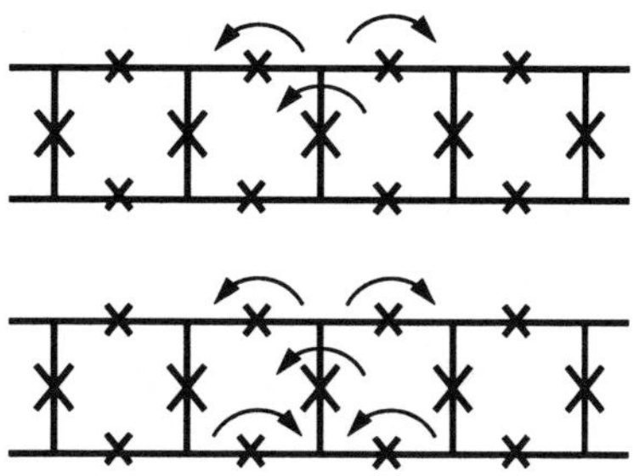

Fig. 23. Two different voltage patterns for a one-site rotobreather in a Josephson ladder. Arrows represent rotating junctions, junctions with dc voltage different from zero.

to this area and the resistance is inversely proportional to it. Thus, also $h = C_h/C_v = R_v/R_h$ and the damping Γ and the plasma frequency ω_p are the same for all the junctions.

We want to recall the equations for the system, introduced in Sec. 3.5:

$$\begin{aligned}\mathcal{N}(\varphi_j^t) &= -\frac{\lambda}{h}\xi_j \\ \mathcal{N}(\varphi_j^v) &= \lambda(\xi_{j-1} - \xi_j) + i_{\text{ext}} \\ \mathcal{N}(\varphi_j^b) &= \frac{\lambda}{h}\xi_j.\end{aligned} \tag{47}$$

Where we have defined

$$\xi_j = -2\pi f_j^{\text{ind}} = \varphi_j^v + \varphi_j^t - \varphi_{j+1}^v - \varphi_j^b + 2\pi f_0, \tag{48}$$

and normalized with respect to the parameters of the vertical junctions, thus $h = I_{ch}/I_{cv} = C_h/C_v = R_h/R_v$ and $\lambda = \lambda_v = \Phi_0/2\pi I_{cv}L$ and $\lambda/h = \lambda_h = \Phi_0/2\pi I_{ch}L$.

5.4. *Rotobreathers in a dc biased ladder*

The simplest rotobreather solutions in a dc biased ladder array are shown in Fig. 23. One vertical junction and some of its neighbors are excited at the gap voltage meanwhile the other oscillate around the superconducting state. Such solutions can be continued from the uncoupled limit or excited by adding local dc currents to the central junction.[60]

The rotobreathers in the ladder can be unambiguously detected by measuring local dc voltages and have been found in numerical simulations of the model and experimentally observed. Other more complex patterns and multi-site DB solutions have been also excited and detected.[61–63] Multi-site

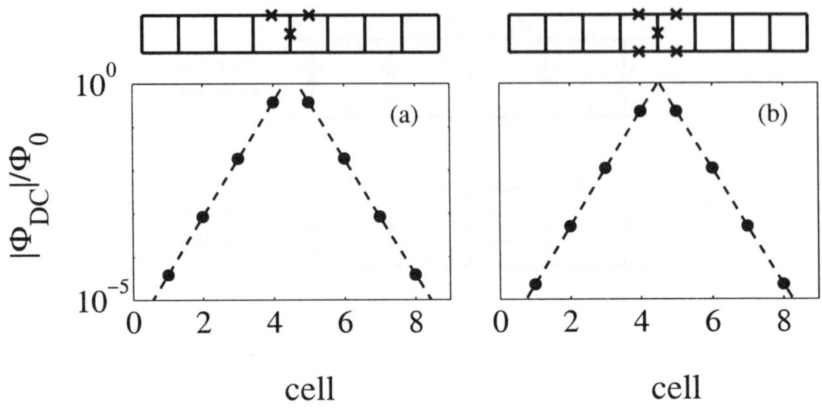

Fig. 24. Simulation of 9x1 array with $\lambda = 0.05, \Gamma = 0.1$ and $h = 0.25$. We have plotted the absolute value of the DC flux per unit cell at $I = 0.7$. The flux decays exponentially with a decay length of 0.32 for both solutions.

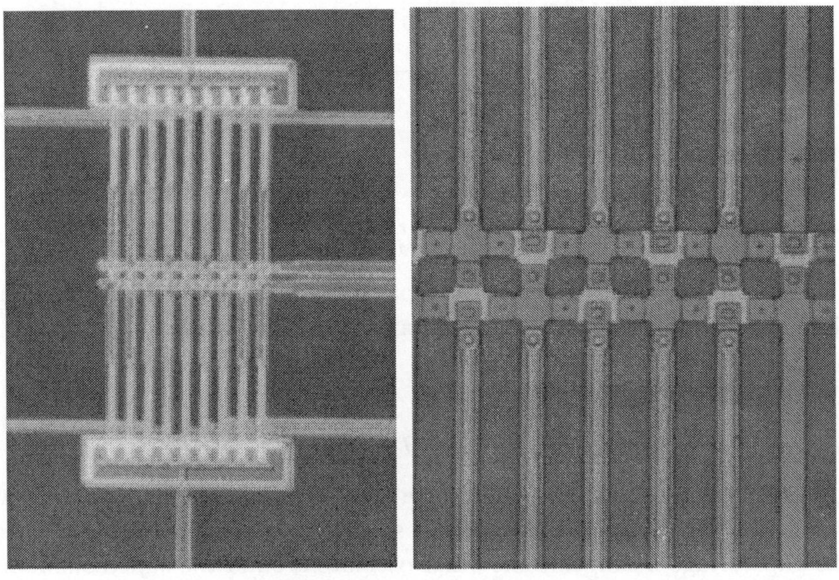

Fig. 25. Picture of the ladder.

solutions are characterized by a breather core with more than one rotating vertical junction.

Figure 24 shows the absolute value of the dc flux per unit cell for a type-

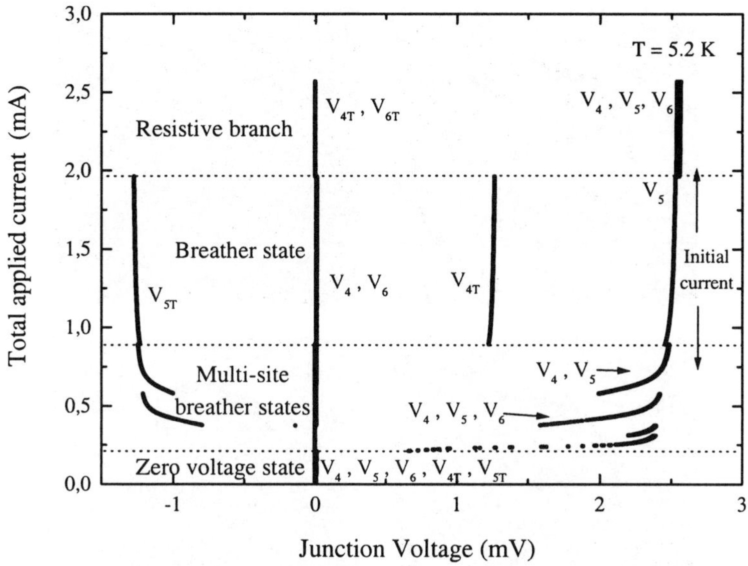

Fig. 26. Measured time-averaged voltages of five junction in the center of the array as the applied current is varied. V_4, V_5, V_6 stand for the voltage in vertical junctions number 4,5 and 6 in a 9 junction ladder. $V_{4\mathrm{T}}$ and $V_{5\mathrm{T}}$ for horizontal junctions in the top branches connecting the three previously mentioned vertical junctions. The breather state was excited at a current close to 1.4mA. When the current was increased the breather solution became unstable at a current close to 2mA and the array switches to a uniform resistive state, where all the vertical junctions rotate with a same voltage. We can see that for the breather state $V_4 = V_6 = 0$ and $V_{4\mathrm{T}} = -V_{5\mathrm{T}} = V_5/2$ which corresponds to the breather state that we have coined as type-B state. In a second experiment, a new breather was excited but now the current was decreased. We observed that the breather becomes unstable at a current close to 0.8mA and a new breather state is excited. This is a multi-site breather state. New instabilities between different multi-site states occur and finally the array reaches its zero-voltage state at about 0.2mA.

A and type-B (Fig. 23) breather solution. It can be seen that, as expected, the flux decays exponentially.

Figure 25 shows a picture of one of the anisotropic arrays where DB were excited and detected.[55,62,40] It is a ladder array with nine vertical junctions biased by an external current source. The array was designed with voltage probes in different vertical junctions (junction 4,5,6, and 9) to measure local voltage at different points of the array (we can measure for instance V_4, V_5, V_6, V_9, V_{4T}, V_{5T} and any other combination of the terminals). Sometimes, when we swept the applied current we found that DB solutions appear spontaneously. However for the experiments, we developed a simple

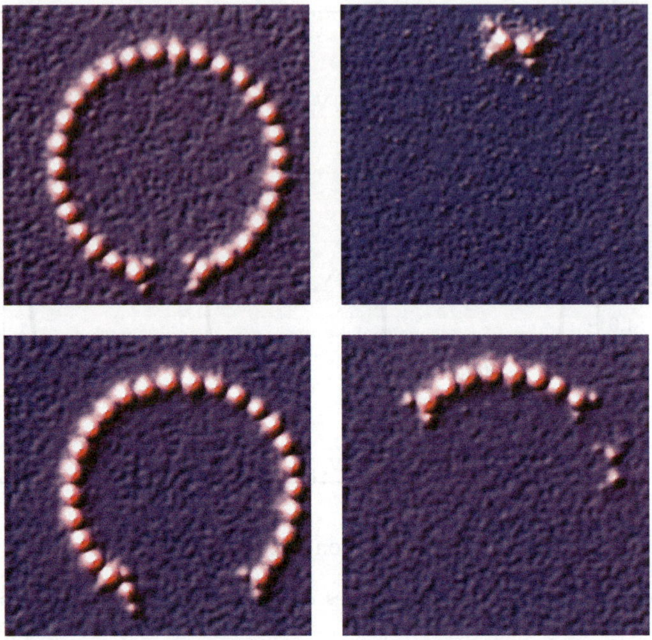

Fig. 27. LTSM image of four different multi-site breather states in an annular Josephson ladder array. Figure from http://www.pi3.physik.uni-erlangen.de/ustinov/.

reproducible method of exciting a breather: (i) Bias the array uniformly to a current below depinning current; (ii) increase the current injected into the middle vertical junction (V_5) until its voltage switches to the gap; (iii) reduce this extra current in the middle junction to zero. Other procedures are also possible.

Fig. 26 shows the result after we have excited the breather and we have increased or decreased the array current. The breather was excited at $I_a \approx 1.4\,\mathrm{mA}$ and then the junction voltages were measured as the applied current was increased or decreased. The DB is initially in the fifth vertical junction (V_4=V_6=0 and $V_5 \neq 0$). When increasing the external current, the breather exists until a maximum current where all the junctions switch to the gap. When decreasing the external current different scenarios were found but typically the breather enlarges from one-site to multi-site breather solutions and then at small enough current the array decays to

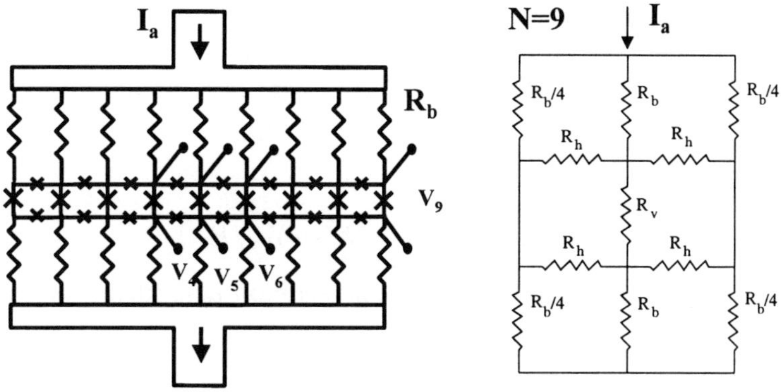

Fig. 28. Ladder array biased by an external current and equivalent circuit for the dc model of a one-site type-B breather.

the uniform superconducting state. More details about the experiment can be found in Ref. 62

Figure 27 shows a low temperature scanning microscopy image of an annular ladder with different multi-site breather solutions. Such pictures were obtained by the group of Alexey Ustinov at Erlangen[39]. The experimental results confirmed the existence of many different breather states in open and annular ladders. In the figure we can observe breathers where two vertical junction are rotating and large multi-site breather states where most of the vertical junctions rotate.

5.4.1. *Analysis of the breather solutions using a dc model*

The simplest theoretical approach to the study of the dynamics of breathers in Josephson arrays is the use of a dc model for the circuit. In this model rotating vertical junctions have a resistance of R_v and rotating horizontal junctions have a resistance of R_h. Oscillating junction will be modeled as shorts. Then we reduce the array to a simple network of resistors and calculate DC properties. The equivalent resistor network for a single-site symmetric breather (type-B breather) located on junction 5 in our 9 junction array is shown in Fig. 28. In addition to its simplicity this model allows to include the effect of the bias resistors (used to distribute uniformly the current through the array).

We will also make the following assumptions: $V_v = sV_h$, $R_v = hR_h$; where s=1 for type-A solutions and s=2 for type-B solutions and h is the parameter which describes the anisotropy of the array.

Using this model we make the following predictions for type-A (s=1) or type-B (s=2) m-site breathers.[62] We present here results for large bias resistances $R_b \gg R_h$ (uniform driving condition).

- The IV curve is given by
$$I_a/N = (1 + 2h/sm)V_v/R_v \tag{49}$$
- The minimun current for the breather solution I_{min} (defined from junction retrapping) is given by
$$i_{min} = I_{min}/NI_{cv} = (2h/m + s)(4/\pi)\Gamma \tag{50}$$
- The maximum current for the breather solution I_{max} (defined from junction switching) is given by
$$i_{max} = I_{max}/NI_{cv} = (2h + sm)/[(2 + m)h + sm] \tag{51}$$
- The effect of the bias resistors in the current distribution is
$$I_5 = [1 + mh/(2h + sm)(1 - m/N)R_h/R_b]^{-1}(I_a/N) \tag{52}$$
For N=9, m=1, s=2, $R_h/R_b \sim 0.8$ and h=0.25, we have $I_5 = 0.934 I_a/N$ and $I_j = 1.008 Ia/N$ ($j \neq 5$).

Figure 29 shows a comparison of the theoretical predictions obtained from the dc model with the experimental results. The different values of Γ in the figure correspond with experiments done at different temperatures from 4.2 to 6.7 K

Although the analysis based on the dc model has been found to be helpful, it presents some important limitations: The model can not account for any λ dependence. It can not explain resonances between breather dynamics and normal modes of the ladder. Such resonances can drive the breather to destabilizate. The model allows for an estimation of parameter values where the breather solution ceases to exist but it gives no information on the dynamical state after the destabilization of the localized solution.

5.4.2. *Simulations*

The breather dynamics has been extensively studied by numerical integration of the Eqs. (47) using an standard 4th order RK scheme. Such integrations are complemented with Floquet stability analysis[64,65] of periodic solutions and the study of the robustness of the breather solutions against

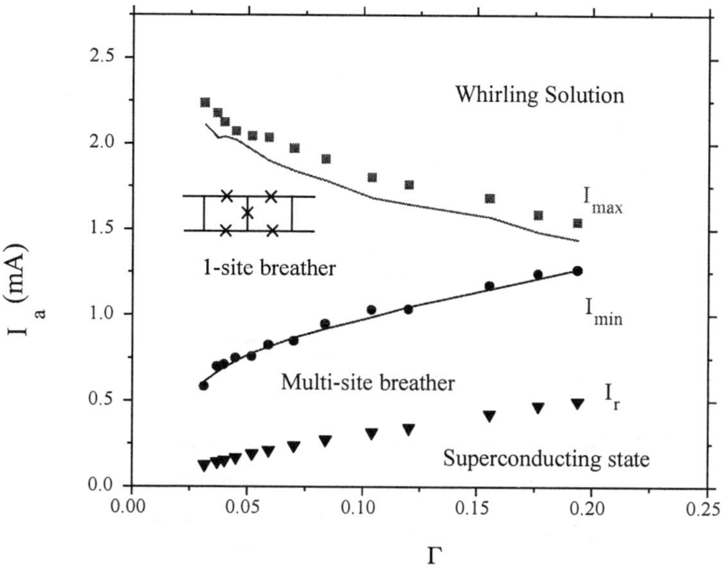

Fig. 29. Comparison between results based on the dc model and experimental results.

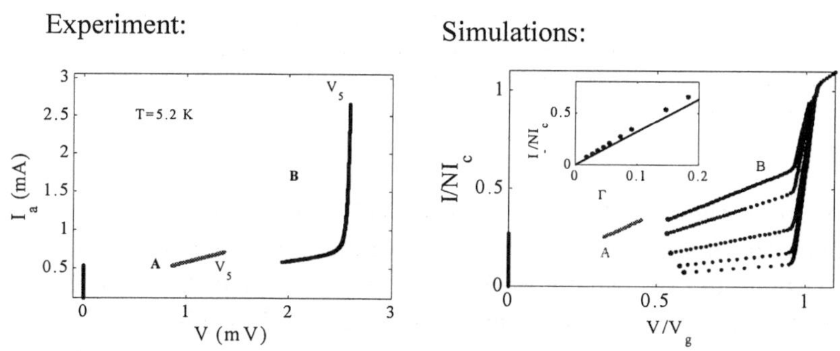

Fig. 30. Experiment (left) and simulation (right) of a type-B breather. The simulations have been done considering different subgap resistances. The experiment, and one of the simulations (upper curve), when decreasing the current show that the type-B breather destabilizes to a type-A one.

thermal fluctuations, modeled by including a noise term in the junction current (see Eq. (8)).

In addition to the noise, in order to get a more detailed description of

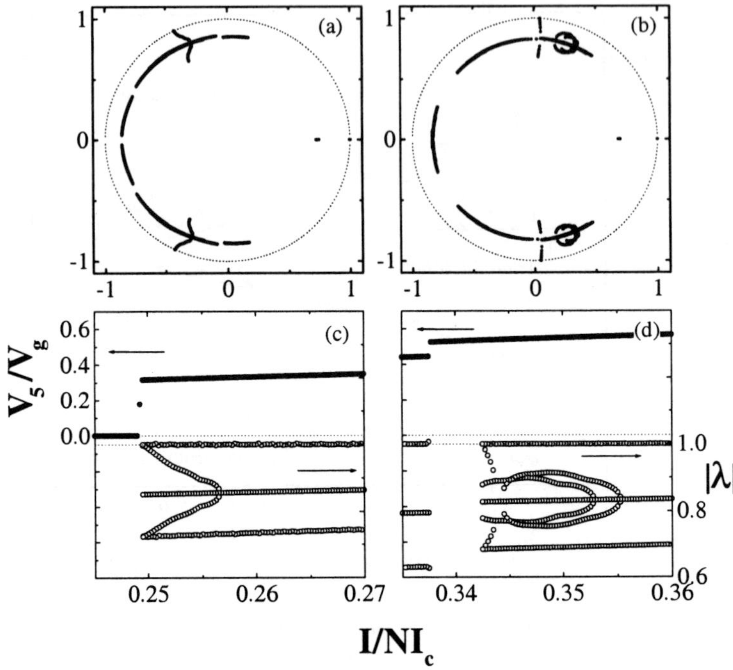

Fig. 31. Floquet multipliers of the type-A (a) and type-B (b) periodic DB showed in Fig. 30 (upper simulation). Figures (c) and (d) show as a function of the current the value of voltage (solid circles) and the modulus of the Floquet multipliers whenever the solution is periodic (open circles).

the system, the model defined in Eqs. (47) can be extended to take into account the nonlinear character of the junction resistance or to include the bias circuit. We usually work at zero external field, however the model allows for the inclusion of other external magnetic fields. Finally, we can also study the effect of considering the full-inductance matrix of the circuit and incorporate disorder, randomizing the junctions critical currents.

In this section we will present simulations made at $\Gamma \simeq 0.1$, $h \simeq 0.25$ and $\lambda \simeq 0.05$. Such values of the parameters are close to the expected parameter values for the experiments reported above.

Figure 30 shows an experimental IV curve of a breather solution that decreasing the current destabilizes from a 1-site type-B solution to a 1-site type-A one. The figure also shows the result of a simulations done including in the model the subgap resistance (as explained in Sec. 2.1) and for different values of this subgap resistance.

Figure 31 shows numerically computed Floquet multipliers for the simulation presented in the previous figure. Fig. 31(b) shows the distribution of the Floquet multipliers at several values of the applied current close to the destabilization current for the type-B breather. Fig. 31(d) shows the voltage and the modulus of the Floquet multipliers (only for periodic solutions) when we decrease the current. This is the typically observed bifurcation scenario for small λ. It seems that for small λ and underdamped junctions, this instability introduces more frequencies in the solution. When the periodic type-B breather losses stability the solution becomes a quasi-periodic type-B breather. This quasi-periodic type-B solution persists up to a smaller current when the array jumps to a periodic type-A solution. For large λ, however, we usually observe a period-doubling bifurcation where a multiplier crosses the unit circle at -1, though the behavior also depends on the damping.

Fig. 31(a) shows the Floquet multipliers for a type-A breather at different current values. In Fig. 31(c) we decrease the current and show the value of the voltage and the modulus of the Floquet multipliers. Below $I \sim 0.25$ the periodic breather is unstable and the solution switches to the superconducting state.

5.4.3. *Breather existence diagrams*

Now we are going to show a series of figures with breather existence diagrams calculated from the results of the dc model and computed with a numerical integration of the system equations.

The equations found in Sec. 5.4.1 allow for a calculation of the IV curves and the maximum and minimum values of external currents supporting DBs. If we write the equations for the single-breather state and use normalized parameters we find

$$\begin{aligned} i &= (1 + 2h/s)\Gamma v_v, \\ i_- &= (2h + s)4\Gamma/\pi, \\ i_+ &= (2h + s)/(3h + s), \end{aligned} \tag{53}$$

where currents are normalized by NI_{cv} and $v_v = \langle\dot{\varphi}_v\rangle$.

Figs. 32 and 33 show the predictions given by the circuit model. The size of the existence regions decreases rapidly when the damping or the anisotropy increase. On the other hand, if the damping is small enough the equations predict the existence of localized solutions even at large values of

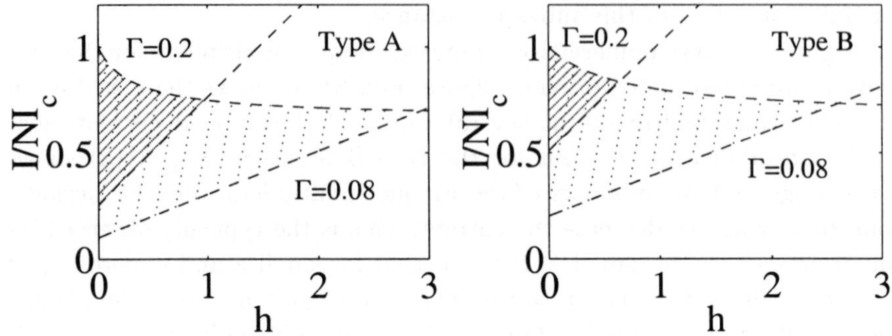

Fig. 32. Prediction of Eqs. (53) for i_+ and i_- as a function of h for single-site type-A (left) and type-B (right) solutions and two values of the damping. Lightly hatched region corresponds to $\Gamma = 0.08$ and the densely one to $\Gamma = 0.2$.

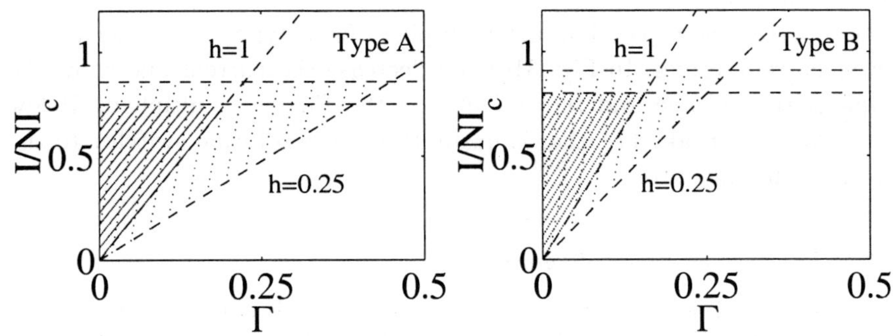

Fig. 33. Prediction of Eqs. (53) for i_+ and i_- as a function of Γ for single-site type-A (left) and type-B (right) solutions and two values of the anisotropy. Lightly hatched region corresponds to $h = 0.25$ and the densely one to $h = 1.0$.

h. We also see that the existence regions are larger for type-A solutions.

This simple model, however, does not account for any dependence of the curves with the parameter λ. This is an important limitation of the dc model and we have confirmed in the numerical simulations that λ affects our predictions in two important ways. First, it affects the value of the array retrapping current. The value used in our circuit models has been calculated from a single junction and should be corrected by λ in the case of the array. Second, it governs the values of the voltage at which resonances between the breather and the normal modes of the array play an important role. Roughly speaking, the resonances split the diagrams in three different

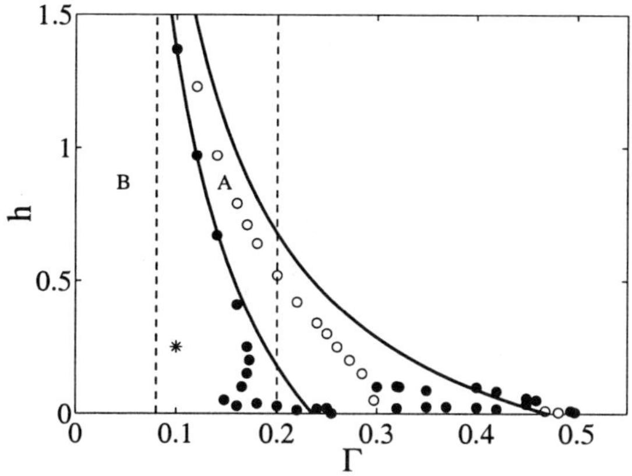

Fig. 34. Numerical calculation of the existence region of single-site DBs when $\lambda = 0.04$ and $I = 0.6$. Open circles correspond to type-A and solid circles to type-B solutions. Vertical lines correspond to cuts shown in Fig. 36 and the asterisk to the experiments.

regions: The small, the moderate and the large λ regions. When λ is small, the resonance frequency is smaller than the breather frequency, and when λ is large the resonance frequency is larger. Thereby, complications of damped resonances between the DB and the lattice eigenmodes are avoided in these limits.

Far from the resonance values the effect of λ is a small correction to our IV curves. This is shown by the numerical simulations. See for instance Figs. 37 and 41, where can be seen that the IV curves numerically integrated agree quite well with the predictions of the dc model.

We have also done numerical simulations based on Eqs. (47) with $f_0 = 0$ in order to study the λ dependence of the breather existence region. The results are presented in Figs. 34, 35, and 36. In these diagrams we show the maximum and minimum values of the parameters for which a localized solution has been numerically found. In some cases the characterization of the solutions inside the existence regions is quite complex and several resonances and transitions between periodic and aperiodic localized states appear.

Figure 34 shows the existence regions in the anisotropy versus damping plane when $\lambda = 0.04$ and $I = 0.6$. The figure confirms that DBs exist at large values of h if Γ is small enough. Fig. 35 shows the existence regions in the current versus λ plane when $\Gamma = 0.08$ and $h = 0.25$. These are the estimated values of h and Γ in our experiments. Fig. 36 shows the diagram

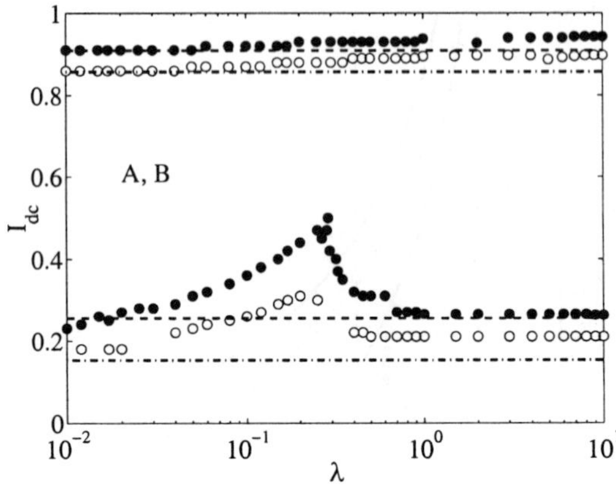

Fig. 35. Numerical calculation of the existence region of single-site DB when $\Gamma = 0.08$ and $h = 0.25$. Open circles correspond to type-A and solid circles to type-B solutions. Horizontal lines correspond to the predictions of the circuit model.

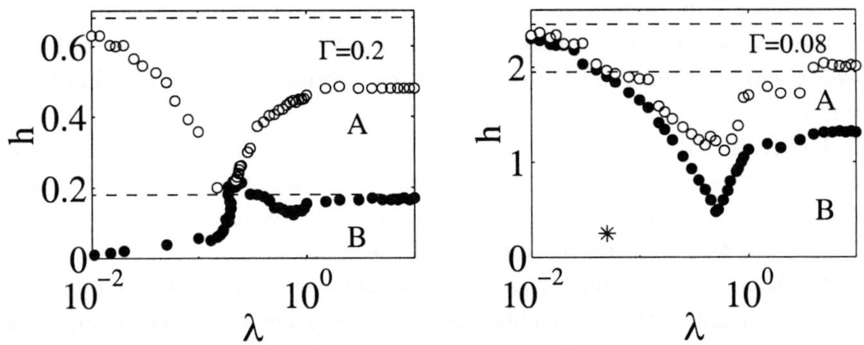

Fig. 36. Numerical calculation of the existence region of single-site DB when $I = 0.6$ and $\Gamma = 0.2$ (left) and $\Gamma = 0.08$ (right). Open circles correspond to type-A and solid circles to type-B solutions. Horizontal lines correspond to the predictions of the circuit model and the asterisk to the experiments.

in the anisotropy versus λ plane for $I = 0.6$ and $\Gamma = 0.2$ (left) and 0.08 (right). The asterisk in the $\Gamma = 0.08$ figure approximately corresponds to the value of the parameter where our experiments were done. We can see in the figures that at moderate λ there is a substantial deviation from the predictions of the dc model. This deviation is caused resonances and other dynamical effects.

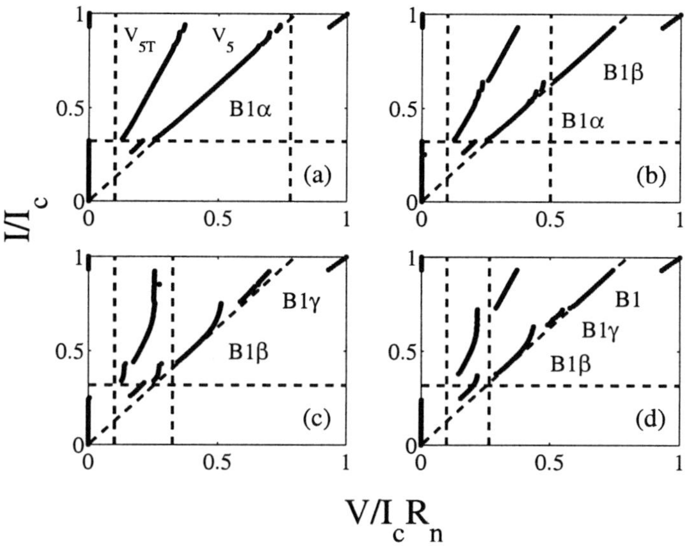

Fig. 37. Simulated IV's for 9×1 ladder of type-B breather as a function of λ. $h = 0.25$, $\Gamma = 0.1$, $f = 0$ and (a) $\lambda = 5$, (b) $\lambda = 2$, (c) $\lambda = 0.8$, and (d) $\lambda = 0.5$. The labels indicate different type-B breather solutions. The vertical dashed lines are $\omega_+(\lambda)$ and $\omega_- = 1$ from Eq. (54). The horizontal dashed line is I_- and the diagonal dashed line is the IV curve, both from Eq. (53).

5.4.4. *Different λ regimes*

In order to understand the role of λ in the simulations we need to study the basic linearized excitations that can occur. Complex behavior appears when the frequency of the rotating junction resonates with the ladder eigenmodes. To calculate the resonant frequencies, we linearize Eq. (47) around a solution. The linear analysis allows to compute such frequencies and the decay length.[62,66,59] The dispersion relation has two branches defined by

$$\begin{aligned} \omega_+ &= \sqrt{1 + \frac{2\lambda}{h} + 4\lambda \sin^2 \frac{q}{2}} \\ \omega_- &= 1, \end{aligned} \tag{54}$$

and we can also estimate the decay length for our waves,

$$\zeta = \cosh^{-1} \left| \frac{2\lambda(h+1)+h}{2\lambda h} - \frac{\omega^2}{2\lambda} \right|. \tag{55}$$

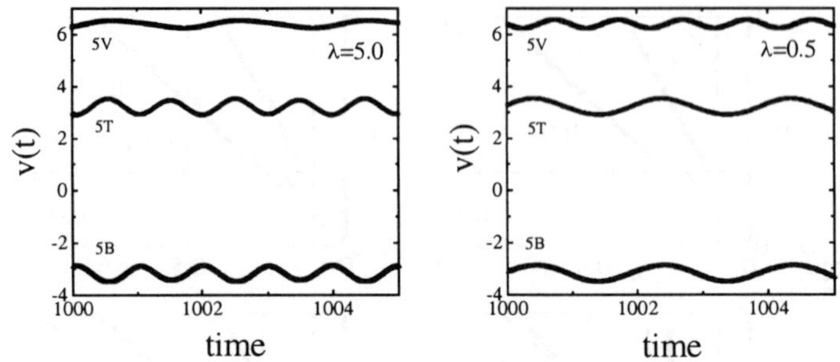

Fig. 38. Time evolution of the time derivative of the phase, $v(t) = d\varphi/dt$, for the large λ type-B solution labeled B1α in Fig. 37(a) and the small λ type-B solution depicted as B1 in Fig. 37(d). We plot $v(t)$ for the rotating vertical junction (5V) and two neighbor horizontal junctions (5T and 5B). Here, $h = 0.25$, $\Gamma = 0.1$ and $I = 0.8$.

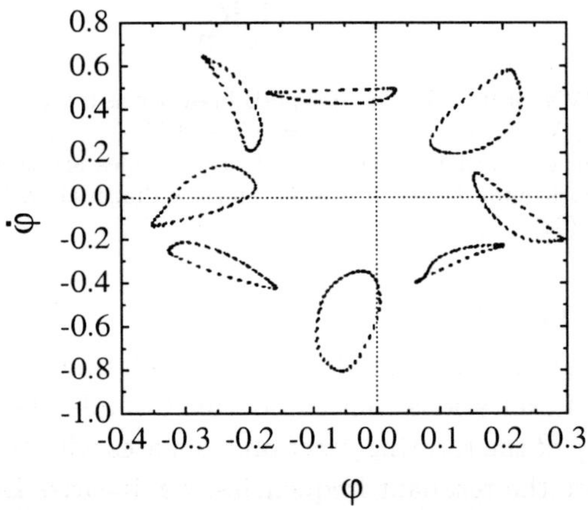

Fig. 39. Poincaré sections of the third top junction (3T) for a B1γ solution. The phases are shown at times $t = t_0 + n\tau$ where $\tau = 2\pi/V_5$ and $h = 0.25$, $\Gamma = 0.1$, $\lambda = 0.5$, and $I = 0.7$.

Equation (54) defines three physical regimes in the system:

$$\text{(i)}\ \ \omega > \omega_R \quad (\text{small}\ \lambda)$$
$$\text{(ii)}\ \ \omega \sim \omega_R \quad (\text{moderate}\ \lambda)$$
$$\text{(iii)}\ \ \omega < \omega_R \quad (\text{large}\ \lambda)$$

The existence of such regimes has been confirmed by the numerical simu-

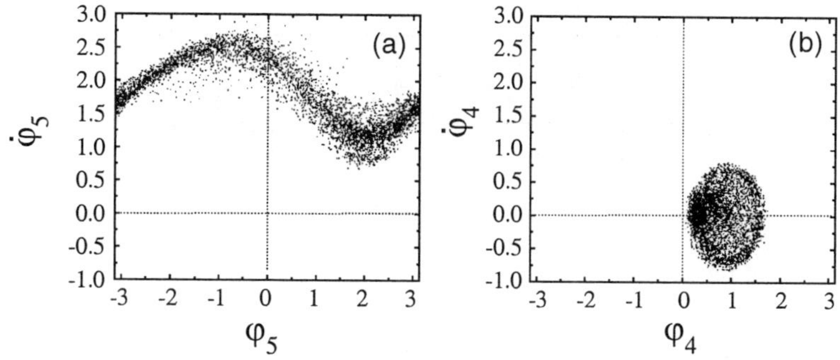

Fig. 40. Poincaré sections of the fourth and fifth vertical junctions for a type-B chaotic solution. The phases are shown at times $t = t_0 + n\tau$ where $\tau = 2\pi/V_5$ and $h = 0.15$, $\Gamma = 0.2$ and $\lambda = 0.2$ and $I = 0.6$. (a) Shows the sections for the rotating vertical junction 5 and (b) for its first neighbor, 4.

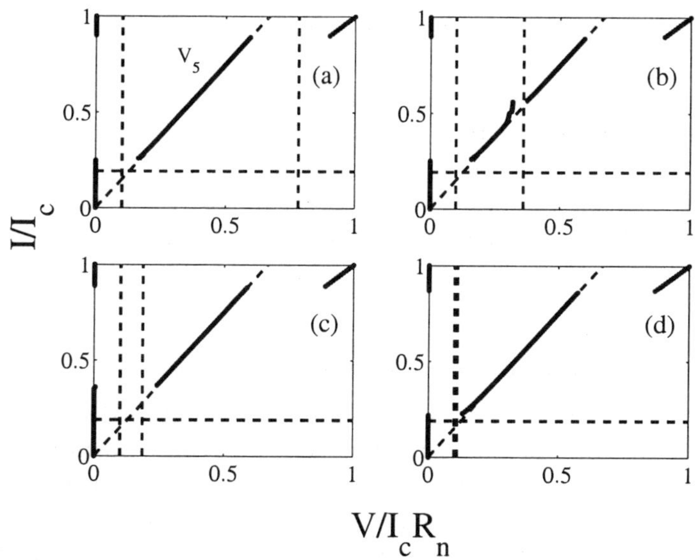

Fig. 41. Simulated IV's for 9×1 ladder of type-A breather as a function of λ. Here, $\Gamma = 0.1$, $f = 0$ and (a) $\lambda = 5$, (b) $\lambda = 1$, (c) $\lambda = 0.2$, and (d) $\lambda = 0.02$. The vertical dashed lines are $\omega_+(\lambda)$ and $\omega_- = 1$ from Eq. (54). The horizontal dashed line is I_- and the diagonal dashed line is the IV curve, both from Eq. (53).

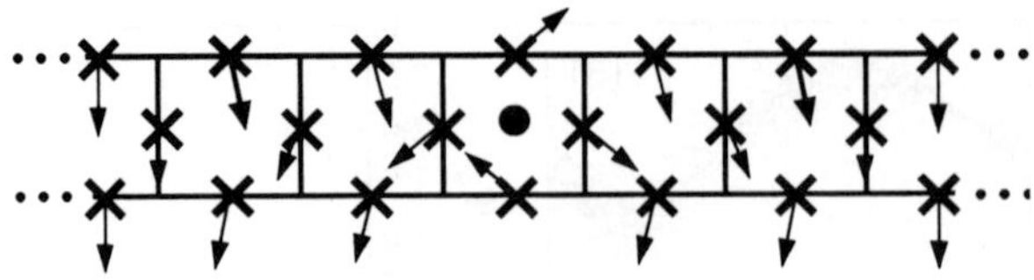

Fig. 42. A vortex configuration in the ladder ($I = 0$). Arrows represent phase differences. The phases of the vertical junctions change from 0 to 2π as we move from one edge to other of the ladder.

lation of IV curves at different values of λ.

Figure 37 shows IV curves at different values of λ of a ladder with a type-B breather (h=0.25, Γ=0.1). From (a) to (d), λ =5, 2, 0.8, 0.5. Labels indicate different types of solutions. Dashed lines are voltage resonances and prediction for I_{min}. We have observed that for large λ the solution is up-down symmetric (see Fig. 38 at $\lambda = 5.0$). However, at small values of λ top and bottom junctions show a dephase of half period (Fig. 38 at $\lambda = 0.5$). Intermediate values of λ show more complex solutions: for instance quasiperiodic (Fig. 39) and chaotic (Fig. 40) solutions. Stable resonant breather solutions appear at moderate values of λ. Such solutions have been also experimentally observed.[67]

Figure 41 shows IV curves at different values of λ of a ladder with a type-A breather (h=0.25, Γ=0.1). From (a) to (d), λ =5, 1, 0.2, 0.02. Dashed lines are voltage resonances and prediction for I_{min}. Again, we can see the existence of resonant breather solutions at intermediate values of λ.

5.4.5. *Breather-vortex collision in the Josephson ladder*

In the same way that kinks exist in parallel arrays and vortices in 2D arrays, the nonlinear static excitations in the ladder are called vortices.

The static properties of a vortex in the ladder are in many aspects similar to those of a kink in the Frenkel-Kontorova or discrete sine-Gordon system.[68–70] There are however some differences, the most important of which is the existence of a critical magnetic field $f_c(\lambda)$ for which if $f < f_c$ a single vortex is not stable in the ladder. Below this critical field the vortex is expelled from the ladder through the horizontal junctions. Thus, vortices are stable static solutions of the array at adequate parameter values. In the absence of external currents, a static vortex in the ladder correspond to a solution for which the phase of the vertical junctions φ_j^v go from 0 to 2π (see Fig. 42).

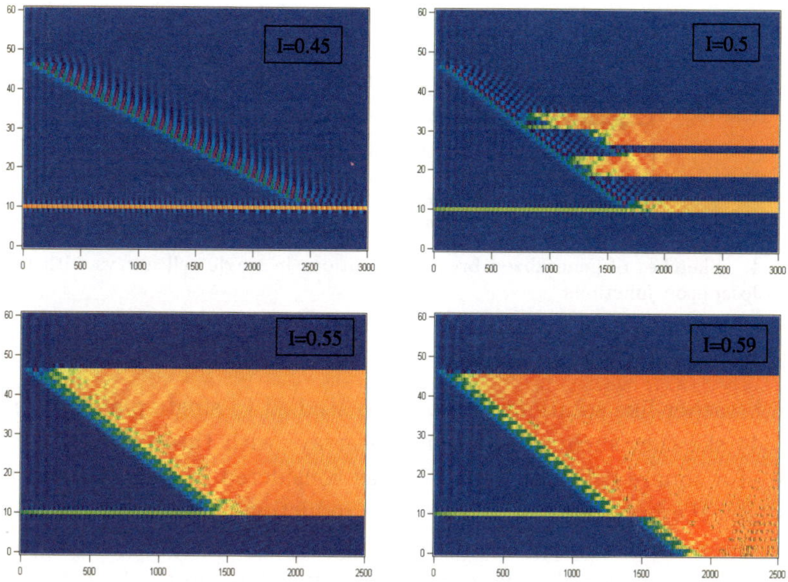

Fig. 43. Simulations a vortex-breather collisions in a Josephson ladder.[71] For all the cases the initial condition is a 1-site rotobreather in junction 11 of the ladder and a vortex in junction 47. The figures plot the value of the instantaneous voltage at every vertical junction.

We have studied vortex-breather collision in a Josephson-junction ladder array.[71] We have computed parameters values of the system for which both types of structures coexist in the ladder. Then, by increasing the circuit bias current, we have found different possible scenarios for vortex-breather collisions (Fig. 43). (i) The breather acts as a pinning center for a single vortex. (ii) Increasing the current, the vortex excites multi-site breathers on its way and is finally pinned by the breather. (iii) Now a whirling mode front is excited by the vortex. However the breather still acts as a pinning center, now for the front. (iv) At higher values of the bias, the front is able to destroy the breather.

For scenario (i), we have also studied thermal activation properties associated with the presence of the vortex-breather pair in the array. We have seen that noise causes the depinning of the vortex and the breather decays into a 2-site breather. The escape time for this process showed an exponential dependence in temperature with an activation energy of 22K.

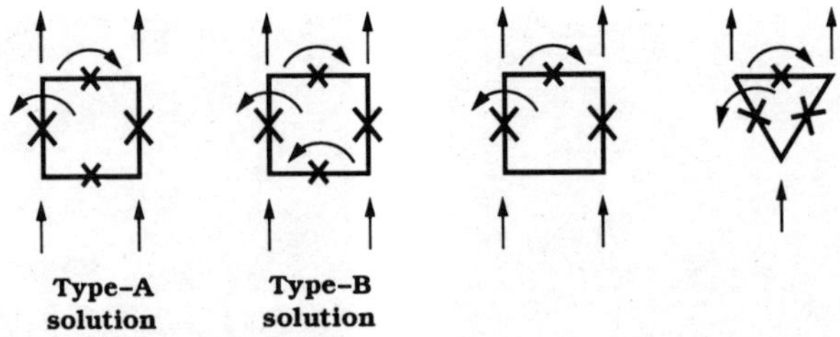

Fig. 44. Sketches of generalized breather solutions in single cell arrays with four and three Josephson junctions.

5.5. *Single-plaquette arrays*

Rotobreather-like solutions can be also studied in single-plaquette arrays. Fig. 44 sketches such solutions in cell arrays with three or four JJs. The case of a square cell with four JJs was considered in Refs. 60 and 62. In this case the breather can be seen as a reduction of the solutions found in the ladder when we impose mirror symmetry with respect to the rotating junction and neglect the dynamics of junctions beyond the first neighbor of the rotating ones. Doing so, we are left with a square plaquette with four junctions. The equations for the ladder can be mapped onto the equations of the plaquette with $h_{\rm plaq.} = 2h_{\rm ladd.}$ and $\lambda_{\rm plaq.} = 2\lambda_{\rm ladd.}$. Then it was found that many of the main aspects concerning DB solutions in the ladder can be studied in single cell arrays.

The analytical and numerical study of the rotobreather states in a single plaquette with three junctions[72,73] showed that this system is complex enough to present most of the phenomena observed in Josephson ladders. The experimental study of the array[74] confirmed the theoretical and numerical results.

5.6. *DBs in two-dimensional Josephson junction arrays*

Two-dimensional Josephson-junction arrays have already been studied in Sec. 4.1 in the context of vortices. Here we are going to present numerical results on the existence of rotobreathers in such arrays.[75]

Based on theoretical arguments from the uncoupled limit of the lattice and on numerical simulations, rotobreather solutions have been predicted to exist in 2DJJA biased by ac currents. The rotobreathers in 2D arrays correspond to voltage localized solutions sketched in Fig. 45. There, four junctions sited in the bulk of the array are in the resistive state, two with voltages $+V$ and two with $-V$, while the rest follow the ac field in a oscil-

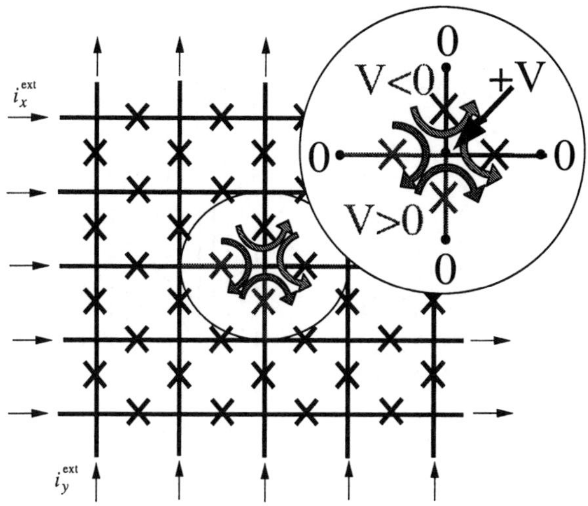

Fig. 45. Sketch of the 2DJJA with breather.

lating state of mean voltage $V = 0$. From the pictures of the dynamics of a single junction (Fig. 21) we see that such scenario of coexistence of three different attractors (0, $+V$, and $-V$) is only possible under ac bias [Figs. 21(c) and (d)]. Thus in the case of 2D arrays we will study rotobreather solutions biased by ac fields.

It can be seen that uncoupled limits can be obtained when $\lambda \to 0$ or when $h \to 0$ in the array (Eqs. (29) and (30)). In any of these limits the junctions behave as independent oscillators where -trivial- localized solutions can be obtained when the array is biased by ac external fields $i_{\text{ext}} = i_{ac} \sin \omega t$.

We have been able to excite and study DBs in many different regions of the parameter space and under diagonal or vertical bias. We have tried large and small values of λ, large and small values of h, different values of the frequency, damping, field amplitude, different array sizes,...

Figure 46 shows four snapshots along one period of the phase dynamics for two rotobreather solutions simulated in two very different situations (see caption).

In addition to the numerical integration of Eqs. (29) we have found that the obtained breather solutions are linearly stable in the framework of the Floquet stability theory and persist in the presence of thermal fluctuations (current noise). We also checked that the solutions also exist when applying

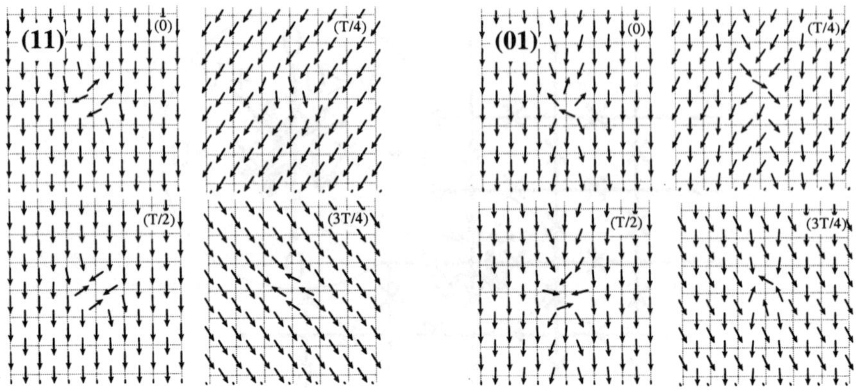

Fig. 46. Four snapshots of the phase evolution for two different DB solutions in a 2DJJA with different bias and very different parameter values. Arrows represent phases of the junctions and we show the central part of a 11×11 array. The solutions are time periodic (period T) and time increases from 0 (arbitrary) to $3T/4$ as labeled. The rotobreather solution is localized in the four central junctions which rotate. Left: current is biased in the diagonal or (11) direction ($\lambda = 0.1$, $h = 1.0$, $i_{\rm ac} = 5.0$, $\Gamma = 0.003$ and $\omega = \pi$). Right: current is biased along the y or (01) direction ($\lambda = 5.0$, $h = 0.05$, $i_{\rm ac} = 0.7$, $\Gamma = 0.02$ and $\omega = \pi/2$).

an external magnetic field and studied the mechanisms involved in the destabilization of the localized solution when changing some of the system parameters.

In relation to the issue of a possible experimental observation of DB's in 2DJJA, the simulations were done at experimentally accessible parameter values and it was found a numerical protocol to excite DB's in the array. The protocol is based in the possibility of adding a local dc current that should be injected in one central island of the array and extracted from the four neighboring islands. With respect to a detection, it can be done for instance by measuring local voltages in different points of the array.

To finish we want to mention that we have also studied and obtained rotobreather solutions in the dynamics of the array in the infinite λ (no induced magnetic flux) limit.[76] In this limit the equations correspond to a realization of the dynamics of the two-dimensional XY model.

Acknowledgments

This lecture is the result of years of work in collaboration with many people. Special thanks are given to F. Falo, L. M. Floría and T. P. Orlando for their guide and collaboration and hospitality. I acknowledge P. J. Martínez and

E. Trías for many collaborations and discussions in these topics. Thanks to M. V. Fistul, S. Flach, J. L. García-Palacios, J. L. Marín, F. Pignatelli, M. Schuster, K. Segall and A. V. Ustinov for insightful discussions. Financial support is acknowledged to FEDER program through BFM2002-00113 project and European Network LOCNET HPRN-CT-1999-0016.

References

1. A. Scott, *Nonlinear Science: Emergence & Dynamics of Coherent Structures.* Oxford University Press, 1999.
2. B. D. Josephson, "Possible new effect in superconductive tunneling". *Phys. Lett.* **1**, 251-253 (1962).
3. J. R. Waldram, "The Josephson effect in weakly coupled superconductors". *Rep. Prog. Phys.* **39**, 751-821 (1976).
4. A. Barone and G. Paternò, *Physics and applications of the Josephson effect.* John Wiley, 1982.
5. K. K. Likharev, *Dynamics of Josephson junctions and circuits.* Gordon and Breach Science, 1984.
6. T. P. Orlando and K. A. Delin, *Foundations of applied superconductivity.* Addison Wesley, 1991.
7. M. Tinkham, *Introduction to superconductivity.* Mc Graw Hill, 1996.
8. A. Barone, "Weakly coupled macroscopic quantum systems: likeness with difference". In *Quantum Mesoscopic Phenomena and Mesoscopic Devices in Microelectronics*, pp. 301-320. Edited by I. O. Kulik, Kluwer Academic, 2000.
9. O. Avenel and E. Varoquax, "Josephson effect and quantum phase slippage in superfluids". *Phys. Rev. Lett.* **60**, 416-419 (1988).
10. A. J. Legget "Bose-Einstein condensation in the alkali gases: some fundamental concepts". *Rev. Mod. Phys.* **73**, 307-356 (2001).
11. V. Ambegaokar and A. Baratoff, "Tunneling between superconductors". *Phys. Rev. Lett.* **10**, 486-489 (1963) and erratum **11**, 104 (1963).
12. W. C. Stewart, "Current-voltage characteristics of Josephson junctions". *Appl. Phys. Lett.* **12**, 277-280 (1968).
13. D. E. McCumber, "Effect of ac impedance on dc voltage-current characteristics of superconductor weak-link junctions". *J. Appl. Phys.* **39**, 3113-3118 (1968)
14. T. Van Duzer and C. W. Turner, *Principles of Superconductive Devices and Circuits.* Pearson Education POD, 1998.
15. P. Hänggi, P. Talkner and M. Borkovec, "Reaction-rate theory: fifty years after Kramers". *Rev. Mod. Phys.* **62**, 251-341 (1990)
16. V. I. Mel'nikov, "The Kramers problem: fifty years of development". *Phys. Rep.* **209**, 1-71 (1991)
17. A. V. Ustinov, "Solitons in Josephson junctions". *Physica D* **123**, 315-329 (1998).
18. J. Clarke, A. N. Cleland, M. H. Devoret, D. Esteve and J. M. Martinis,

"Quantum mechanics of a macroscopic variable: the phase difference of a Josephson junction". *Science* **239**, 992-997 (1988).

19. L. J. Geerlings, V. F. Anderegg, J. Romijn and J. E. Mooij, "Single Copper-pair tunneling". *Phys. Rev. Lett.* **65**, 377-380 (1990).
20. W. J. Elion, M. Matters, U. Geigenmüller and J. E. Mooij, "Demonstration of Heisenberg's uncertainty principle in a superconductor". *Nature* **371**, 594-595 (1994).
21. Penttilä, Ü. Parts, P. J. Hakonen, M. A. Paalanen and E. B. Sonin, "Superconductor-Insulator Transition in a Single Josephson Junction". *Phys. Rev. Lett.* **82**, 1004-1007 (1999).
22. A. Wallraff, A. Lukashenko, J. Lisenfeld, A. Kemp, M. V. Fistul, Y.Koval, and A. V. Ustinov, "Quantum dynamics of a single vortex". *Nature***425**, 155-158 (2003)
23. Y. Nakamura, Yu. A. Pashkin and J. S. Tsai, "Coherent control of macroscopic quantum states in a single-Cooper-pair box". *Nature* **398**, 786-788 (1999).
24. J. R. Friedman, V. Patel, W. Chen, S. K. Tolpygo and J. E. Lukens, "Quantum superposition of distinct macroscopic states". *Nature* **406**, 43-46 (2000)
25. C. H. Van der Waal, A. C. J. ter Haar, F. K. Wilhelm, R. N. Schouten, C. J. P. M. Harmans, T. P. Orlando, S. Lloyd and J. E. Mooij, "Quantum superposition of macroscopic persistent-current states". *Science* **290**, 773-777 (2000).
26. D. Vion, A. Aassime, A. Cottet, P. Joyez, H. Pothier, C. Urbina, D. Esteve and M. H. Devoret, "Manipulating the quantum state of an electrical circuit". *Science* **296**, 886-889 (2002).
27. Y. Yu, S. Han, X. Chu, S. Chu and Z. Wang, "Coherent temporal oscillations of macroscopic quantum states in a Josephson junction". *Science* **296**, 889-892 (2002)
28. I. Chiorescu, Y. Nakamura, C. J. P. M. Harmans and J. E. Mooij, "Coherent quantum dynamics of a superconducting flux qubit". *Science* **299**, 1869-1871 (2003).
29. Yu. A. Pashkin, T. Yamamoto, O. Astafiev, Y. Nakamura, D. V. Averin and J. S. Tsai, "Quantum oscillations in two coupled charge qubits". *Nature* **421** 823-826 (2003).
30. C. A. Hamilton, C. J Burroughs and S. P. Benz, "Josephson voltage standard: A review". *IEEE Trans. Appl. Supercond.* **7**, 3756-3761 (1997).
31. J. Clarke, "Principles and applications of SQUIDs". *Proc. IEEE* **77**, 1208-1223 (1989).
32. S. T. Ruggiero and D. A. Rudman (eds.), *Superconducting Devices.* Academic Press, 1990.
33. D. Koelle, R. Kleiner, F. Ludwig, E. Dantsker and J. Clarke, "High-transition-temperature superconducting quantum interference devices". *Rev. Mod. Phys.* **71**, 631-686 (1999).
34. L. M. Floría and J. J. Mazo, "Dissipative dynamics of the Frenkel-Kontorova model". *Adv. Phys.* **45** 505-598 (1996).
35. O. Braun and Y.Kivshar, "Nonlinear dynamics of the Frenkel-Kontorova

model". *Phys. Rep.* **306**, 1-108 (1998).
36. O. Braun and Y.Kivshar, *The Frenkel-Kontorova model. Concepts, methods and applications.* To be published by Springer.
37. A. V. Ustinov, M. Cirillo and B. A. Malomed, "Fluxon dynamics in one-dimensional Josephson-junction arrays". *Phys. Rev. B* **47**, 8357-8360 (1993).
38. S. Watanabe, H. S. J. van der Zant, S. H. Strogatz and T. P. Orlando, "Dynamics of circular arrays of Josephson junctions and the discrete sine-Gordon equation". *Physica D* **97**, 429-470 (1996).
39. A. V. Ustinov, "Imaging of discrete breathers". *Chaos*, **13**, 716-724 (2003).
40. J. J. Mazo and T. P. Orlando, "Discrete breathers in Josephson arrays", *Chaos*, **13**, 733-743 (2003).
41. R. S. Newrock, C. J. Lobb, U. Geigenmüller and M. Octavio, "The two-dimensional physics of Josephson-junction arrays". *Sol. State Phys.* **54**, 263-512 (2000).
42. P. Hadley, M. R. Beasley and K. Wiesenfeld, "Phase-locking of Josephson-junction series arrays". *Phys. Rev. B* **38**, 8712-8719 (1988).
43. H. J. Mooij and G. Schön, "Single charges in 2-dimensional junction arrays". In *Single Charge Tunneling*, pp. 275-310. Ed. by H. Grabert and M. H. Devoret, Plenum Press, 1992.
44. R. Fazio and H. S. J. van der Zant, "Quantum phase transitions and vortex dynamics in superconducting networks". *Phys. Rep.* **355**, 235-334 (2001).
45. P. Delsing, "One-dimensional arrays of small tunnel junctions". In *Single Charge Tunneling*, pp. 249-274. Ed. by H. Grabert and M. H. Devoret, Plenum Press, 1992.
46. D. B. Haviland and P. Delsing, "Cooper-pair charge solitons: the electrodynamics of localized charge in a superconductor". *Phys. Rev. B* **54**. R6857-R6860 (1996).
47. F. Falo, P. J. Martínez, J. J. Mazo and S. Cilla, "Ratchet potential for fluxons in Josephson-junction arrays". *Europhys. Lett.* **45**, 700-706 (1999).
48. E. Trías, J. J. Mazo, F. Falo and T. P. Orlando, "Depinning of kinks in a Josephson-junction ratchet array". *Phys. Rev. E* **61**, 2257-2266 (2000).
49. F. Falo, P. J. Martínez, J. J. Mazo, T. P. Orlando, K. Segall and E. Trías, "Fluxon ratchet potentials in superconducting circuits". *Appl. Phys. A* **75**, 263-269 (2002).
50. A. J. Sievers and S. Takeno, "Intrinsic localized modes in anharmonic lattices". *Phys. Rev. Lett.* **61**, 970-973 (1988)
51. S. Takeno and M. Peyrard, "Nonlinear modes in coupled rotator models". *Physica D* **92**, 140-163 (1996).
52. S. Flach and C. R. Willis, "Discrete breathers". *Phys. Rep.* **295**, 181-264 (1998).
53. S. Aubry, "Breathers in nonlinear lattices: Existence, linear stability and quantization". *Physica D* **103**, 201-250 (1997).
54. Focus Issue on "Nonlinear localized modes: Physics and applications". *Chaos* **13(2)** (2003).
55. E. Trías, J. J. Mazo and T. P. Orlando, "Discrete breathers in nonlinear lattices: experimental detection in a Josephson array". *Phys. Rev. Lett.* **84**,

741-744 (2000).
56. P. Binder, D. Abraimov, A. V. Ustinov, S. Flach, and Y. Zolotaryuk, "Observation of breathers in Josephson ladders". *Phys. Rev. Lett.* **84**, 745-748 (2000).
57. E. Trías, *Vortex motion and dynamical states in Josephson arrays.* Ph.D. thesis, Massachusetts Institute of Technology, 2000.
58. P. Binder, *Nonlinear localized modes in Josephson ladders.* Ph.D. thesis, Universität Erlangen-Nürnberg, 2001.
59. M. V. Fistul, "Resonant breather states in Josephson coupled systems". *Chaos*, **13**, 725-732 (2003).
60. J. J. Mazo, E. Trías and T. P. Orlando, "Discrete breathers in dc-biased Josephson-junction arrays". *Phys. Rev. B* **59**, 13604-13607 (1999).
61. P. Binder, D. Abraimov and A. V. Ustinov, "Diversity of discrete breathers observed in a Josephson ladder". *Phys. Rev. E* **62**, 2858-2862 (2000).
62. E. Trías, J. J. Mazo, A. Brinkman and T. P. Orlando, "Discrete breathers in Josephson ladders". *Physica D* **156**, 98-138 (2001).
63. P. Binder and A. V. Ustinov, "Exploration of a rich variety of breather modes in Josephson ladders". *Phys. Rev. E* **66**, 016603-1(9) (2002).
64. D. W. Jordan and P. Smith, *Nonlinear ordinary differential equations.* Ofxord Univ. Press, 1999.
65. S. Flach, chapter 1, *Energy localization and transfer*, T. Dauxois, A. Litvak-Hinenzon, R. S. Mackay, A. Spanoudaki Eds., Advanced Series on Nonlinear Dynamics, World Scientific (2003).
66. A. E. Miroshnichenko, S. Flach, M. V. Fistul, Y. Zolotaryuk and J. B. Page, "Breathers in Josephson junction ladders: Resonances and electromagnetic wave spectroscopy". *Phys. Rev. E* **64**, 066601-1(14) (2001).
67. M. Schuster, P. Binder and A V. Ustinov, "Observation of breather resonances in Josephson ladders". *Phys. Rev. E* **65**, 016606-1(6) (2001).
68. J. J. Mazo, F. Falo and Luis M. Floría, "Josephson-junction ladders: ground state and relaxation phenomena". *Phys. Rev. B* **52**, 10433-10440 (1995).
69. J. J. Mazo and J. C. Ciria, "Equilibrium properties of a Josephson-junction ladder with screening effects". *Phys. Rev. B* **54**, 16068-16076 (1996).
70. M. Barahona, S. H. Strogatz and T. P. Orlando, "Superconducting states and depinning transitions of Josephson ladders". *Phys. Rev. B* **57**, 1181-1199 (1998).
71. E. Trías, J. J. Mazo and T. P. Orlando, "Interactions between Josephson vortices and breathers". *Phys. Rev. B* **65**, 054517-1(10) (2002).
72. A. Benabdallah, M. V. Fistul and S. Flach, "Breathers in a single plaquette of Josephson junctions: existence, stability and resonances". *Physica D* **159**, 202-214 (2001).
73. M. V. Fistul, S. Flach and A. Benabdallah, "Magnetic-field-induced control of breather dynamics in a single plaquette of Josephson junctions". *Phys. Rev. E* **65**, 046616-1(4) (2002).
74. F. Pignatelli and A. V. Ustinov, "Observation of breather-like states in a single Josephson cell". *Phys. Rev. E* **67**, 036607-1(7) (2003).
75. J. J. Mazo, "Discrete breathers in two-dimensional Josephson-junction ar-

rays", Phys. Rev. Lett. **89**, 234101-1(4) (2002).
76. D. Zueco and J. J. Mazo, private communication.

CHAPTER 5

LOCALIZED EXCITATIONS IN JOSEPHSON ARRAYS. PART II: EXPERIMENTS

Alexey V. Ustinov

Physikalisches Institut III, Universität Erlangen-Nürnberg
Erwin-Rommel-Str. 1, 91058 Erlangen, Germany
E-mail: ustinov@physik.uni-erlangen.de

The purpose of this review is to give an introduction into experiments with Josephson junction arrays. For many years this experimental field has been the test ground of nonlinear science. Josephson junctions and arrays are unique hardware to study solitons, synchronization, chaos, pattern formation, and localized excitations. After introducing the reader into fabrication and measurement techniques, I will discuss several experiments with Josephson arrays in both classical and quantum regimes.

1. Introduction

Josephson junction arrays are small thin-film devices which can be prepared using superconducting materials. Since the superconducting state can be reached by decreasing temperature towards absolute zero, measurements of these structures require special cooling techniques and are typically performed at liquid helium temperatures (a few Kelvin). Using modern fabrication technology, Josephson junctions can be prepared in a very clean and controlled way, almost without any noticeable imperfections. The arrays of such junctions can be easily tailored for studying specific problems.

Besides fundamental studies of nonlinear dynamics, there are several useful applications of Josephson junctions and arrays. The most common are superconducting quantum interference devices (SQUIDs), which allow to measure extremely low magnetics fields, currents and voltages. Major high-frequency applications of Josephson arrays are voltage standards, sub-millimeter wave oscillators (approximately up to 1 THz) and superconducting digital electronics. Due to the macroscopic quantum nature of superconductivity, Josephson junctions are currently widely discussed and studied in experiments as candidates for solid-state qubits to build a scalable quantum computer.

This review is an introduction into experiments with Josephson junction arrays. The theoretical background on localized excitations in these systems can be learned from the preceding paper by Juan Mazo[1] in this book.

2. Fabrication of Josephson arrays

Josephson junction arrays are usually fabricated using the superconductive integrated circuit technology. It includes evaporation or sputtering of several metallic and insulating layers on a dielectric substrate. These thin films are then patterned one after another using such micro-fabrication tools as photolithography (for structure sizes of about 1 micrometer or larger) and/or electron beam lithography (for sub-micrometer structures). Below we briefly discuss the fabrication aspects starting from materials to layout. The attainable physical parameters for modern fabrication technology are also outlined.

The fabrication procedure for integrated circuits containing Josephson junctions usually consists of the following steps:

(1) Making a circuit layout using computer aided design (CAD);
(2) Fabrication of photo masks;
(3) Deposition of superconducting and insulating layers on a wafer;
(4) Photolithography and/or electron beam lithography;
(5) Dicing the wafer into individual chips.

The steps (3) and (4) may be repeated several times, depending on the number of superconducting layers and the complexity of the circuit.

Josephson tunnel junctions are formed by a thin oxide tunnel barrier between two superconducting electrodes. They are characterized by large normal resistance R and large capacitance C. These junctions are typically strongly underdamped (with McCumber parameter $\beta_c \gg 1$) and have a hysteretic I-V curve.

There exists nowadays a possibility of fabrication of Josephson junction circuits at the so called foundries. In this case the CAD layout is provided by a customer and the full fabrication procedure is performed at a foundry. The foundry service is provided, *e.g.*, by Hypres[2] in the USA and by IPHT Jena[3] in Germany. These foundries have specially developed design rules and formats which should be obeyed while designing the circuits.

2.1. *Materials*

Out of many known superconducting materials (pure metals, many alloys, complex ceramic compounds) there are a few which proved to be most convenient for fabrication of Josephson tunnel junctions and arrays of these junctions. The highest superconducting transition temperatures T_c between pure metals are known for Nb and Pb, of about 9.2 K and 7.2 K, respectively. Since the discovery of superconductivity in 1911, there have been found many superconducting alloys and compounds, but the highest T_c did not climb above 24 K. A striking breakthrough came in 1987 with the discovery of high-temperature superconductivity in the complex ceramic oxide $YBa_2Cu_3O_7$, which has a critical temperature of about 92 K. More recently, several further compounds based on copper oxides have shown T_c's of up to 130 K. The major refrigeration technique for experiments with low-temperature (low-T_c) superconductors is based on liquifying helium, which boils at 4.2 K at the atmospheric pressure. The high-temperature (high-T_c) superconductivity occurs well above the 77 K boiling point of nitrogen, a cheaper and more convenient cryogenic liquid than helium.

2.1.1. *Low-temperature superconducting technology*

The superconducting integrated circuit technology is usually based on using Nb as the superconducting material. The circuits combine Nb films and the co called Nb-trilayer Josephson junctions. Niobium has a critical temperature of 9.2 K which allows operation of circuits at liquid helium temperature (4.2 K at atmospheric pressure).

The Nb-trilayer Josephson junctions have Al-oxide as the tunnel barrier between two Nb superconductors. The formation of this oxide barrier layer involves depositing an ultra-thin layer of metallic Al (about 5 nm) on top of freshly sputtered Nb film, without breaking the vacuum between both deposition processes. Then a controlled amount of oxygen is introduced into the system to oxidize the Al film. This forms a thin (2-3 nm) Al-oxide tunnel barrier. Finally, the Nb-counter electrode is sputtered on top of the barrier layer. The whole Nb/Al-AlO_x/Nb trilayer is patterned to form Josephson junctions by using a reactive ion etching process. Typical critical current density j_c for Nb trilayer junctions ranges from from 10^2 to 10^4 A/cm^2. The uniformity of the junction parameters over the chip is usually very high, better than few percent.

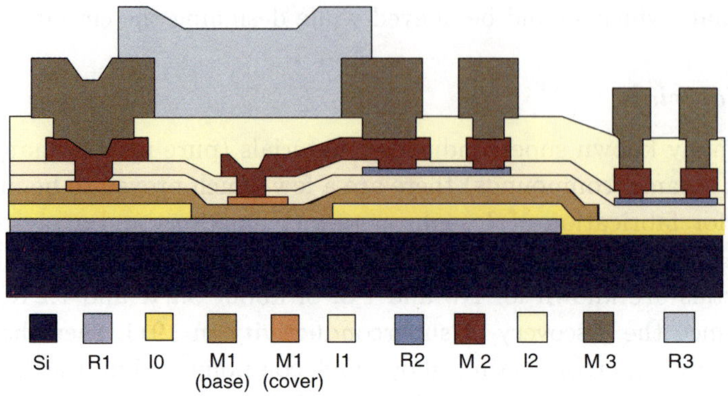

Fig. 1. An example of low-temperature superconducting technology for fabrication of Josephson junction arrays. The figure illustrates layer sequence used for the fabrication process at Hypres[2] (dimensions are not to scale). The bottom layer is a Si substrate, **R** stands for resistive metallic layers, **I** layers are insulators (SiO_2), **M** are the superconducting Nb layers (**M1** is the trilayer).

Typically, the whole fabrication process requires more than 10 layers. These may include several auxiliary superconducting and insulating layers (forming so called ground planes), Nb trilayer, Nb-wiring and resistor layers. As an insulator between superconducting and resistor layers one can use Si-oxide (SiO or SiO_2). The optional resistor layer should be made of a metal which remains non-superconductive down to very low temperatures (*e.g.*, AuPd) and has a specified sheet resistance. Figure 1 shows a cross section of a typical Nb-based superconducting circuit.

An alternative junction fabrication technology is all-Al shadow evaporation process. This process is usually required for making very small (less than 1 μm in size) Josephson tunnel junctions. There is an extra complication with these junctions as the critical temperature T_c of Al is rather low (about 1.2 K) and experiments with them imply using liquid ^{3}He or a dilution cryostat in order to reach these low temperatures.

The shadow evaporation fabrication process[4] uses two layers of electron beam resist that can be developed selectively. Through proper timing of the development a large undercut in the lower layer can be achieved and the top layer forms a resist mask hanging above the substrate surface. The $Al/Al_2O_3/Al$ junctions are then formed using the technique of shadow evaporation. That is, aluminium is evaporated in a vacuum chamber at two different angles, with an oxidation step in between, see Fig. 2. The mask is lifted off after evaporation. Typical critical current density j_c for all-Al

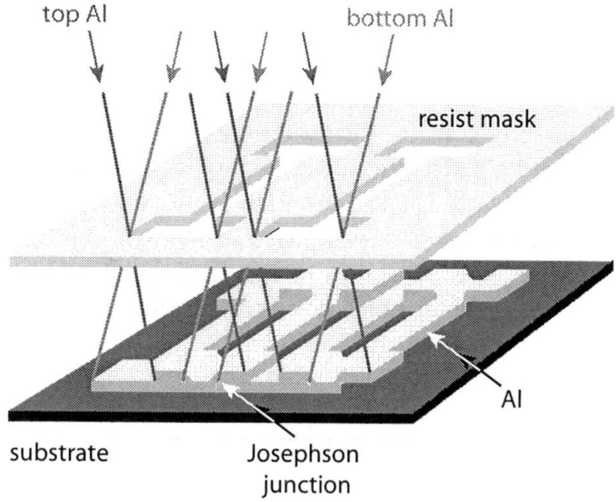

Fig. 2. A schematic sketch of the shadow evaporation technique (dimensions are not to scale).

junctions ranges from 1 to 10^2 A/cm^2.

2.1.2. *High-temperature superconducting technology*

As compared to the low-temperature superconducting technology, the technology based on high-temperature superconductors is still in its infancy. Ramp-type and grain boundary junctions can be made using $YBa_2Cu_3O_7$ as the superconducting material, but their uniformity is usually at least order of magnitude worse than that of typical low-T_c junctions. Moreover, the effective dissipation in these junctions is rather high with $\beta_c \sim 1$, *i.e.* the junctions are overdamped.

Very unique Josephson junction arrays made by nature were discovered about 10 years ago by R. Kleiner and P. Müller.[5] These series arrays of intrinsic junctions occur in some high-temperature superconductors such as $Bi_2Sr_2CaCu_2O_{8+x}$ and are formed naturally on the atomic scale due to the highly anisotropic crystal structure of these materials. These junctions have a hysteretic I-V curve ($\beta_c \gg 1$), very similar to standard Josephson tunnel junctions. Due to their atomic-scale nature intrinsic junctions are still very difficult to tailor in more complex geometries.

2.2. *Layout*

There are many different ways to combine Josephson junctions into an array. For example, one may connect them in series as shown in Fig. 3(a). In this case the junctions are mutually decoupled and may only interact via, *e.g.*, an external load. To study localized excitations we are interested to make arrays of *mutually coupled* oscillators. An array of coupled nonlinear oscillators can be created by connecting many Josephson junctions with superconducting leads, as shown in Fig. 3(b)-(f). The junctions here are coupled via mesh currents, which flow through their common superconducting leads.

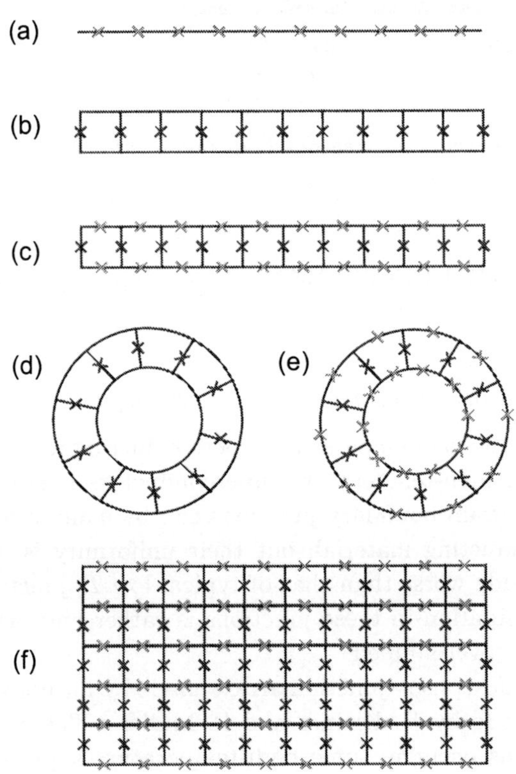

Fig. 3. Schematic view of different Josephson array geometries: (a) series array; (b) linear one-dimensional (1-D) parallel array; (c) linear ladder; (d) annular 1-D array: (e) annular ladder; (f) two-dimensional (2-D) array. Josephson junctions are shown by crosses and straight segments are the superconducting electrodes which also serve as inductors.

Major types of Josephson junction arrays, in which one may study localized excitations, are sketched in Fig. 3(b)-(f). The simplest one is the one-dimensional (1-D) parallel array, which can be viewed as the discrete version of a long Josephson junction. It can be made in the linear version (Fig. 3(b)) with open boundary conditions or in the annular geometry (Fig. 3(d)) with periodic boundary conditions. By adding one or two horizontal junctions per cell we obtain the Josephson ladder, which can also be fabricated in either linear (Fig. 3(c)) or annular (Fig. 3(e)) versions. The latter is characterized by periodic boundary conditions, supplemented by an additional equation for the central cell.[6] Finally, a two-dimensional Josephson junction array is sketched in Fig. 3(f).

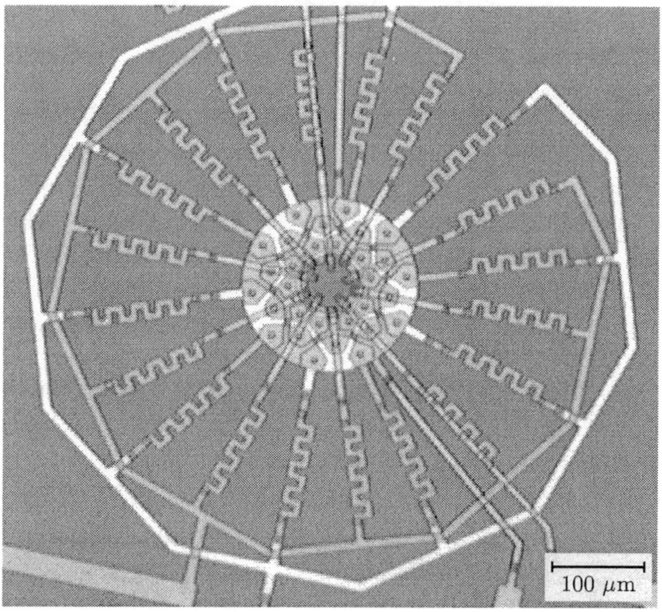

Fig. 4. A photograph of a 10-cell annular Josephson ladder.[7]

The first step in preparing an experiment with a Josephson array is preparation of a CAD design file using the standard format and the foundry rules (see, *e.g.*, Ref. 2). The designed layout can be then fabricated at a foundry. As an example, a photograph of an annular Josephson ladder is shown in Fig. 4. The ladder itself is located in the central part of the picture. The meandering radial strips are thin-film resistors R_{B} which provide uniform distribution of the bias current through the ladder. A magnified

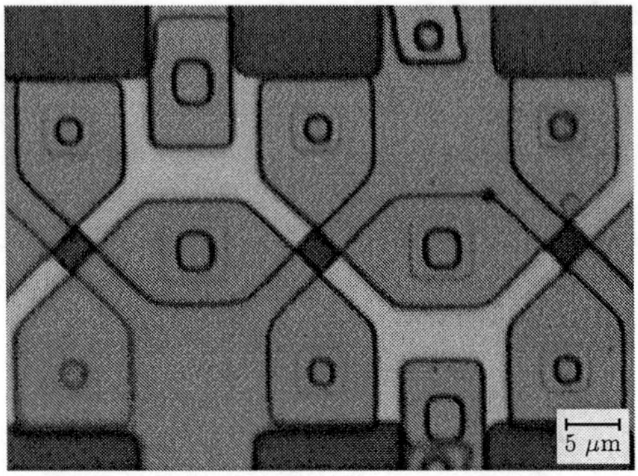

Fig. 5. A zoomed photo of few cells of a linear Josephson ladder.[7]

image showing few cells of another (linear) ladder is shown in Fig. 5. Each cell of the ladder contains four small Josephson junctions. The distance between the Josephson junctions is about a few μm.

2.3. *Junction parameters*

To illustrate the role of various parameters we quote the equations of motion for a Josephson junction ladder (see, *e.g.*, Flach and Spicci paper[8] for details; note that in the preceding review[1] the notations are slightly different):

$$\begin{aligned} \ddot{\varphi}_l^{\mathrm{V}} + \alpha\dot{\varphi}_l^{\mathrm{V}} + \sin\varphi_l^{\mathrm{V}} &= \gamma + \tfrac{1}{\beta_{\mathrm{L}}}(\Delta\varphi_l^{\mathrm{V}} - \nabla\tilde{\varphi}_{l-1}^{\mathrm{H}} + \nabla\varphi_{l-1}^{\mathrm{H}}) \, ; \\ \ddot{\varphi}_l^{\mathrm{H}} + \alpha\dot{\varphi}_l^{\mathrm{H}} + \sin\varphi_l^{\mathrm{H}} &= -\tfrac{1}{\eta\beta_{\mathrm{L}}}(\varphi_l^{\mathrm{H}} - \tilde{\varphi}_l^{\mathrm{H}} + \nabla\varphi_l^{\mathrm{V}}) \, ; \\ \ddot{\tilde{\varphi}}_l^{\mathrm{H}} + \alpha\dot{\tilde{\varphi}}_l^{\mathrm{H}} + \sin\tilde{\varphi}_l^{\mathrm{H}} &= \tfrac{1}{\eta\beta_{\mathrm{L}}}(\varphi_l^{\mathrm{H}} - \tilde{\varphi}_l^{\mathrm{H}} + \nabla\varphi_l^{\mathrm{V}}) \, , \end{aligned} \tag{1}$$

where l is the junction index, $l = 0, 1, ..., N$. Here $\varphi_l^{\mathrm{V}}, \varphi_l^{\mathrm{H}}, \tilde{\varphi}_l^{\mathrm{H}}$ are the phase differences across the l-th vertical junction and its right upper and lower horizontal neighbours, $\nabla\varphi_l = \varphi_{l+1} - \varphi_l$, $\Delta\varphi_l = \varphi_{l+1} - 2\varphi_l + \varphi_{l-1}$, and $\gamma = I_{\mathrm{B}}/I_{\mathrm{cV}}$ is the normalized bias current. There are two mechanisms of coupling between cells in a ladder. The first is the inductive coupling between the cells that is expressed by the self-inductance parameter $\beta_{\mathrm{L}} = 2\pi L I_{\mathrm{cV}}/\Phi_0$, where L is the inductance of the elementary cell. The second is the nonlinear Josephson coupling via horizontal junctions. The ratio of the horizontal and vertical junction areas is called the anisotropy factor and can be expressed in terms of the junction critical currents $\eta = I_{\rm cH}/I_{\rm cV}$. In the limit of $\eta \to 0$ the vertical junctions are decoupled and operate independently one from another. On the other hand, if η goes to infinity the ladder behaves like a parallel 1-D array, in which horizontal junctions are replaced by superconducting shorts.

The experimental parameters from sample to sample can be varied by changing the layout of arrays. The hole between the superconducting electrodes which form the cell determines the discreteness of the ladder. For a given value of the junction's critical current, the parameter $\beta_{\rm L}$ can be varied by changing the cell size a, as $L \approx 1.25\mu_0 a$. The anisotropy η can be chosen by varying the size of vertical junctions relative to that of horizontal ones. Conveniently, the dissipation constant α can be varied from less than 0.01 to 0.3 or even more by varying the temperature of the experiment.

3. Measurement techniques

Measurements of Josephson junction arrays are performed at liquid helium temperatures, typically around 4 K or higher. The sample is put in a cryostat or transport dewar in good thermal contact with liquid helium. As the simplest measurement, one applies a dc bias current to the array and measures a dc voltage proportional to the Josephson phase rotation frequency. The bias current $I_{\rm B}$ can be uniformly injected into an array by using thin-film resistors $R_{\rm B}$ fabricated on chip. To avoid extra noise and electromagnetic interference, the biasing is usually provided by a battery-powered current source. The sample should be well shielded against external magnetic fields, usually using a screen made of cryopermalloy.

Direct detection of the Josephson electromagnetic radiation (ac voltages and currents) from the arrays is a much more complex task. The typical radiation wavelengths lie in the millimeter and submillimeter range, which translates into frequencies from several 10 GHz to almost 1 THz. Due to unavoidable impedance mismatch between the array and its electromagnetic environment, the array radiation power is rather small (in the range of 10^{-10} - 10^{-7} W) and its detection and spectral resolution require a sensitive receiver. An example of radiation detection data is presented in Fig. 6. Such experiments are usually rather difficult and complicated, but of course not impossible. The same applies to direct time-resolved measurements of Josephson voltage of a chosen Josephson junction in the array. Thus, in con-

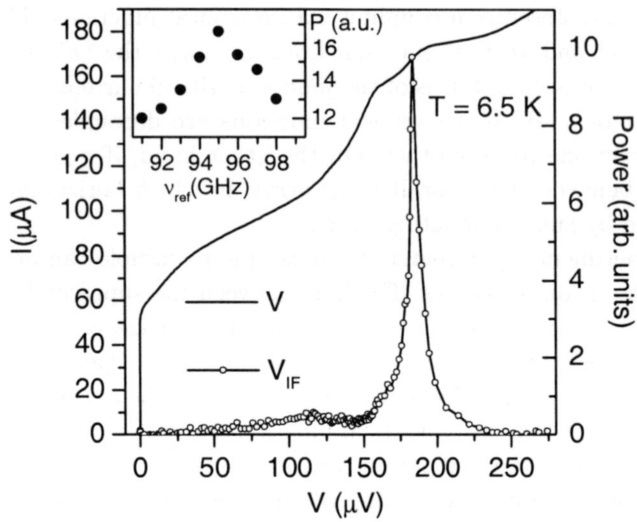

Fig. 6. Current-voltage (I-V) characteristics of a ladder array presented along with the detected array radiation power at a fixed receiver frequency $\nu_{\text{ref}} = 88\,\text{GHz}$. The inset shows the power measured at the front end of the receiver in the frequency range $\nu_{\text{ref}} = 92 - 98\,\text{GHz}$.[9]

trast to rotobreathers, experimental detection of oscillobreathers in Josephson arrays is a formidable task which has not yet been solved.

There have been theoretical proposals to support persisting rotobreathers in Josephson arrays by ac bias current. Since the characteristic frequencies of lattice modes (Josephson plasmons) typically fall in the range around 100 GHz, ac currents should have even high frequency and thus require applying millimeter-wave irradiation to Josephson arrays. The major uncertainty here is that the spatial distribution of the ac bias density is generally unknown, as it is hard to control and verify experimentally. Nevertheless, some preliminary experiments with small systems have been recently done[10] and we may expect more of these coming out in the near future.

3.1. *Generation of localized excitations*

The most common type of localized excitations in Josephson arrays are vortices or fluxons. A fluxon moving in an array generates the dc voltage V proportional to the fluxon's velocity u. In experiments with parallel 1-D arrays, the signatures of fluxon motion are the resonant steps which appear in the current-voltage characteristics (I-V curve) of the junction. The shape of these steps is given by the dependence of the average fluxon velocity $u \propto V$ on the driving force $\gamma \propto I_B$. Similarly to long Josephson junctions, a fluxon arriving at the open boundary of an array may undergo reflection into an anti-fluxon which is driven then back into the array and the process repeats again at the opposite boundary. This process sets in as an instability point of uniformly rotating state at zero magnetic field. Alternatively, an external magnetic field may generate vortices in the array, which can be then be driven by a bias current through the array. This motion in magnetic field is usually noted as flux flow. Vortices are continuously generated at one boundary of the array and annihilated at another boundary. Reversing of either the magnetic field of the bias current direction reverses the direction of vortex motion.

Intrinsic localized modes in Josephson ladders such as discrete rotobreathers are dynamic states with a few resistive (rotating) junctions. Rotobreathers persist under the influence of a spatially-uniform bias current, but usually they first have to be generated by preparing the ladder in a proper initial condition. In order to form such conditions, it is usually possible to introduce a local bias current and then reduce it while increasing the uniform bias current, which will finally support the breather.[13–15] This procedure allows to generate discrete breathers in a controlled way. In addition, experiments show that rotobreather states often occur spontaneously under biasing with a uniform current, due to the influence of both intrinsic (thermal) and extrinsic noise.[11,12]

3.2. *Hot probe imaging techniques*

The simplest and very straightforward way to detect a rotobreather in an array is to measure local dc voltages on different junctions in the array. This method works well for simple localized states in which only a few resistive junctions are involved. Having many voltage leads, even if technically possible, is rather inconvenient and makes the data analysis complicated. Moreover, every pair of leads unavoidably generates extra noise due to the external electromagnetic interference. Having in mind these limitations, it is much more convenient and spectacular to use an *imaging technique* that is described below.

In order to visualize localized excitations in arrays one can use the so called hot-probe imaging technique. The idea of this technique is to apply a focused low-power laser beam or electron beam to the *current-biased* array.

The beam locally heats the sample and changes the dissipation in an area of few micrometers in diameter. If the junction illuminated by the beam is in the resistive (rotating) state, a small change of its dc voltage drop is induced. This voltage change occurs due to the temperature dependence of the damping constant α. Junctions that are in the superconducting (non-rotating) state do not show any voltage response, as their dc voltage remains zero. By scanning the laser beam over the whole ladder one can detect the rotating junctions and thus visualize of the dynamic state in the ladder.

In the studies of rotobreathers in Josephson ladders we have used the hot-spot method of low temperature laser scanning microscopy (LSM). Details about the method of scanning laser microscopy can be found elsewhere.[16–18] The diameter of the laser beam is around $1 - 2\,\mu$m.

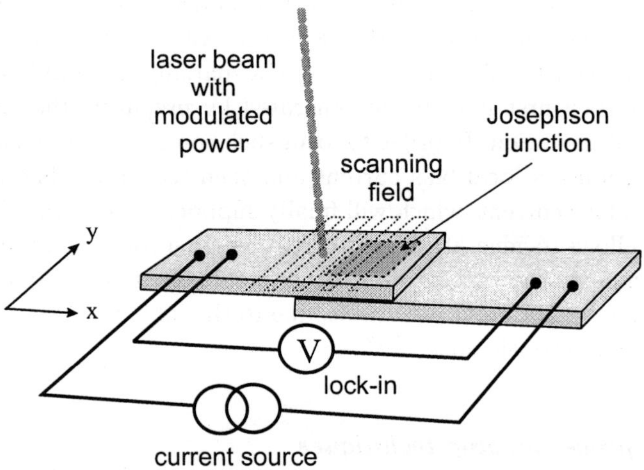

Fig. 7. The principle of low temperature laser scanning microscopy (LSM): A modulated laser beam locally heats the sample and increases the dissipation in a current-biased Josephson junction, which voltage is measured as a contrast response for the imaging.

Figure 7 shows the principle of low temperature LSM. Resistive junctions of the ladder contribute to the voltage response, while junctions in the superconducting state show no response. The laser beam locally heats the sample and therefore introduces extra dissipation in the area of few micrometers in diameter. Such a dissipative spot is scanned over the sample. The voltage variation at a given bias current is recorded as a function of the beam coordinates (x, y). The power of the laser beam is modulated at a frequency of several kHz while the sample voltage response is measured using a lock-in technique. Note, that the modulation frequency of the laser beam is much smaller (by a factor of at least 10^6) than the Josephson junction oscillation frequency. The laser beam is scanned by mirrors or moved by step motors. The typical time, which is needed to make an electrical image, is between 3 and 15 minutes depending on the chosen scanning area an the resolution. During the measurement the sample resides in vacuum and is placed on a cold finger in an optical cryostat.

4. Experiments in the classical regime

4.1. *Fluxons in Josephson arrays*

The first experiments detecting the motion of fluxons (Josephson vortices) in spatially extended Josephson junctions were done by Fulton and Dynes[19] in 1973. Since then there have been a lot of experimental studies of the dynamics of these classical solitons in long Josephson junctions and related systems, see Refs. 20-22 for reviews. Here we will only briefly review experiments with discrete arrays of Josephson junctions (1-D parallel arrays, ladders and 2-D arrays) performed in the classical regime with relatively large junctions, for which the Josephson energy E_J is much larger than the charging energy E_C.

4.1.1. *Parallel 1-D arrays*

As compared to soliton motion in long Josephson junctions,[22] array discreteness introduces a number of aspects into the fluxon dynamics that have no counterparts in the continuum systems. A parallel biased array of small Josephson junctions schematically shown in Fig. 3(a), represents an experimental realization of the spatially discrete sine-Gordon lattice. Although the discrete sine-Gordon equation is much simpler for numerical studies, its non-integrability is probably a major reason for which relatively small number of theoretical studies and thus also very few experiments have been done on fluxon dynamics in parallel 1-D arrays and even less with ladders.

In general, a kink moving through a discrete lattice generates small-amplitude oscillations of the lattice. The dispersion relation for such linear waves $\varphi_j = \varphi^{(0)}\exp[\mathrm{i}(\omega t - kaj)]$ is

$$\omega^2 = 1 + \frac{4}{a^2}\sin^2\left(\frac{ka}{2}\right), \tag{2}$$

where ω is the frequency of oscillations and k is their wave number. Peyrard and Kruskal[23] pointed out that the phase velocity $v_{\text{ph}} = \omega/k$ of the linear waves may coincide with the velocity of the kink v. Considering the kink as a quasi-particle, this radiation emission mechanism is equivalent to the Cherenkov radiation, which leads to resonances at some kink velocities.[24] Experimentally, these discreteness-induced resonances were observed by van der Zant *et al.*[25] using 8–junction annular 1-D array shown in Fig. 3(c) with a single trapped fluxon. Later, similar and also high-order resonances have been investigated in experiments by Caputo *et al.*[26] and Duwel *et al.*[27]

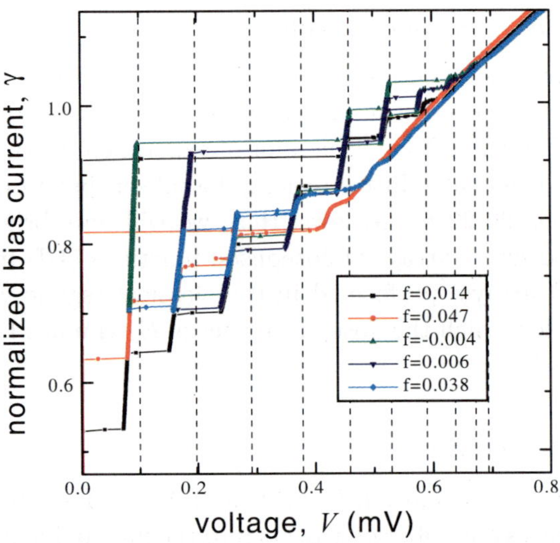

Fig. 8. Measurements[6] of ZFSs in 10-cell parallel array. The parameter f is the frustration which was slightly varied around zero. The dashed lines indicate the calculated asymptotic positions of the steps.

In contrast to ladders and 2-D arrays, in 1-D parallel array a moving fluxon may get reflected from a boundary into an antifluxon. The antifluxon then travels back to another boundary and reflects there as a fluxon, and the process repeats again. This type of shuttle-like fluxon motion is responsible for zero-field steps (ZFSs) in the current-voltage characteristics (I-V curve) of long Josephson junctions[19] and it was also observed experimentally[28] in discrete 1-D arrays. More recently Marcus Schuster[6] has done very clean and systematic measurements of ZFSs in 1-D parallel arrays illustrated in Fig. 8. These experimental data are also in good accord with numerical simulations.[6]

4.1.2. *Ladders*

In difference to parallel arrays, closing a ladder in a ring, as shown in Fig. 3(d), does not prevent a vortex from escaping the array. This happens due to the presence of horizontal junctions. These junctions increase the effective discreteness of the array and fluxons pinning by ladder cells gets stronger. It can be shown[29] that for a ladder with 4 junctions per cell and anisotropy parameter η the effective discreteness parameter becomes $\beta_L' \approx \beta_L + 2/\eta$. Thus, even for isotropic ladder with $\eta = 1$ and very small cell size the fluxons pinning is very strong and, thus, their motion is rather difficult to observe.

Fluxons in Josephson ladders can be studied experimentally in the presence of external magnetic field. Schuster[6] has done experiments with annular ladders placed in a small magnetic field and observed motion of one and more individual fluxons along the ladder. Interestingly, the observed fluxon resonances have shown sometimes switching back to the superconducting branch of I-V curve. In this case, the fluxon moving at high velocity excited large amplitude oscillations in horizontal junctions, which facilitated the escape of the fluxon from the ladder.

In sufficiently high magnetic field, the fluxons may not any more be properly treated as localized excitations as they strongly overlap and interact with each other. In this case the phase distribution along the ladder is nearly linear and a more simple quasi-linear cavity mode approach can be used. Experiments in this regime have been performed and good accord with analysis has been found.[30]

4.1.3. *2-D arrays*

2-D Josephson junction arrays have attracted a lot of interest and vortex motion in these systems has been studied as well. Even at zero dissipation, fluxons moving in a 2-D array continuously lose their energy emitted in the form of small-amplitude waves (Josephson plasmons). The extent to which a concept of ballistic quasiparticle can be applied to vortices in 2-D arrays is often still debated. There has been experiments[31] and simulations[32] which have shown some features of ballistic motion for vortices. Still, there have been also critical analysis concluding the opposite.[33] In general, the fluxon dynamics of underdamped 2-D arrays is very complicated and, thus, requires further studies.

4.2. *Rotobreathers in Josephson ladders*

Laser scanning experiments revealed a rich breather "zoo" in Josephson ladders.[13–15] Various rotobreather states distinguish from one another by the number of whirling horizontal junctions at their edges and interior. For many breathers we observed similar states of single site, two sites, three sites and so on. The number of sites is defined as the number of rotating vertical junctions. Very detailed accounts of the performed experiments can be found in published papers[13–15] and recent review.[34] Since the nonlinear localized modes cease to exist in the continuum case, it is interesting to increase the anisotropy η to the largest possible values at which these modes may still occur. We observed rotobreathers in Josephson ladders with the anisotropy parameter η as large as unity.

In order to briefly illustrate the above mentioned experimental observations of rotobreathers, here we present several laser microscope images of annular ladders. Figure 3(d) shows a schematic view of annular ladder. The photograph of the device that we studied is presented in Fig. 4. A bias current is applied uniformly in each outer node and extracted from each inner node. In annular ladders the magnetic flux cannot be accumulated in the interior of the ladder. This implies that the sum of voltages on inner horizontal junctions should be zero. The annular ladder has an outer diameter of 140 μm, anisotropy $\eta = 0.73$ and the parameter $\beta_L = 1.5$. A homogeneous whirling state with all vertical junctions in the resistive state is shown in Fig. 9(a). The observed symmetric rotobreathers are shown in Figs. 9(b)–9(d). In Fig. 9(b) a two-site breather and in Fig. 9(c) a three-site breather can be seen.

Around the vertical junctions two horizontal junctions on every side are whirling. The symmetric breathers were very stable and it was possible to take all their images starting from the two-site and up to the eight-site breather. The latter one is presented in Fig. 9(d). In these measurements the voltage of the horizontal junction was always half of the voltage of the vertical junctions. Several examples of asymmetric breathers are shown in Fig. 10. There we find a single-site, a two-site, a three-site, and a nine-site breather in Figs. 10(a), 10(b), 10(c), and 10(d), respectively. It was also possible to obtain images of all the other asymmetric multi-site breather states. The inner horizontal junctions were whirling in most cases as in Figs. 10(a)–(c), but sometimes also the outer horizontal junctions were excited as in Fig. 10(d). A possible reason for this different occurrence can be the annular geometry of the sample. There is a circulating current

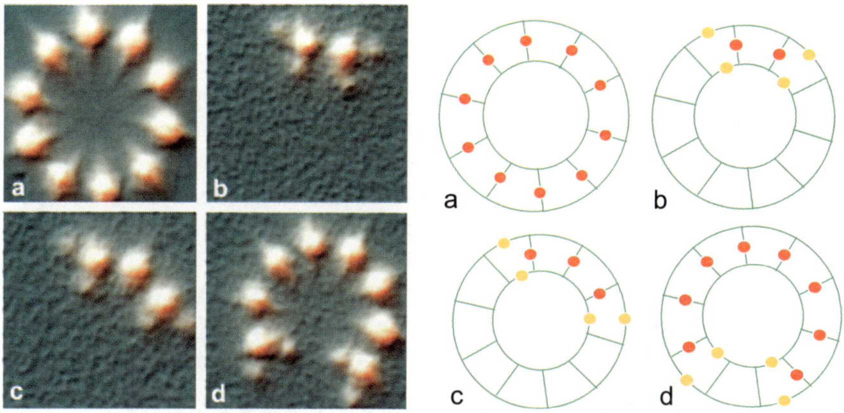

Fig. 9. Top four pictures are LSM images of the homogeneous whirling mode (a) and of two-site (b), three-site (c), and eight-site (d) symmetric rotobreathers. The four plots sketch the corresponding states; dots denote resistive (rotating) junctions.

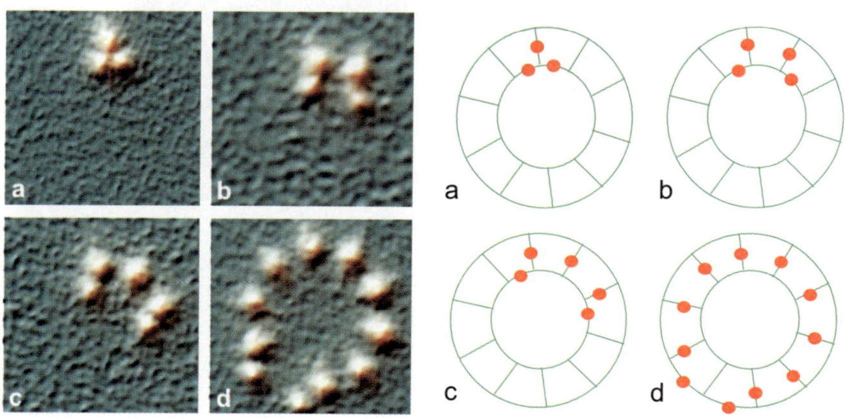

Fig. 10. LSM images of single-site (a), two-site (b), three-site (c), and nine-site (d) asymmetric rotobreathers.

through the inner horizontal junctions due to magnetic flux trapped in the ring, which facilitates to their switching to the whirling mode.

4.3. *Meandered states in 2-D Josephson arrays*

Using laser scanning technique interesting rotobreather-related states with broken symmetry in 2-D Josephson arrays have been found in experiments by Abraimov *et al.*[11] While switching the arrays by a uniform bias current into resistive states, a violation of the uniform row switching accompanied by the meandering of the resistive rows of junctions has been observed.

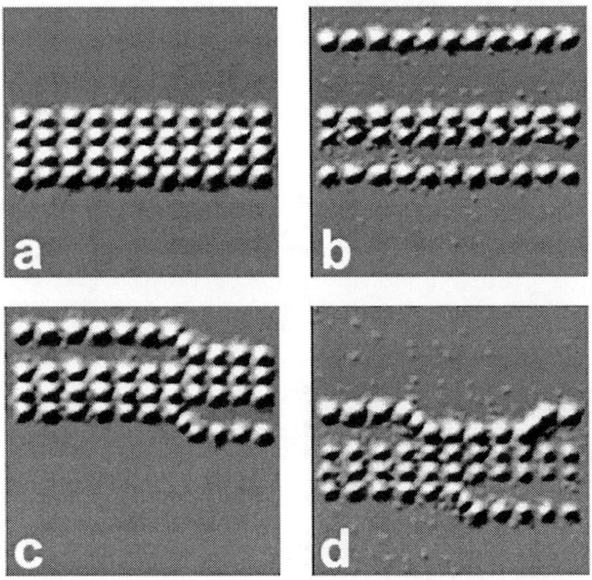

Fig. 11. Laser scanning microscope images of a simple row-switching (a,b) and row meandering states (c,d) in a 2-D Josephson junction array.[11]

The junctions of the studied 2-D arrays are placed at the crossings of the superconducting lines, which are arranged in a square lattice with 4 junctions per elementary cell. An equivalent circuit of the square 2-D array is shown in Fig. 3(f). By choosing a bias point of the array at a multiple of gap voltage V_g, in experiment one can select the number of rows of junctions switched to the resistive state, while the other rows remain in the superconducting state. Typical images of a square 2-D array are shown in Fig. 11 at the bias points close to the $4V_g$ voltage region. The light spots correspond to junctions which are in the resistive state. The junctions that are in the superconducting state do not appear on the image. The images (a) and (b) show four rows in the resistive state; the number of switched rows is equal to the number n of gap voltages selected by the bias point. The major part of taken images shows various combinations of ***straight*** resistive rows. The striking feature seen in Fig. 11(c),(d) was that the resistive lines are not straight, but show *meandering* towards the neighboring row involving one of the horizontal junctions in the resistive state.[11] We have found that such a broken symmetry in row switching systematically appears in all studied arrays, independently of the topology (square or triangular) and array parameters (large or small self-inductance, different temperatures).[18]

The meandering of the row switching can be understood as a spontaneous (*e.g.*, induced by thermal fluctuations) generation of rotobreathers in 2-D array. It is clear that a single "step" of the row can be considered as a superposition of two truncated left- and right-side rotobreathers that occupy the n–th and $(n+1)$–th rows of the array. Thus, in this experiment underdamped 2-D arrays driven by dc bias current show spontaneous generation of quasi-(1-D) rotobreathers. Undoubtedly, this complex behavior of 2-D arrays deserves further investigation.

5. Experiments in the quantum regime

At low temperatures and small damping the dynamics of a current-biased Josephson junction is governed by the macroscopic quantum mechanics of the superconducting phase-difference across the junction (see, *e.g.* Refs. 35, 36 and references therein). Macroscopic quantum tunneling of the phase, energy level quantization and the effect of dissipation have been studied in detail in these systems.[36] Josephson junction parameters can be adjusted in a wide range and can be well controlled. Josephson junction circuits have been proposed and recently successfully tested as qubits in quantum information processing.[37–41]

5.1. *Single Josephson junction*

The macroscopic quantum tunneling in Josephson junctions has been already studied in great detail in the past.[35,36] This very appealing signature of quantum nature of the Josephson phase can be observed in a statistical measurement of the current at which the junction switches from the zero-voltage state to a finite voltage state. In terms of the junction energy dependence on the phase φ, it means an escape form a metastable well in

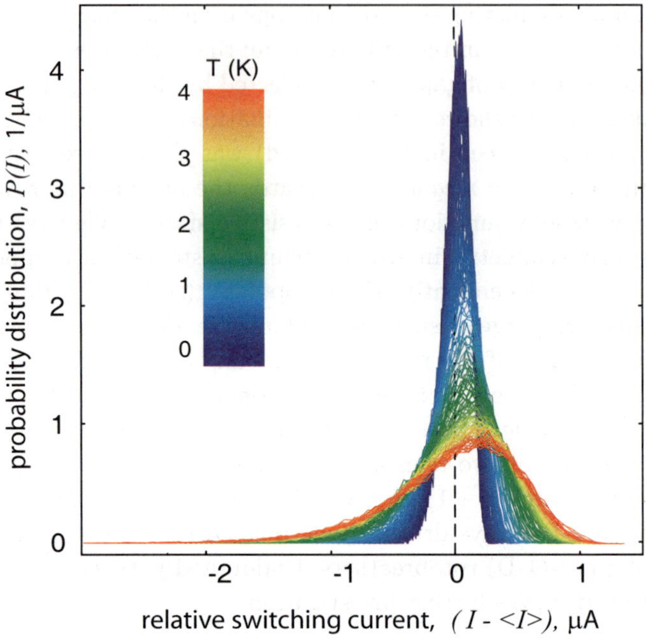

Fig. 12. Switching current distributions $P(I - \langle I \rangle)$ at bath temperatures between $T = 4.2\,\mathrm{K}$ and $T = 25\,\mathrm{mK}$. The temperature is color coded, red corresponds to high and blue to low temperatures.[44]

a tilted washboard potential $U(\phi)$. This potential is given by

$$U(\phi) = E_J\left(-\gamma\phi - \cos\phi\right), \tag{3}$$

where the Josephson energy $E_J = \Phi_0 I_c/2\pi$ and $\gamma = I/I_c$ is the normalized bias current. $U(\phi)$ is a cosinusoidal potential with a depth proportional to E_J, which is tilted proportionally to the applied bias current I. In the absence of thermal or quantum fluctuations and for $\gamma < 1$, the junction is in the zero voltage state, similarly to the particle being localized in the potential well. At finite temperature, the particle may escape from the well at bias currents $\gamma < 1$ by thermally activated processes or by quantum tunneling through the barrier.

Figure 12 shows experimental data[44] obtained using a high quality $5 \times 5\,\mu\mathrm{m}^2$ tunnel junction. The switching current distribution of the sample was measured at bath temperatures T between 4.2 K and 20 mK. The current was ramped up at a constant rate of $\dot{I} = 0.245\,\mathrm{A/s}$ with a repetition rate of $\nu_I = 500\,\mathrm{Hz}$. Typically 10^4 switching currents were recorded at each

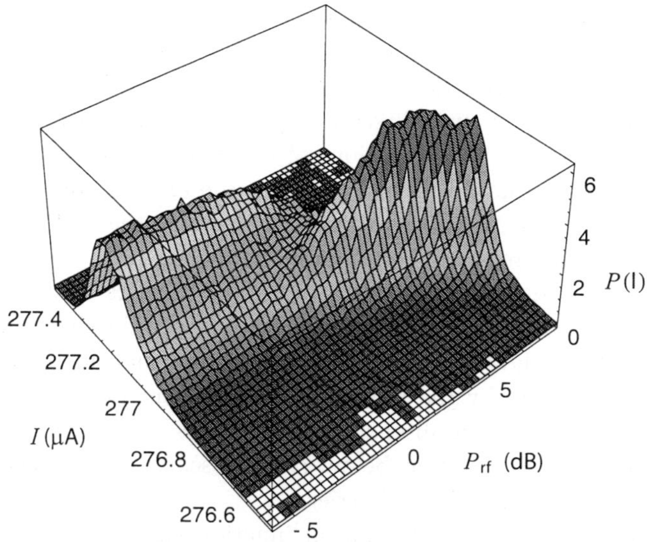

Fig. 13. 3-D plot of the measured[45] $P(I)$ distribution versus the applied microwave power $P_{\rm rf}$ at frequency of 36.6 GHz and temperature $T = 100\,{\rm mK}$.

temperature. Histograms of the switching currents were calculated with bin widths of approximately 10 nA in order to determine the $P(I)$ distributions. It is observed that the width σ_I of the $P(I)$ distributions decreases with temperature and then saturates at low temperatures. At temperatures above approximately 300 mK, the distribution width σ_I decreases approximately linearly with T, indicating the temperature dependent thermal activation of the phase across the barrier. At the characteristic temperature $T^\star \approx 300\,{\rm mK}$, σ_I saturates as the escape of the phase occurs predominantly by quantum tunneling through the barrier. The $P(I)$ distributions in both the thermal and the quantum regime can be analyzed quantitatively[44] and agree well with theory.

Another clear signature of quantum behavior of the Josephson phase is the energy level quantization. The level spectroscopy can be performed by applying microwaves to the junction. While monitoring the $P(I)$ distributions of the junction, the microwave power $P_{\mu{\rm w}}$ is swept from low values, at which the $P(I)$ distribution is not changed by the microwaves, to higher values for each chosen frequency. At negligibly small microwave powers the $P(I)$ distribution is essentially determined by the unperturbed quantum tunneling of the phase from the ground state of the well. If the microwave power is increased to substantially populate the excited level, the $P(I)$ distribution becomes double-peaked.[45] This double-peak structure smoothly varies with $P_{\mu w}$, as shown for $\nu = 36.6\,\mathrm{GHz}$ in Fig. 13. Further increasing the power, only the pronounced resonant peak is visible in the distribution. At this level of power the populations of the ground and the first excited state are equal, but the tunneling rate from the excited state is exponentially larger than that from the ground state. Thus, the $P(I)$ distribution is dominated by the resonant peak due to tunneling from the first excited state and the initial peak in the distribution disappears. Due to the resonance excitation of transitions between the two levels, the switching current distribution at this level of power is *more narrow* than in the absence of microwaves. This fact, by the way, is an independent proof that the measured $P(I)$ distribution in absence of microwaves is not limited by noise in experimental setup and that below $T^{\star}$ the escape indeed occurs due to quantum tunneling through the barrier.

The quantum state of a Josephson junction can be prepared and its coherent evolution is controlled by short (few nanosecond long) microwave pulses. Two experimental groups[42,43] have already demonstrated such a controlled time-resolved *manipulation by a quantum state* of a single current biased Josephson junction. It can be operated as a qubit by coupling its two lowest energy quantum states by microwaves, in a manner similar to the way it is usually done for nuclear spins and atoms. The measured Rabi oscillation decay time is at least of the order of 100 ns[42] and was even claimed to be as large as 5 μ.[43] The major difficulty of this approach lies in the fact that the quantum superposition is extremely fragile to interactions with the environment and tends to decay to one of the two classical states, an effect known as decoherence. Therefore the characteristic time scale for microwave pulses (between the initial state preparation and final readout) should be shorter than decoherence, limited by the energy relaxation time T_1 and dephasing time T_2.

5.2. *Coupled Josephson junctions*

In the view of great interest to study quantum localized excitations (*quantum discrete breathers*) in Josephson junction arrays, we may now think of coupling several junctions in the quantum regime. The first experiment demonstrating quantum energy level spectroscopy of a pair of capacitively coupled quantum Josephson oscillators has been reported this year.[46]

In the near future, it should be possible to perform pulse-probe experiments with Josephson arrays in order to read out quantum states at the upper levels of an anharmonic potential. The idea here is to use two resonant microwave frequencies ω_{01} and ω_{12} to work with individual oscillators as 3-level systems, having the level separation $\hbar\omega_{01} \neq \hbar\omega_{12}$. In a coupled system (of two or more anharmonic quantum oscillators) the quantum behavior should be manifested by a very long relaxation time of the second excited state for one of the oscillators. Such a state should neither relax nor propagate, providing thus a signature of quantum localization by nonlinearity in a discrete system.

6. Conclusions and outlook

This review has focussed almost exclusively on experiments with discrete arrays of Josephson junctions. Localized excitations in the form of *fluxons* (Josephson vortices) in these structures have been studied already for many years and their dynamics, in particular for discrete 1-D arrays, is to a large extent well understood. Fluxons in 2-D arrays and ladders still require further studies as their interaction with lattice oscillations (Josephson plasmons) make the dynamics very rich and complex. In the past few years another type of localized excitations, namely discrete *rotobreathers* has been also observed and thoroughly investigated. Experiments with Josephson arrays are unique as they allow for direct visualization of discrete breathers in nonlinear lattices.

Recent progress in the experiments with Josephson junctions in the quantum regime opens the possibility to experimentally verify a number of theoretical predictions concerning *quantum discrete breathers*. These experimental results are extremely encouraging and point towards a possibility of direct time-resolved experiments with quantum breathers in arrays of current-biased Josephson junctions at millikelvin temperatures.

Acknowledgments

This work was supported by LocNet under the EC contract number HPRN-CT-1999-00163 and Deutsche Forschungsgemeinschaft (DFG). I would like to thank the Erlangen breather team of D. Abraimov, P. Binder, F. Pignatelli and M. Schuster and also our qubit crew of A. Wallraff, A. Lukashenko, A. Kemp, J. Lisenfeld, and C. Coqui for their encouragement and effort. Stimulating discussions with M. Fistul, S. Flach, and J. J. Mazo are gratefully acknowledged.

References

1. J. J. Mazo, chapter 4, *Energy localization and transfer*, T. Dauxois, A. Litvak-Hinenzon, R. S. Mackay, A. Spanoudaki Eds., Advanced Series on Nonlinear Dynamics, World Scientific (2003).
2. Hypres Inc., Elmsford, NY 10523, USA (http://www.hypres.com).
3. Institute for Physical High Technology, P.O.B. 100239, D-07702 Jena, Germany (http://www.ipht-jena.de).
4. G. J. Dolan, *Appl. Phys. Lett.* **31**, 337 (1977).
5. R. Kleiner, F. Steinmeyer, G. Kunkel, and P. Müller. *Phys. Rev. Lett.* **68**, 2394 (1992); R. Kleiner and P. Müller, *Phys. Rev. B* **49**, 1327 (1994).
6. M. Schuster, *Dipl. Thesis*, Universität Erlangen-Nürnberg, 2000. Electronic version available at `http://www.pi3.physik.uni-erlangen.de/ustinov`.
7. P. Binder, *Ph.D. Thesis*, Universität Erlangen-Nürnberg, 2002. ISBN: 3-932392-36-1. Electronic version available at `http://www.pi3.physik.uni-erlangen.de/ustinov`.
8. S. Flach and M. Spicci, *J. Phys.: Condens. Matter* **11**, 321 (1999).
9. P. Caputo, A.V. Ustinov, and S.P. Yukon. *IEEE Trans. Appl. Supercond.* **11**, 454 (2001).
10. F. Pignatelli and A. V. Ustinov, *Phys. Rev. E* **67**, 036607 (2003).
11. D. Abraimov, P. Caputo, G. Filatrella, M. V. Fistul, G. Yu. Logvenov, and A. V. Ustinov, *Phys. Rev. Lett.* **83**, 5354 (1999).
12. M. Schuster, F. Pignatelli, and A. V. Ustinov, unpublished (2003).
13. P. Binder, D. Abraimov, A. V. Ustinov, S. Flach, and Y. Zolotaryuk, *Phys. Rev. Lett.* **84**, 745 (2000).
14. P. Binder, D. Abraimov, and A. V. Ustinov, *Phys. Rev. E* **62**, 2858 (2000).
15. P. Binder, and A. V. Ustinov, *Phys. Rev. E* **66**, 016603 (2002).
16. A. G. Sivakov, A. P. Zhuravel', O. G. Turutanov, I. M. Dmitrenko, *Appl. Surf. Sci.* **106**, 390 (1996).
17. P. M. Shadrin, Y. Y. Divin, S. Keil, J. Martin, and R. P. Huebener, *IEEE Trans. Appl. Supercond.* **9**, 3925 (1998).
18. D. Abraimov, *Ph.D. Thesis*, Universität Erlangen-Nürnberg, 2002. ISBN: 3-932392-34-5. Electronic version available at `http://www.pi3.physik.uni-erlangen.de/ustinov`.
19. T. A. Fulton and R. C. Dynes, *Sol. St. Commun.* **12**, 57 (1973).
20. N. F. Pedersen, in: *Solitons*, S. E. Trullinger, V. E. Zakharov, V. L. Pokrovsky, eds. (Elsevier, Amsterdam, 1986), p. 469.
21. R. D. Parmentier, in: *The New Superconducting Electronics*, H. Weinstock, R. W. Ralston, eds. (Kluwer, Dordrecht, 1993), p. 221.
22. A. V. Ustinov. *Physica D* **123**, 315-329 (1998).
23. M. Peyrard and M. D. Kruskal, *Physica D* **14**, 88 (1984).
24. A. V. Ustinov, M. Cirillo, and B. A. Malomed, *Phys. Rev. B* **47**, 8357 (1993).
25. H. S. J. van der Zant, T. P. Orlando, S. Watanabe, and S. H. Strogatz, *Phys. Rev. Lett.* **74**, 174 (1995).
26. P. Caputo, A. V. Ustinov, N. Iosad, and H. Kohlstedt. *J. Low Temp. Phys.* **106**, 353 (1997).

27. A. E. Duwel, S. Watanabe, E. Trías, T. P. Orlando, H. S. J. van der Zant, and S. H. Strogatz, *J. Appl. Phys.* **79**, 7864 (1997).
28. A. V. Ustinov, M. Cirillo, B. H. Larsen, V. A. Oboznov, R. Leoni, G. Rotoli, and I. Modena. *Phys. Rev. B* **51**, 3081 (1995).
29. P. Binder, P. Caputo, G. Filatrella, M. V. Fistul, and A. V. Ustinov, *Phys. Rev. B* **62**, 8679 (2000).
30. P. Caputo, M. V. Fistul, A. V. Ustinov, B. A. Malomed, and S. Flach. *Phys. Rev. B* **59**, 14050 (1999).
31. H. S. J. van der Zant, F. C. Fritschy, T. P. Orlando, and J. E. Mooij, *Europhys. Lett.* **18**, 343 (1992).
32. T. J. Hagenaars, J. E. van Himbergen, J. V. Josè, and P. H. E. Tiesinga, *Phys. Rev. B* **53**, 2719 (1996).
33. R. S. Newrock, C. J. Lobb, U. Geigenmuller, and M. Octavio, *Solid State Physics* **54**, 263 (2000).
34. A. V. Ustinov, *Chaos* **13**, 716 (2003).
35. A. J. Leggett, In: *Percolation, localization, and superconductivity*, Plenum Press, 1984.
36. M. H. Devoret, D. Esteve, C. Urbina, J. Martinis, A. Cleland, and J. Clarke, In: *Quantum Tunneling in Condensed Media*, North-Holland, 1992.
37. Y. Nakamura, Y. A. Pashkin, and J. S. Tsai, *Nature* **198**, 786 (1999).
38. J. R. Friedman, V. Patel, W. Chen, S. K. Tolpygo, and J. E. Lukens, *Nature* **6791**, 44 (2000).
39. C. H. van der Wal, A. C. J. ter Haar, F. K. Wilhelm, R. N. Schouten, C. J. P. M. Harmans, T. P. Orlando, S. Lloyd, and J. E. Mooij, *Science* **290**, 773 (2000).
40. I. Chiorescu, Y. Nakamura, C. J. P. M. Harmans, and J. E. Mooij, *Science* **299**, 186 (2003).
41. D. Vion, A. Aassime, A. Cottet, P. Joyez, H. Pothier, C. Urbina, D. Esteve, M. H. Devoret, *Science* **296**, 886 (2002).
42. J. M. Martinis, S. Nam, J. Aumentado, and C. Urbina, *Phys. Rev. Lett.* **89**, 117901 (2002).
43. Yang Yu, S. Han, X. Chu, S. Chu, and Z. Wang, *Science* **296**, 889 (2002).
44. A. Wallraff, A. Lukashenko, C. Coqui, T. Duty, and A.V.Ustinov, *Rev. Sci. Instr.* **74**, 3740 (2003).
45. A. Wallraff, T. Duty, A. Lukashenko, and A. V. Ustinov, Phys. Rev. Lett. **90**, 037003 (2003).
46. A. J. Berkley, H. Xu, R. C. Ramos, M. A. Gubrud, F. W. Strauch, P. R. Johnson, J. R. Anderson, A. J. Dragt, C. J. Lobb, and F. C. Wellstood, *Science* **300**, 1548 (2003).

CHAPTER 6

PROTEIN FUNCTIONAL DYNAMICS: COMPUTATIONAL APPROACHES

Yves-Henri Sanejouand
Laboratoire de Physique, École Normale Supérieure de Lyon
46 allée d'Italie, 69364 Lyon Cedex 07, France
E-mail: Yves-Henri.Sanejouand@ens-lyon.fr

As polymers, proteins have very unusual properties. The main one is that they are able to fold, that is, under physiological conditions, to adopt a very well defined three-dimensional structure. Such a folding process has been studied using simple lattice models that were shown to share many properties with natural proteins. Simple models, where a protein is represented as a network of harmonic springs, have also recently proved useful for studying the other functionally important motions of proteins, namely, their conformational changes. Both kinds of models could be used in order to test the hypothesis that non-linear energy localisation phenomena may occur in proteins and play a role in their functional dynamics.

1. Introduction

A question of interest is whether or not phenomena of non-linear energy localisation play a role in the way biological macromolecules achieve their function. As far as I know, in the case of proteins elements for discussing this question are still missing. The goal of this course is just to introduce the functional dynamics of proteins, focusing on the description of a few approaches and models that could be used in order to address this question.

2. Protein structure

Proteins are natural, unbranched hetero-polymers made of twenty different building blocks, the aminoacids, linked together according to a precisely defined sequence, which is encoded in the corresponding gene. For instance, the ten first aminoacids of human myoglobin (a protein responsible for oxygen transport in muscles) are: glycine-leucine-serine-aspartate-glycine-glutamate-tryptophan-glutamine-leucine-valine (those are usually given using an one-letter code, namely, here: GLSDGEWQLV, where G stands for glycine, L for leucine *etc*). Human myoglobin is a 153 aminoacids protein while typical proteins have sequences two or three times longer, though there are some with less than fifty aminoacids and a few with more than ten thousand of them.

As exemplified in Fig. 1 for the case of the LAO binding protein, the most remarkable property of these polymers is that they have very well defined three-dimensional (3-D) structures. These are quite compact, as one can see by looking at an all-atoms representation (top of Fig. 1, where each atom is represented as a van der Waals sphere), though they are often substructured into "domains" (two of them, in the LAO binding protein case).

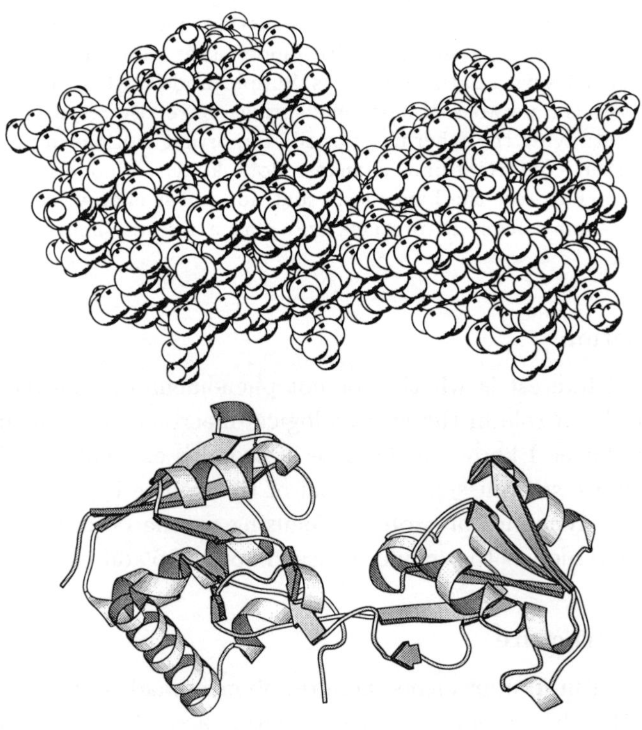

Fig. 1. Two standard representations of the LAO binding protein.

The all-atoms representation is quite realistic but it is rarely used. The reason is that the path followed by the polypeptidic chain, which is the information that is useful in order to discriminate the various kinds of protein structures, is hidden. Such a path is shown in the bottom of Fig. 1, drawn with the widely used Molscript program.[1] This representation gives also information concerning the series of hydrogen bonds within the polypeptidic chain that determine the local conformation: the polypeptidic chain is either in an helical configuration (the famous α-helix) or extended in side-by-side stretches (the β-sheets). In this later case, represented as arrows in Fig. 1, the hydrogen bonds are formed between atoms that can be far apart in the protein sequence.

Protein structures like the one shown in Fig. 1 are obtained using x-ray crystallography or nuclear magnetic resonance. While less than a dozen structures were known in 1975, more than 20000 are now available in the protein databank (PDB), a public repository for the processing and distribution of 3-D biological macromolecular structure data. These 20000 structures can be classified into nearly 1000 different folds, that is, 1000 clearly different paths of the polypeptidic chain.

Note that most proteins whose structure is presently known are soluble proteins, that is, non-membrane ones. Although the latter are much more difficult to produce in large amounts and to crystallize (*sic*), huge progress has recently been made in this field. For instance, in year 2000 seven new structures of membrane proteins were released in the PDB.

3. Energetics of protein stabilisation

When viewed at the level of description of the protein fold, that is, of the path followed by the polypeptidic chain within the structure, it seems that hydrogen bonding plays a major role in protein stability, through the formation of the secondary structure elements, namely, α-helices and β-sheets. Moreover, because prediction of secondary structure elements from the knowledge of the protein sequence in the vicinity of a given sequence site is quite efficient (more than 75% of the aminoacids are correctly predicted by current methods to be in one of the three following states: $\alpha-$helix, $\beta-$strand or coil[2]), it appears that interactions between neighbouring aminoacids in the sequence also play an important role. Indeed, during the folding process, secondary structure elements are often found to be formed quite rapidly.

However, the key aspect of the protein folding process is related to the question of how the protein is able to "choose" between more than 1000 different folds, given the fact that many of these folds are alternative packings with similar content of each of the secondary structure elements. For instance, in the last release (May 2003) of the SCOP protein classification database, there are 171 different folds of all-alpha proteins, that is, of proteins with no β-sheet in their structure.

At this level there are strong indications that hydrophobic interactions play a key role. Indeed, when all known soluble protein structures are analyzed, most aminoacids buried in their cores are found to be hydrophobic ones. In other words, the folding process seems analogous to the formation of an oil droplet in water, the hydrophobic parts being clustered, so as to minimize their surface accessible to the water environment. Moreover, when the genetic code is analyzed, it is found that its structure helps protecting aminoacid hydrophobicity, by restricting the impact of errors (that is, of mutations) on this property.[3] Correspondingly, when massive mutational experiments are performed, it is observed that a majority of sites along the protein sequence are generally tolerant to aminoacid substitutions and that they belong to segments acting as spacers between hydrophobic aminoacids.[4] Some of these hydrophobic aminoacids seem to be more important than others. For instance, when proteins of same fold but with very different sequences are compared, some buried positions, coined "topohydrophobic", are found to be always occupied by strong hydrophobic aminoacids.[5]

4. Protein folding

Because a protein is a long and flexible heteropolymer, it is not obvious how it is able to fold, that is, how starting from a more or less random coil it can reach its structure among the huge number of its possible conformations. Indeed, let us consider that each aminoacid along the polypeptidic chain can choose between at least two conformations: either the one it has when it belongs to an α-helix or the one it has when it belongs to a β-strand. Then for a small one hundred aminoacids protein, a low estimate for the number of its possible conformations is: $2^{100} \backsimeq 10^{30}$. As underlined by Cyrus Levinthal in 1969, a few years after the first protein structure was determined, this is a huge number (the so-called "Levithal's paradox"). Indeed, if the folding process were a straightforward enumeration of all these conformations one after the other, and if one conformation were examined every, say, ten femtoseconds (one period of the fastest bond-length vibration in a protein), the whole enumeration would last nearly 400 million years, while small proteins typically fold within milliseconds.

It would be easy to get rid of such a paradox if there were a clear path in the conformational space towards the native state of the protein. For instance, in the case of a long α-helix Levinthal's argumentation also holds. Nevertheless, it is well known that the limiting step is the formation of the first turn of the helix, which is stabilised by a single hydrogen bond. Then, for each neighbouring aminoacid adopting the helical configuration, another hydrogen bond is gained, allowing for the fast formation of the rest of the helix. So, following Levinthal's conclusion, such paths were seeked for, through the search of transient intermediaries during the folding process. There was however little, if any, success, as far as the way a given fold is reached by a given protein is concerned.

Another way to get rid of Levinthal's paradox is to suppose that there are some biological mechanisms able to guide the protein towards its native state. Though such mechanisms were indeed found, namely, an ensemble of proteins called chaperonins which facilitate folding of proteins that are otherwise destined to aggregate, the fact that they are not necessary for the proper folding of at least small proteins was most convincingly shown by total chemical synthesis of some of them. Moreover, in the case of a viral protease, the enantiomer of the natural protein, made with the same aminoacid sequence but with aminoacids of the non-natural D-type, was also synthesised. As expected, it exhibits reciprocal chiral specificity on peptide substrates, that is, each enzyme enantiomer cuts only the corresponding substrate enantiomer. This implies that the folded forms of the chemically synthesized D- and L-enzyme molecules are mirror images of one another in all elements of their 3-D structure.[6]

4.1. *On-lattice models*

In order to gain a deeper understanding of Levinthal's paradox, lattice protein models were studied in the late seventies,[7–9] starting with the case of the 49-mer on a square lattice. In these models, the protein is represented by a chain of beads occupying the sites of a lattice in a self-avoiding way (see Fig. 2). The main advantage of lattice models over more detailed ones is that in many cases their whole conformational space can be examined. However, even for such simple models the number of possible conformations is growing very quickly as the size of the polymer increases. For instance, on the square lattice, a 18-mer has 5808335 different conformations unrelated by symmetries except reverse-labeling.[10] Simply enumerating them is tricky in the above case, while in the 49-mer case it is out of reach (there are $\simeq 10^{20}$ of them). However, as shown by Nobuhiro Go and his collaborators, starting from a random conformation, the 49-mer can reach its ground state, that is, its lowest-energy configuration, within a few thousands steps of a Monte Carlo simulation, as long as the energy surface is defined as follows. First, the lowest-energy, compact 7×7 conformation, is chosen *a priori.* Then, for all pairs of monomers which are close neighbours in this conformation, the contact energy is assumed to be attractive, while for all others it is not.

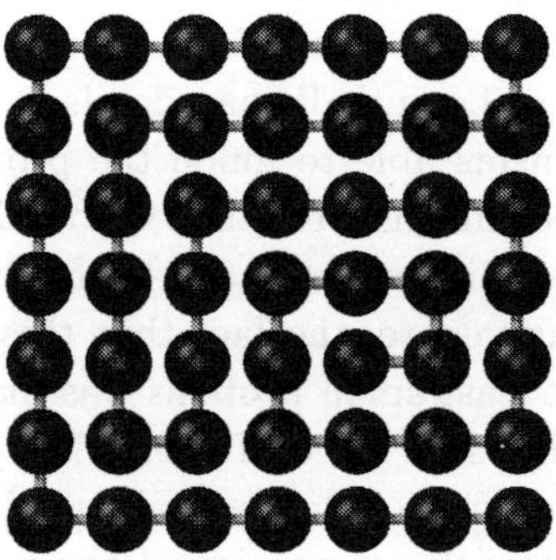

Fig. 2. A compact conformation of the 49-mer on the square lattice.

In other words, when the ground-state is at the bottom of a deep funnel on the energy surface, then it is quite easy for a flexible polymer to find its way and reach it, through a random search biased by the average energy gradient. However, though the funnel picture is nowadays the preferred view for understanding the folding process,[24] there is no indication that protein energy surfaces are as funneled and as deep as in a Go model. On the contrary, when the probabilities for two aminoacids of types k and l (*e.g.*, glycine and leucine, leucine and leucine) to be found in contact in proteins of known 3-D structures are computed, it turns out that the effective interaction energy between them can be assumed to be nearly additive,[11] that is:

$$E_{kl} \simeq E_k + E_l \,, \tag{1}$$

where E_k is a quantity correlated with the hydrophobicity of the aminoacid of type k. Later on, studies were performed with more realistic energy functions and, though models on the square lattice are still being considered, models on the cubic lattice were more and more often preferred, like the popular 27-mer.[12,13] However, even in this rather simple case, the number of possible conformations is so large that only estimates can be given (nearly 10^{16}).[14] This is why most studies were performed with energy functions including a "compaction term", that is, a bias towards the most compact conformations of the polymer.

For the 27-mer case, the most compact geometry is the $3\times3\times3$ sublattice, in which there are only 103346 self-avoiding possible conformations, unrelated by symmetry except reverse-labeling,[12,13] that is, 103346 different ways for a chain to go through all 27 lattice sites, going from a site to a neighbouring one at each step. One of these so-called hamiltonian paths is shown in Fig. 3, for the following sequence: $PHP_4HPHPHP_{15}H$, where only two kinds of monomers, either polar (P) or hydrophobic (H), have been considered.

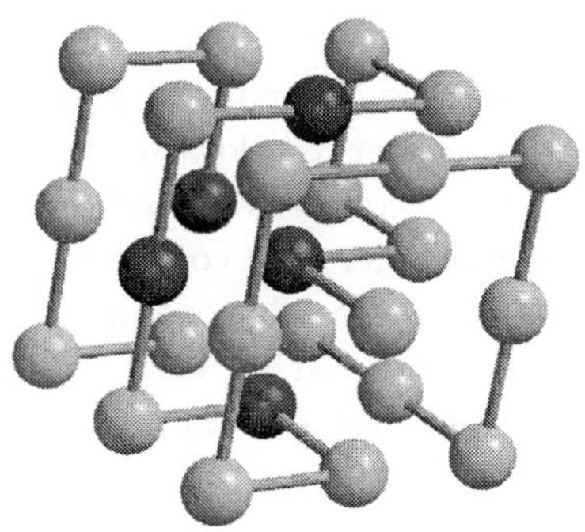

Fig. 3. One of the 120 remarkable compact conformations of the 27-mer.

Of course, Fig. 3 can also be viewed as showing another possible conformation, the reverse-labeled one, for the corresponding reverse-labeled sequence, namely, $HP_{15}HPHPHP_4HP$. If such pairs of conformations are assumed to be identical, the number of different compact conformations drops to 51704.[15]

The fact that a sequence can be "threaded" in a given structure in two different ways, the position of the first aminoacid in the first way being the position of the last aminoacid in the other way, is certainly not a property of natural proteins, which are made with asymmetric building blocks, namely, aminoacids of the L-series. As a consequence, the α−helix is right-handed, and left-handed helices have not yet been observed in structures of natural proteins. Note that the reverse-labeled sequence of a given protein could quite well be synthesized. If built with aminoacids of the D-series, its 3-D structure should be the same as that of the natural protein, as far as the positioning of the aminoacid sidechains is concerned. However, to our knowledge, up to now, following the seminal work of Shemyakin and its collaborators,[16] only small reverse-labeled all-D peptides, now called "retro-inverso" peptides,[17] have been synthesized.

Because the number of compact conformations of the 27-mer is small enough so that the energy, for a given sequence, can be calculated for all them, the lowest-energy compact conformation can be determined. Then, starting from any non-compact conformation on the cubic lattice, the folding of 27-mers can be studied, using Monte Carlo simulations.[14] Doing so, it was shown that for certain sequences the lowest-energy conformation can be reached within $5 \cdot 10^7$ steps, during which only a few dozen compact conformations are sampled, among the 103346 possible ones, while the choice of the initial conformation has little consequence, if any, on the outcome of the simulation.[14] Such *special* sequences, which are able to fold rapidly and, as such, appear to behave like natural proteins, all proved to bear a pronounced energy gap between the lowest-energy and the first excited compact state,[14] that is, the second lowest-energy compact conformation.

When only two kinds of monomers are considered, either hydrophobic or polar ones, the sequence-structure relationship for the model can be exhaustively examined, and compared to what it is in the case of actual proteins.[15,18] The usual choice for the energy function has the following form:

$$H = \sum_{i<j} E_{ij} \Delta(r_i - r_j) , \qquad (2)$$

where $\Delta(r_i - r_j) = 1$ if monomers i and j are close neighbours in the lattice, $\Delta(r_i - r_j) = 0$ otherwise and E_{ij} depends on the nature of the interacting monomers. A popular choice for the E_{ij} values has been $E_{ij} = E_{HH} = -\epsilon$, when monomers i and j are both hydrophobic and $E_{ij} = 0$ otherwise.[12] When the additive, more realistic case is considered, that is, when:

$$E_{ij} \backsimeq E_i + E_j , \qquad (3)$$

with for instance $E_i = E_H = -1$ and $E_i = E_P = 0$, the sequence-structure relationship in the model exhibits remarkable features. Noteworthy, if the energy is determined for all sequences for each of the 103346 conformations, it is found that only 122750 of them (0.09%) have non-degenerate ground states, that is, for each of these sequences a given conformation is the lowest-energy one, while for all of them the energy gap Δ between this conformation and the second lowest-energy one is: $\Delta = 2$. Strikingly, out of the 103346 possible conformations only 120 (0.11%) are found to be possible, non-degenerate, ground states. All these so-called "remarkable structures" are characterized by a large "designability"[15] as measured by the number N_s of sequences of which they are the ground state (N_s ranges between 513 and 2306). The fact that the number of possible folds in the above 27-mer model is limited is consistent with the current belief that the number of protein folds also seems to be limited. Though nearly a thousand folds are now known, new folds are discovered less and less often and it is usually estimated that there may not exist more than 10000 protein folds in nature.[19]

Moreover, when the N_s sequences with a given fold are analyzed, properties similar as that of ensembles of protein sequences with the same fold are recovered. For instance, in half of the sequence sites hydrophobic and polar monomers are equiprobable,[15] while, as a consequence, pairs of sequences as different as pairs of random sequences are found to adopt the same fold.[18] In Fig. 3 the "top structure"[15] is shown, that is, the conformation which is the ground state of the largest number of sequences. Among these 2306 sequences, $PHP_4HPHPHP_{15}H$ is the one with the smallest number of hydrophobic monomers. As a matter of fact, each remarkable structure is the ground state conformation of a *single* five-hydrophobic monomers sequence and the corresponding five monomers are always located as shown in Fig. 3, that is, one of them being at the cube center, the four other ones being at the center of facets which are not bonded to the monomer at the cube center. Furthermore, in all sequences sharing a given ground state, these five monomers are all hydrophobic.[18] This later point helps to clarify why there are 120 remarkable conformations. Indeed, if a sequence like $PHP_4HPHPHP_{15}H$ has a non-degenerate ground state, this means that, out of the 103346 possible ones, there is only one way to bring its five hydrophobic monomers close together, so that each of them can interact with at least another hydrophobic one.[18] In other words, it is from a topological point of view that the 120 remarkable conformations of the $3 \times 3 \times 3$ cubic lattice model are "atypical".[20]

This is a quite satisfactory property of the model, since it means that the same 120 conformations are also expected to be remarkable when different E_{ij} values are chosen. For instance, with $E_{ij} = E_{HH} = -2 - \gamma, E_{ij} = E_{HP} = -1, E_{ij} = E_{PP} = 0$, and $\gamma = 0.3$, the conformation shown in Fig. 3

is still the top structure, but with $N_s = 3794$ (see Ref. 15). The picture is now more complicated than in the case of the additive potential. The difference between additive and nearly additive cases comes from the fact that the departure from additivity lifts the degeneracy of many sequences. Now energy gaps of $\Delta = 0 + n\gamma$ and $\Delta = 2 \pm n\gamma$ are observed, with $n = 1, 2,$ ... (see Ref. 21). For instance, with $\gamma = 0.3$, nearly half of the sequences whose ground state is one of the remarkable conformations have energy gaps lower than 1.0 (see Ref. 18). Moreover, while 4.75% of the sequences have a unique ground state, for the majority of them it is not one of the remarkable conformations but the corresponding energy gap is small, being on average close to 0.3, the γ value.[15] When only sequences with energy gaps larger than 1.0 are considered, the picture obtained with the additive potential is restored, despite a few minor differences arising from the overlap of the energy gaps splitting around the $\Delta = 0$ and $\Delta = 2$ cases.[18]

An interesting protein-like property of the model allows for the determination of *all* large gap sequences (*i.e.*, with $\Delta = 2$, in the case of the additive potential) whose ground state is a given remarkable conformation, without the need of any huge enumeration.[18] To do so, starting from the corresponding remarkable five hydrophobic monomers sequence, all 27 singly-mutated sequences are generated. Then, those among them with the same single ground state are retained. Next, for each sequence of this subset all 27 singly-mutated sequences are generated and so on, until no new sequence can be retained. What is shown through the success of such a protocol is that the N_s sequences of each of the 120 remarkable structures belong to a "neutral island" of the sequence space.[18] Note that such a property has also been found in the case of square lattice models,[22] and that it is expected to be shared by ensembles of protein sequences corresponding to a given fold. Thus, the sequence-structure relationship for the above cubic lattice model seems to have much in common with the sequence-structure relationship observed in the case of natural proteins. This suggests that topology may play an important role in the folding process, allowing for a sequence to choose between significantly different folds, each characterized by a topologically unique "folding nucleus".[23]

As far as Levinthal's paradox is concerned, studies of lattice models have yielded a so-called "new view" of the protein folding process in which folding is seen as a parallel microscopic multi-pathway diffusion-like process. In other words, it is nowadays assumed that folding occurs through funneling to a single stable state by multiple routes in the conformational space.[24]

4.2. *Off-lattice models*

However, despite their many successes, cubic lattice protein models have several drawbacks. First, from a chemical point of view, the cubic geometry is a very unlikely one for a polymer, since on this lattice *par définition* the angle between three consecutive monomers is either 90 or 180 degrees. Second, using the Monte Carlo Metropolis algorithm for studying dynamical processes, like the folding process, has clear limits. Indeed, in order to study such processes using molecular dynamics methods, that is, standard step by step integration of the equations of motion, would be a more obvious choice. But then, since the conformational space in which dynamics take place is continuous, if the energy function is not chosen in an *ad hoc* way, knowing the lowest-energy conformation of a given polymer becomes difficult as soon as it is not a very short or highly constrained one.
Hereafter, a class of 3-D, off-lattice protein models is briefly described, for which this can be done almost as easily as in the case of cubic lattice models, while the folding process of the polymer towards its lowest-energy conformation can be followed as a function of time using molecular dynamics methods.

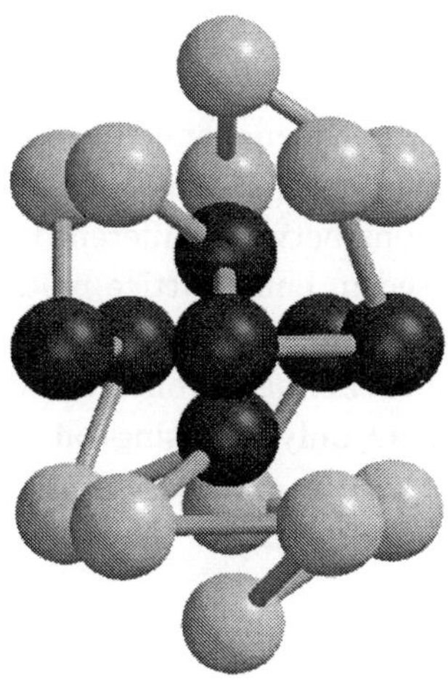

Fig. 4. A compact conformation of the off-lattice 19-mer.

Like in the case of lattice protein models, in the following a N-mer is modeled as a chain of N beads linked together. Moreover, in order to retain some of the major advantages of lattice models, the diameter D of the bead is chosen to be equal to the average length of the bonds between two linked beads. Together with a compaction term in the energy function, such a choice allows to know *a priori* which is the approximate geometry of the lowest-energy conformations of the polymer, as long as N is well chosen. For instance, on a flat surface, the most compact conformation of a 7-mer has the shape of an hexagon, the hexagonal packing being the most compact possible one for a set of disks of same size. For the corresponding 3-D models, though the most compact packing of a set of N spheres is not that obvious, it is often very well defined.[25] For instance, a maximally compact set of thirteen spheres is known to have the shape of an icosahedron, while for nineteen spheres a compact conformation of the corresponding polymer is shown in Fig. 4. So, for such a family of cases also, all hamiltonian paths as well as the energy gap of any given sequence can be determined. Then, starting from random conformations, folding simulations can be performed using molecular dynamics methods.

From a practical point of view, the energy function associated to such models can be a simplified version of standard empirical energy functions used when protein dynamics is studied at the atomic level, like that implemented in the CHARMM program,[26] namely:

$$H = \frac{1}{2}k_b \sum_{bonds} (l_{ij} - D)^2 + \sum_{i>j} (\epsilon_{ij} + E_c)\left(\frac{D^{12}}{r_{ij}^{12}} - 2\frac{D^6}{r_{ij}^6}\right), \tag{4}$$

where l_{ij} is the distance between bonded monomers i and j, r_{ij} the distance between non-bonded ones, E_c the compaction term, while ϵ_{ij} allows for defining the interactions between different types of monomers. In practice, values like those used in cubic lattice model studies could be chosen, assuming for instance an additive description (see Eq. 3).

However, up to now, such Lennard-Jones chains have been considered in the case of homopolymers only, focusing on the collapse transition,[27,28] eventually as a function of an external stretching constraint[29] and often still using Monte Carlo methods for the sampling of the configurational space.[30]

4.3. *More detailed models*

Of course, one could think of trying to study the protein folding process using brute force, by solving the equations of motion for a protein model at an all-atoms level of description, surrounded by explicit water molecules, as done in standard molecular dynamics studies of biological macromolecules. Here, the main limitation is the time step required for solving accurately enough the equations of atomic motion, which is in the femtosecond range. In practice, this limits the time scale accessible with such approaches to nanoseconds, though a few microseconds long simulations have been performed in the case of very small, fast-folding proteins. Indeed, recently, interesting results were obtained in the case of a 36 aminoacids protein, using an "implicit solvent" model and a worldwide distributed computing network of tens of thousands of PCs.[31] However, for such small proteins, results of similar quality were also obtained using a much simpler protein description, Monte Carlo methods and a single PC.[32]

5. Protein conformational changes

5.1. *Functional motions*

The precise positioning of the aminoacids in space is vital for the ability of proteins to function and, noteworthy, for the specificity of the kind of function they perform. In particular, if enzymes are efficient catalysts, their main property, as far as their role in the very existence of the life phenomenon is concerned, is to be able to catalyze very narrowly defined chemical reactions.

However, in order to perform their function, most, if not all, protein structures have to be flexible. For instance, when the first structure of a protein (the myoglobin) was obtained at a level of resolution allowing for the identification of the aminoacid side chains,[33] it was found that the ligand (the oxygen molecule) is buried deeply into the structure and that it would have no possibility to escape the binding pocket, if the protein were a perfectly rigid object.

More recently, using hydrogen/deuterium exchange as a tool for evaluating protein flexibility, it was found that the 3-isopropylmalate dehydrogenase enzyme from a thermophilic bacterium which is hardly active at room temperature, is much more rigid at this temperature than the same enzyme from a normal, mesophilic bacterium, whereas these enzymes have nearly identical flexibilities under their respective optimal working conditions.[34]

5.2. *Collective motions*

In many proteins, large conformational transitions involve the relative movement of almost rigid structural elements: a loop, a helix, a whole

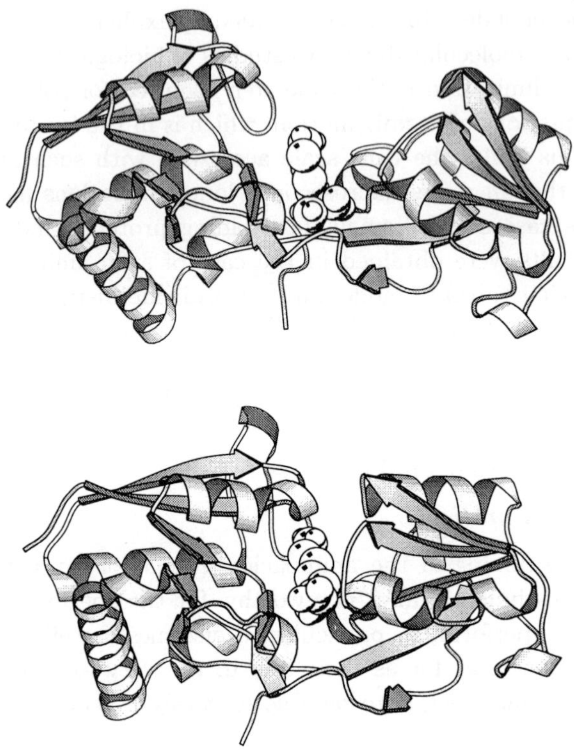

Fig. 5. Closure motion of the LAO binding protein upon lysine binding.

domain. Such motions are important for a variety of protein functions, including catalysis, regulation of activity, *etc.* For example, the binding of the coenzyme A on the citrate synthase enzyme, a two-domain protein, induces a 18° rotation of the smaller domain around an axis close to aminoacid 274, which functions as the hinge.[35] One consequence of this motion is the closure of the cleft between the two domains, in which the binding site lies, allowing for the catalyzed chemical reaction to occur in a non-water, organic (*sic*) environment. As in most other cases (hexokinase,[36] phage T4 lysozyme,[37] *etc*) this motion was probed by x-ray crystallography. Fig. 5 shows one such simple large-amplitude, collective motion, called "hinge-bending", in which two structural domains of similar size move quite rigidly, so as to close a binding pocket. This protein binds charged aminoacids, namely, lysine (represented as van der Waals spheres), arginine or ornithine. Hence its name: the Lysine-Arginine-Ornithine (LAO) binding protein. The two conformations shown can be found in the Brookhaven protein data bank under the codes 2lao (top) and 1lst (bottom). In this case, distances between pairs of aminoacids vary by up to 20 Å.

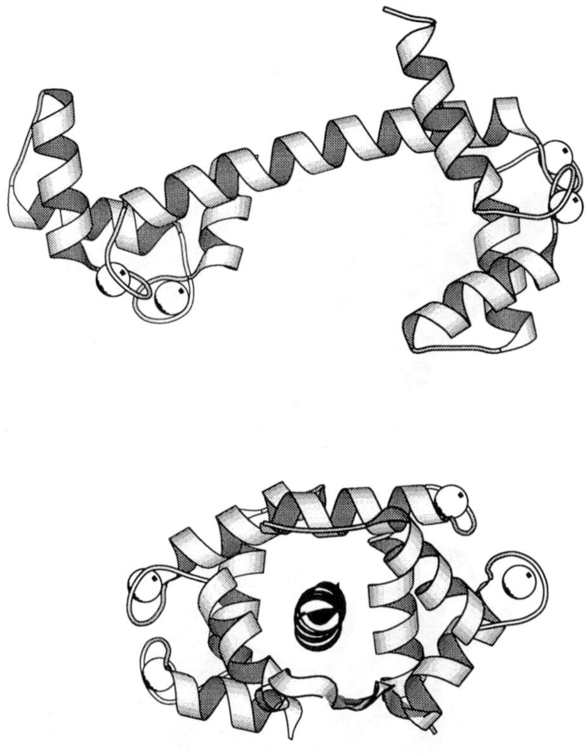

Fig. 6. The conformational change of calmodulin upon peptide binding.

A more spectacular case is that of calmodulin, a protein of the EF-hand family, whose members act as Ca^{2+} sensors. Calmodulin reacts to transient increases in cellular calcium concentration occurring during a variety of cellular activities, ranging from cell cycle control to muscle contraction. As shown in the top part of Fig. 6, its Ca^{2+}-bound conformation has a dumbell shape, with two globular domains linked by a long helix (Here, van der Waals spheres stand for calcium ions). One of the effects of Ca^{2+}

binding is the exposure to the solvent of hydrophobic patches that form the binding site of target proteins. Indeed, Ca^{2+}-bound calmodulin is able to bind to more than one hundred different target proteins, and it does so by breaking its long central helix, allowing for the hydrophobic patches to grasp an hydrophobic part of the target protein, usually an α-helix, as shown in bottom part of Fig. 6 (the black moiety). In this case, distances between pairs of aminoacids vary by up to 50 Å.

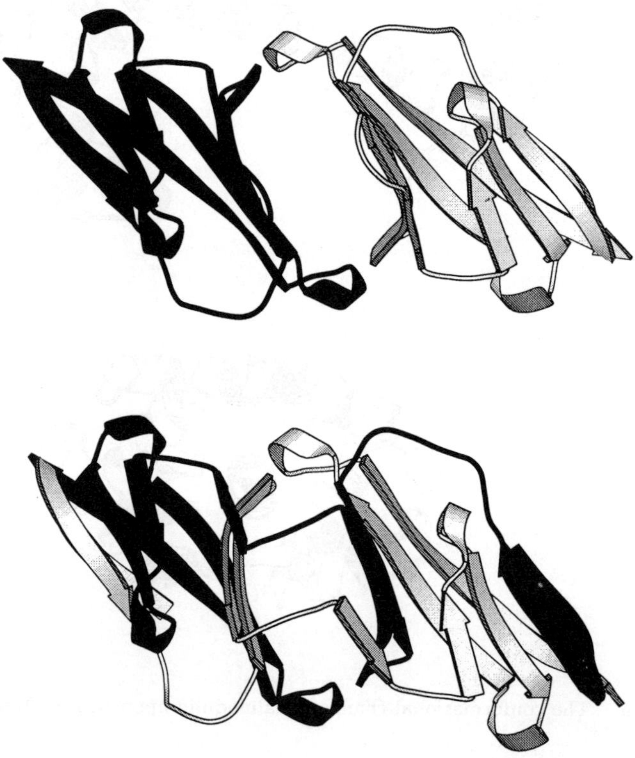

Fig. 7. Domain-swapping motion of the CD2 dimer.

Because it is that large, the conformational change of calmodulin can be observed out of the crystal state, namely, in solution by x-ray scattering, and it is one of the few cases where direct measurements of the kinetics of the conformational change have been performed. For instance, using fluorescence resonance energy transfer (FRET) between two chromophores,

each attached to one domain, it was possible to show that the time scale of domains separation, when the bound peptide is released (the "off-rate"), is in the tens of a second range.[38]
Except in a few cases, conformational changes take place in proteins without any modification of the secondary structural elements: the lengths of the α-helices and of the β-strands are the same before and after the motion. Also, most interactions between secondary structural elements are preserved and the motion can often be described as a relative motion of two "blocks" of structural elements, the "domains", with respect to each other, like in Figs. 5 or 6. An interesting exception is the recently discovered "domain-swapping" conformational change, which can occur between members of a dimer, that is, a pair of identical proteins associated together at physiological concentrations. Here, as shown in Fig. 7 in the case of the CD2 protein, the association can either be a standard, simple contact between the two monomers (top), or the two monomers can be intertwined (bottom). While the packing of the secondary structural elements is the same, it is obtained through contacts between elements belonging to different monomers, the difference between the two conformations being a change in the configuration of a single, flexible loop which in the later case "jumps" from a monomer to the other.

5.3. *Low-frequency normal modes*

One of the best suited theoretical methods for studying collective motions in proteins is the normal mode analysis (NMA), which leads to the expression of the dynamics in terms of a superposition of collective variables, namely the normal mode coordinates. The principles underlying NMA are briefly recalled below.

5.3.1. *Normal mode analysis*

This kind of analysis is based on the following ideas.[39] In the vicinity of a stationary point, the potential energy V of a system of N atoms, can be approximated by:

$$V = \frac{1}{2}\sum_{i=1}^{3N}\sum_{j>i}^{3N} k_{ij}(r_i - r_i^s)(r_j - r_j^s)\,, \tag{5}$$

where k_{ij} is the mass-weighted second derivative of the potential energy with respect to coordinates r_i and r_j, and where r_i^s and r_j^s are the i and

j coordinates of the stationary structure under study. Within the frame of this approximation, the equations of motion can be solved analytically, leading to the following solutions:

$$r_i(t) = r_i^s + \frac{1}{\sqrt{m_i}} \sum_{j=1}^{3N} a_{ij} q_j(t) ,$$

with:

$$q_i(t) = C_j cos(\omega_j t + \phi_j) ,$$

where m_i is the atomic mass, a_{ij} the ith coordinate of normal mode j and $\nu_j = \frac{\omega_j}{2\pi}$ the corresponding frequency. ω_j^2 is the jth eigenvalue of the Hessian, that is, the $3N \times 3N$ matrix built with the k_{ij}'s, while a_{ij} is the ith coordinate of the corresponding eigenvector. C_j and ϕ_j, the amplitude and phase of mode j, are determined once the coordinates and the velocities of the system at $t = 0$ are known. Noteworthy, C_j is such that:

$$C_j = \frac{\sqrt{2E_j}}{\omega_j} , \tag{6}$$

where E_j is the amount of energy in mode j, which according to the equipartition principle is expected to be at a given temperature T: $E_j \backsimeq k_B T$, where k_B is the Boltzmann constant. One can see that at a given temperature the lower the frequency of a mode, the larger its amplitude. Typically, the normal modes whose frequencies lie under 30-100 cm^{-1} are found to be responsible for most of the amplitude of the atomic displacements in proteins.[40,41]

From a practical point a view, the potential energy of the studied protein is first minimized, using standard algorithms that seek minima close to the starting structure. Such are for example the Powell or the Newton-Raphson algorithms. Note that during this process each atom in the structure drifts by $1 - 2$ Å on average. Then, the Hessian is diagonalized. Due to its size, this step used to be the technically limiting one. Indeed, though the normal mode analysis of the small, 58 aminoacids BPTI protein was performed as soon as 1982 (see Ref. 42) ten years later the largest protein studied at the atomic level of description was still the 153 aminoacids myoglobin,[43] while most interesting proteins are much larger. Since then, efficient algorithms were designed (*e.g.* DIMB, Ref. 44) or adapted to the case of macromolecular assemblies (*e.g.* the block Lanczos approach, Ref. 45) in order to compute the lowest-frequency normal modes, which proved to be the most informative ones.

5.3.2. *The RTB approximation*

Instead of diagonalizing the Hessian, $\mathbf{H}$, as in standard NMA, the principle of the RTB method is to diagonalize $\mathbf{H_b}$, a $6n_b \times 6n_b$ matrix built as follows:[46,47]

$$\mathbf{H_b} = \mathbf{P^t H P}\,,$$

where $\mathbf{P}$ is an orthogonal $3N \times 6n_b$ matrix composed of the vectors describing the six rigid-body rotations and translations of each of the n_b blocks the protein is split into. For instance, each block can contain a single aminoacid. $\mathbf{A_p}$, the $3N \times 6n_b$ matrix with the $6n_b$ approximate lowest-frequency normal modes of the protein, is then obtained as follows:

$$\mathbf{A_p} = \mathbf{P A_b}\,,$$

where $\mathbf{A_b}$ is the matrix diagonalizing $\mathbf{H_b}$, $\mathbf{A_b}$ being obtained with standard diagonalization routines.

Of course, the RTB approximation can only be used for calculating modes in which aminoacids behave almost rigidly. Even in that case, calculated frequencies are found to be higher than exact ones, reflecting the fact that aminoacids can not adapt their conformation so as to make the whole motion easier. However, for frequencies lower than 40 cm^{-1} it was shown that, when one aminoacid is put in each block, a linear relationship holds between approximated and exact frequencies, that is:

$$\nu_{rtb} \backsimeq d_P \cdot \nu_s$$

where ν_s and ν_{rtb} are the frequencies obtained using, respectively, standard approaches or the RTB approximation. In the case of a set of proteins of various sizes, using the standard CHARMM force field[26] and a 8.5 Å cutoff for electrostatic interactions, it was found that d_P does not depend upon protein size or fold type ($d_P \backsimeq 1.7$).[47] This allows to get fair estimates for exact frequencies, once the approximated ones are known.

5.3.3. *Comparison with crystallographic B-factors*

Frequencies are necessary in order to obtain atomic fluctuations, which themselves can be compared to crystallographic B-factors (the "temperature factors"). Indeed, according to NMA, B_i, the crystallographic B-factor for coordinate i is as follows:

$$B_i = \frac{8\pi^2 k_B T}{3m_i} \sum_{j=1}^{n} \frac{a_{ij}^2}{\omega_j^2}\,, \tag{7}$$

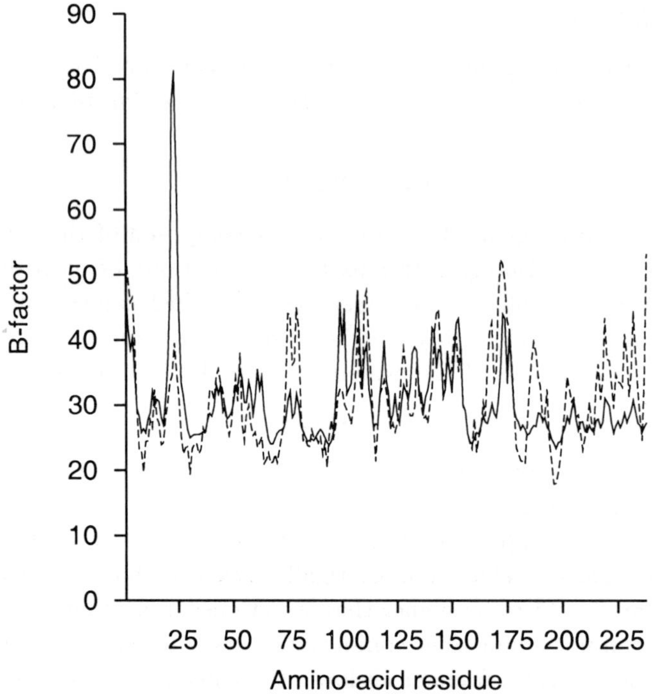

Fig. 8. Calculated and experimental B-factors of the LAO binding protein.

where n is the number of low-frequency normal modes retained for the calculation. Usually, the six rotational and translational rigid-body motions of the whole system are not included in the calculation. However, it is well known that crystal static disorder contributes significantly to the experimental value of B-factors. This contribution can be partly described in terms of rigid-body rotations and translations of the whole system within the crystal cell, assuming for instance the corresponding frequencies to be identical, very low, but non-zero.

However, even without taking static disorder into account, B-factors calculated according to Eq. 7 are often found to be well correlated with experimental values. A typical exemple is shown in Fig. 8, in the case of the open form of the LAO binding protein (see top of Fig. 5), where calculated (dotted line) and experimental (plain line) values are given as a function of the aminoacid number. Here, calculated values were scaled so as to have same average and root-mean-square than experimental ones.

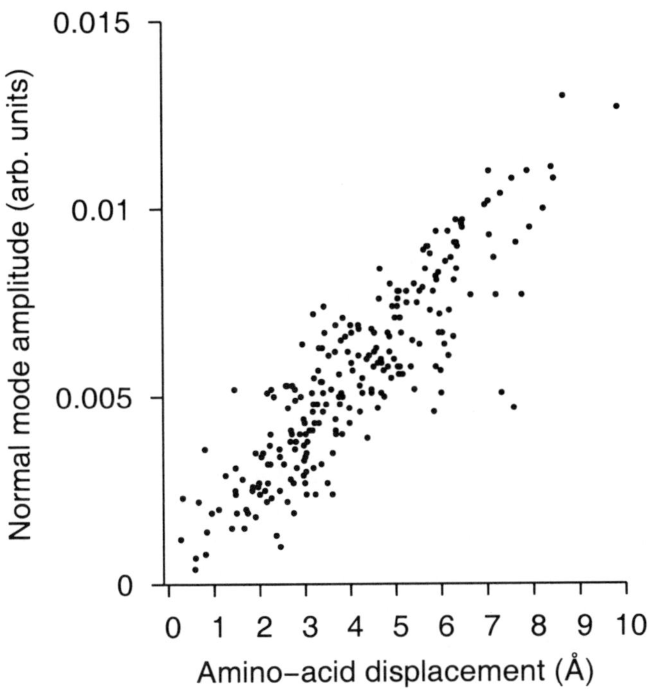

Fig. 9. LAO binding protein: the lowest mode versus the conformational change.

5.3.4. *Comparison with conformational changes*

Normal mode theory has also been shown to give a fair picture of x-ray diffusive scattering of protein crystals.[48] However, the idea that NMA may be an accurate tool for studying protein conformational changes comes from the fact that in several cases, *e.g.*, hexokinase,[49] lysozyme,[50,51] citrate synthase,[45] *etc*, the largest amplitude motion obtained with this theory, that is, the one with the lowest frequency, was found to compare well with the conformational change observed by crystallographers in these proteins upon ligand binding. Indeed, as shown in Fig. 9, in the case of the LAO binding protein conformational change shown in Fig. 5, there is a clear correlation between the observed aminoacid displacements and the amplitude of the aminoacid motions in the lowest-frequency normal mode of this protein, as calculated for the "open" form of the LAO binding protein (top of Fig. 5). The calculation was performed in the absence of the lysine ligand, whose position in the open form is not known, at variance with what suggests

Fig. 5, since it induces the closure motion. Here, C_j, the amplitude of the normal mode (see Eq. 6), was set to one, and data are shown for α-carbons only, as representatives of aminoacid motions.

In order to quantify how well a conformational change is described by a given normal mode, one can calculate I_j, the overlap between $\overrightarrow{\Delta r} = \{\Delta r_1, \cdots, \Delta r_i, \cdots, \Delta r_{3N}\}$, the conformational change observed by crystallographers, and $\overrightarrow{a_j}$, the jth normal mode of the protein. This is a measure of the similarity between the direction of the conformational change and the one given by mode j. It is obtained as follows:[45]

$$I_j = \overrightarrow{\Delta r} \cdot \overrightarrow{a_j} = \frac{\sum^{3N} \Delta r_i a_{ij}}{[\sum^{3N} \Delta r_i^2 \sum^{3N} a_{ij}^2]^{1/2}},$$

where $\Delta r_i = r_i^o - r_i^c$, r_i^o and r_i^c being, respectively, the ith atomic coordinate of the protein in the "open" crystallographic structure and in the "closed" one. A value of one for the overlap means that the direction given by $\overrightarrow{a_j}$ is identical to $\overrightarrow{\Delta r}$. From a practical point of view, $\overrightarrow{\Delta r}$ is calculated after both crystallographic conformations of the protein were superimposed, using standard fitting procedures. These pairs of conformations are often referred to as "open" or "closed", because many known conformational changes involve the closure of a binding pocket site. Note that $O_s(n)$, the cumulative square overlap, calculated as:

$$O_s(n) = \sum_{j=1}^{j=n} I_j^2$$

is equal to one when $n = 3N$, that is, when all modes are taken into account, since the 3N modes form a basis set.[39] In the case of the LAO binding protein conformational change, I_1, the overlap with the lowest-frequency motion is: $I_1 = 0.90$, and $O_s(50) = 0.90$, which means that 90% of the atomic displacements observed during the conformational change can be described with 50 coordinates *only*, namely, the 50 normal coordinates corresponding to the 50 lowest-frequency normal modes. Note that there are 2253 atoms in the system considered here, that is, 6759 coordinates.

The fact that atomic displacements corresponding to a protein motion with a high "collective" character, that is, a motion in which many atoms are involved, can be accurately described with a small subset of low-frequency normal coordinates is not a surprising result, because low-frequency normal coordinate themselves have such a collective character. However, the fact that one, or a few, of them may prove enough for obtaining a fair description of a conformational change was not *a priori* expected. On the contrary, many good reasons can be found in favor of the opinion that it should not be so. For instance, from a physical point of view, the energy function used to compute protein normal modes is a very approximate one, and it is quite sure that frequency values would be significantly different, if it were possible to compute them at an *ab initio* level. Moreover, the low-frequency part of protein normal mode spectra is not characterized by clear gaps, as shown in Fig. 10, in the case of the open form of the LAO binding protein.

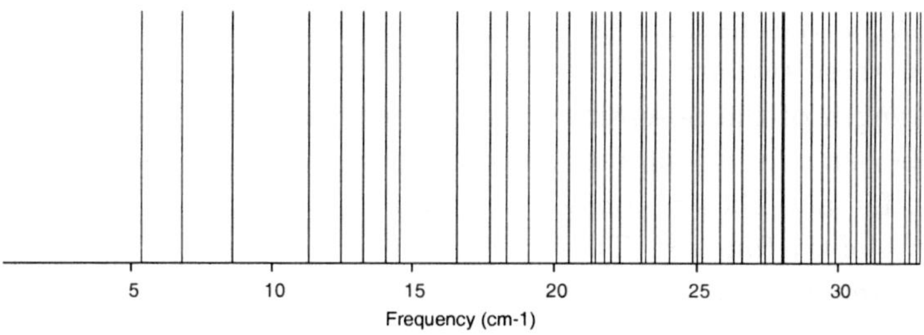

Fig. 10. The low-frequency normal mode spectrum of the LAO binding protein.

Indeed, in the case of such spectra, it was shown that $g(\omega)$, the density of states, follows a characteristic, universal curve with, for the lowest-frequency values:[52]

$$g(\omega) \backsim \omega$$

Furthermore, from a biological point of view, proteins are known to fold and function in a water environment, within a narrow range of pH, temperature, ionic strength, *etc*, while NMA is performed *in vacuo*. Also, standard NMA requires a preliminary energy minimization which drifts the atoms of the protein up to a few Å away from their position in the crystallographic structure. As a consequence, the structure studied with standard NMA is always a distorded one. More generally, NMA is based on a severe small displacements approximation, which amounts to suppose that a protein behaves like a solid does at low temperature, while it is well known that a protein is a somewhat flexible polymer, undergoing many local conformational transitions at room temperature.

5.3.5. *Simplified potentials*

Recent results have shed some light on this paradox. Noteworthy, it was shown that using a single parameter hookean potential for taking into account pairwise interactions between neighbouring atoms yields results in good agreement with those obtained when NMA is performed with standard semi-empirical potentials, as far as low-frequency normal modes are concerned.[53,54] More specifically, within the frame of the approach proposed by M. Tirion, the standard detailed potential energy function is replaced by:

$$E_p = \sum_{d_{ij}^0 < R_c} C(d_{ij} - d_{ij}^0)^2 \,, \tag{8}$$

where d_{ij} is the distance between atoms i and j, d_{ij}^0 being the distance between these two atoms in the given studied crystallographic structure. The strength of the potential C is a phenomenological constant assumed to be the same for all interacting pairs. This energy function was designed so that for any chosen configuration the total potential energy E_p is a minimum of the function. Thus, with such an approach *par définition* NMA does not require any prior energy minimization.

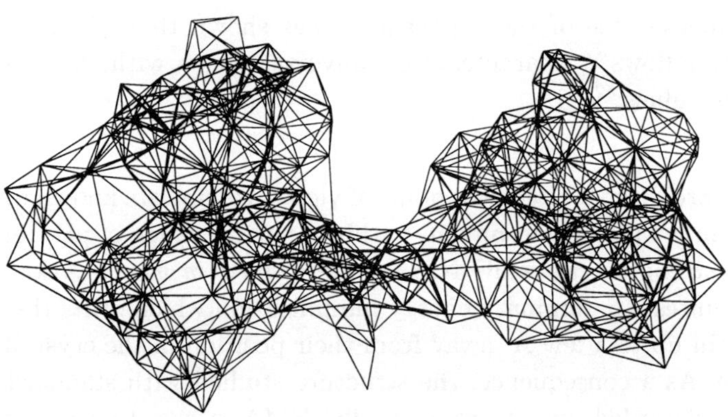

Fig. 11. Tirion's model for the LAO binding protein.

Note that in Eq. 8, the sum is restricted to atom pairs separated by less than R_c, which is an arbitrary cut-off parameter. When, as proposed by Ivet Bahar and her collaborators, only the C_α atoms are taken into account,[55] a cut-off of 8-10 Å can for instance be used. Such a network of harmonic springs, shown for the case of the open form of LAO binding protein in Fig. 11, is enough to study backbone motions, which in turn proves sufficient for characterizing low-frequency normal modes of proteins. Moreover, it allows for studying proteins of large size on common workstations, using small amounts of CPU time, since, with this simple model, the matrix to be diagonalized is a $3N_r \times 3N_r$ one, where N_r is the number of aminoacids of the protein.

Using this kind of highly simplified potential yields low-frequency normal modes in good agreement with those obtained using standard NMA.[54] Moreover, when the interactions between closely located α-carbon pairs are described by a gaussian network model, crystallographic temperature B-factors are also found to be accurately predicted.[55] Again, this means that low-frequency normal modes of proteins are well described with such models.

Indeed, like when detailed semi-empirical energy functions are used, a few low-frequency normal modes are often found to yield a good description of the functional motion of the protein, especially when the corresponding conformational change has a highly collective character.[56–58]

Thus, results obtained with NMA in the field of low-frequency protein dynamics seem to be of a very good quality even when most atomic details are simply ignored. This means that the low-frequency normal modes of a protein depend mainly upon its shape, that is, upon the distribution of its aminoacids in space. Reciprocally, because protein conformational changes are often found to be well described by one or by a few such modes, it seems that during the course of evolution proteins have taken advantage of these solid-like motions for finding paths in configuration space that allow for conformational changes to proceed within reasonable time spans.

6. Dissipation of energy in proteins

Though protein dynamics at room temperature is certainly not harmonic, the harmonic picture has often proved useful for analyzing it. For instance, when temperature jumps are performed during a protein molecular dynamics simulation, if the total kinetic energy of the protein is monitored during the following hundreds of femtoseconds, "echoes" are observed (coined "temperature echoes"), which can be explained in terms of the dynamics of a set of damped oscillators.[59] Unfortunately, the question of how well protein dynamics can be described as the dynamics of a set of coupled harmonic oscillators has not yet been fully worked out, maybe because of a too important focus on slow motions, whose behaviors are known to be highly anharmonic, involving energy barrier crossings, *etc*, or because of a reluc-

tance to study in depth high-frequency protein motions, which are expected to behave quantum-mechanically at room temperature. However, non-linear energy localisation may well have already been observed in proteins, in the course of molecular dynamics simulations. Indeed, when analyzing the very first simulation of a protein at an all-atoms level of description, McCammon, Gelin and Karplus noticed that the kinetic energy of a few aminoacids remained higher than expected from the equipartition principle by 10-20% for more than 9 picoseconds, concluding that "vibrational energy can localise for relatively long time periods in proteins".[60]
Experimentally, though there have been attempts to follow how the energy given to a protein, usually through laser impulses on an aromatic moiety, is redistributed within the protein, probing energy paths within such large and complex systems remains problematic. In this respect, studies of model peptides may prove useful. For instance, recently the kinetics of energy release within a cyclic peptide was followed, taking advantage of the insertion in the peptide of a bi-aromatic bistable ring with clearly different absorption spectra in each of its two possible conformations.[61]

7. Conclusion

Simple models have proved useful for studying the two major kinds of protein functional motions, namely their folding and their conformational changes. Up to now, the folding process has mainly been studied using lattice models. However, Lennard-Jones chains are an obvious generalization for such models if one wishes to characterize dynamical aspects of the process, like the way energy is redistributed in the structure, as a function of its progress towards lower and lower energy configurations.
Models considered for studying conformational changes, in which a protein is figured as a network of harmonic springs, could also be generalized in a straightforward way, by including non-linear terms, so as to explore possible links between non-linear energy localisation phenomena and functional protein motions, which have been shown to be often very well described with a few of their lowest-frequency normal modes.

Acknowledgments

I thank the LOCNET members for their interest as well as for their numerous and stimulating questions. I have tried to answer some of them above. For several of the many remaining ones: work is still in progress...

References

1. P. Kraulis, *Appl. Cryst.* **24**, 946 (1991).
2. D. Przybylski and B. Rost, *Proteins* **46**, 197 (2002).
3. G. Trinquier and Y. H. Sanejouand, *Protein Engineering* **11**, 153 (1998).
4. P. Markiewicz, L. Kleina, C. Cruz, S. Ehret and J. Miller, *J. Mol. Biol.* **240**, 421 (1994).
5. A. Poupon and J. Mornon, *Proteins* **33**, 329 (1998).
6. R. Milton, S. Milton and S. Kent, *Science* **256**, 1445 (1992).
7. H. Taketomi, Y. Ueda and N. Go, *Int. J. Pept. Prot. Res.* **7**, 445 (1975).
8. N. Go and H. Taketomi, *Proc. Natl. Acad. Sci. USA* **75**, 559 (1978).
9. N. Go, *Adv. Biophys.* **18**, 149 (1984).
10. A. Irback and E. Sandelin, *J. Chem. Phys.* **108**, 2245 (1998).
11. H. Li, C. Tang and N. Wingreen, *Phys. Rev. Lett.* **79**, 765 (1997).
12. H. Chan and K. A. Dill, *Macromolecules* **22**, 4559 (1989).
13. E. Shakhnovich and A. Gutin, *J. Chem. Phys.* **93**, 5967 (1990).
14. A. Sali, E. Shakhnovich and M. Karplus, *Nature* **369**, 248 (1994).
15. H. Li, R. Helling, C. Tang and N. Wingreen, *Science* **273**, 666 (1996).
16. M. Shemyakin, Y. Ovchinnikov and V. Ivanov, *Angew. Chem. Int. Ed. Engl.* **8**, 492 (1968).
17. M. Chorev and M. Goodman, *Trends Biotechnol.* **13**, 438 (1995).
18. G. Trinquier and Y. H. Sanejouand, *Phys. Rev.* **E59**, 942 (1999).
19. E. Koonin, Y. Wolf and G. Karev, *Nature* **420**, 218 (1992).
20. H. Li, C. Tang and N. Wingreen, *Proc. Natl. Acad. Sci. USA* **95**, 4987 (1998).
21. M. Ejtehadi, N. Hamedani, H. Seyed-Allaei, V. Shahrezaei and M. Yahyanejad, *Phys. Rev.* **E57**, 3298 (1998).
22. E. Bornberg-Bauer, *Biophys. J.* **73**, 2393 (1997).
23. K. F. Lau and K. A. Dill, *Proc. Natl. Acad. Sci. USA* **87**, 638 (1990).
24. K. A. Dill and H. Chan, *Nat. Struct. Biol.* **4**, 10 (1997).
25. D. Wales and J. Doye, *J. Phys. Chem.* **A101**, 5111 (1997).
26. B. R. Brooks, R. Bruccoleri, B. Olafson, D. States, S. Swaminathan, and M. Karplus, *J. Comp. Chem.* **4**, 187 (1983).
27. M. Taylor, *J. Chem. Phys.* **114**, 6472 (2001).
28. T. Frisch and A. Verga, *Phys. Rev.* **E66**, 041807 (2002).
29. T. Frisch and A. Verga, *Phys. Rev.* **E65**, 041801 (2002).
30. F. Calvo, J. Doye and D. Wales, *J. Chem. Phys.* **116**, 2642 (2002).
31. V. S. Pande, I. Baker, J. Chapman, S. P. Elmer, S. Khaliq, S. M. Larson, Y. M. Rhee, M. R. Shirts, C. D. Snow, E. J. Sorin and B. Zagrovic, *Biopolymers* **68**, 91 (2003).
32. P. Derreumaux, *Phys. Rev. Lett.* **85**, 206 (2000).
33. J. C. Kendrew, R. E. Dickerson, R. G. Strandberg, R. G. Hart, D. R. Davies, D. C. Phillips and V. C. Shore, *Nature* **185**, 422 (1960).
34. P. Zavodszky, J. Kardos, A. Svingor and G. A. Petsko, *Proc. Natl. Acad. Sci. USA* **95**, 7406 (1998).
35. G. Wiegand and S. Remington, *Annu. Rev. Biophys. Biophys. Chem.* **15**, 97 (1986).

36. W. S. Bennet and T. Steitz, *J. Mol. Biol.* **140**, 210 (1980).
37. H. Faber and B. Matthews, *Nature* **348**, 263 (1990).
38. Y. Zhang, M. N. Waxham, A. L. Tsai and J. A. Putkey, *Biophys. J.* **78**, 412P (2000).
39. H. Goldstein, *Classical Mechanics*, Addison-Wesley, Reading, MA, 1950.
40. R. Levy, D. Perahia and M. Karplus, *Proc. Natl. Acad. Sci. USA* **79**, 1346 (1982).
41. S. Swaminathan, T. Ichiye, W. Van Gunsteren and M. Karplus, *Biochemistry* **21**, 5230 (1982).
42. T. Noguti and N. Go, *Nature* **296**, 776 (1982).
43. Y. Seno and N. Go, *J. Mol. Biol.* **216**, 95 (1990).
44. L. Mouawad and D. Perahia, *Biopolymers* **33**, 569 (1993).
45. O. Marques and Y.-H. Sanejouand, *Proteins* **23**, 557 (1995).
46. P. Durand, G. Trinquier and Y. H. Sanejouand, *Biopolymers* **34**, 759 (1994).
47. F. Tama, F.-X. Gadea, O. Marques and Y.-H. Sanejouand, *Proteins* **41**, 1 (2000).
48. P. Faure, A. Micu, D. Perahia, J. Doucet, J. C. Smith and J.P. Benoit, *Nat. Struct. Biol.* **1**, 124 (1994).
49. W. Harrison, *Biopolymers* **23**, 2943 (1984).
50. B. R. Brooks and M. Karplus, *Proc. Natl. Acad. Sci. USA* **82**, 4995 (1985).
51. J. Gibrat and N. Go, *Proteins* **8**, 258 (1990).
52. D. Ben-Avraham, *Phys. Rev.* **B47**, 14559 (1993).
53. M. Tirion, *Phys. Rev. Lett.* **77**, 1905 (1996).
54. K. Hinsen, *Proteins* **33**, 417 (1998).
55. I. Bahar, A. R. Atilgan and B. Erman, *Folding & Design* **2**, 173 (1997).
56. F. Tama and Y.-H. Sanejouand, *Protein Engineering* **14**, 1 (2001).
57. M. Delarue and Y.-H. Sanejouand, *J. Mol. Biol.* **320**, 1011 (2002).
58. W. G. Krebs, V. Alexandrov, C. A. Wilson, N. Echols, H. Yu and M. Gerstein, *Proteins* **48**, 682 (2002).
59. O. M. Becker and M. Karplus, *Phys. Rev. Letters* **70**, 3514 (1993).
60. J. A. McCammon, B. R. Gelin and M. Karplus, *Nature* **267**, 585 (1975).
61. S. Sporlein, H. Carstens, H. Satzger, C. Renner, R. Behrendt, L. Moroder, P. Tavan, W. Zinth and J. Wachtveitl, *Proc. Natl. Acad. Sci. USA* **99**, 7998 (2002).

CHAPTER 7

NONLINEAR VIBRATIONAL SPECTROSCOPY: A METHOD TO STUDY VIBRATIONAL SELF-TRAPPING

Peter Hamm and Julian Edler

Universität Zürich, Physikalisch Chemisches Institut, Winterthurer Strasse 190, CH-8057 Zürich, Switzerland

Emails: phamm@pci.unizh.ch & j.edler@pci.unizh.ch

We review the capability of nonlinear vibrational spectroscopy to study vibrational self-trapping in hydrogen-bonded molecular crystals. For that purpose, the two relevant coupling mechanisms, excitonic coupling and nonlinear exciton-phonon coupling, are first introduced separately using appropriately chosen molecular systems as examples. Both coupling mechanisms are subsequently combined, yielding vibrational self-trapping. The experiments unambiguously prove that both the N-H and the C=O band of crystalline acetanilide (ACN), a model system for proteins, show vibrational self-trapping. The C=O band is self-trapped only at low enough temperature, while thermally induced disorder destroys the mechanism at room temperature. The binding energy of the N-H band, on the other hand, is considerably larger and self-trapping survives thermal fluctuations even at room temperature.

1. Introduction: The Story of Davidov's Soliton

Biological macromolecules, such as proteins and DNA, often form 1-dimensional quasi-crystals with approximate translational symmetry. The most important of these structure motifs in proteins are α-helices (see Fig. 1) and β-sheets. These structures are stabilized by hydrogen bonds, which are weak chemical bonds that are, however, of fundamental importance for bio-macromolecules. The hydrogen bond has the unique property that its potential energy surface contains an anharmonic term which strongly couples the N−H and/or C=O stretching vibration of the hydrogen bond to a lattice phonon (i.e. a N−H$\cdots$C=O stretching motion). This observation motivated Davidov to speculate about nonlinear collec-

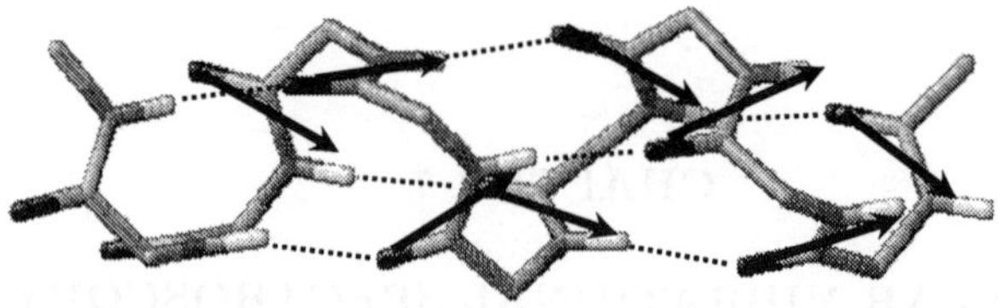

Fig. 1. An α-helix. The dotted lines depict the hydrogen bonds stabilizing the structure and the arrows the transition dipoles of the C=O groups.

tive phenomena in bio-macromolecules.[1] Dipole-dipole interaction between adjacent C=O and N−H vibrations tends to delocalize the excitation along the crystal, forming a vibrational exciton (vibron). Coupling of the vibron to lattice-deformation modes, mediated through the nonlinearity of the hydrogen bonds, self-localizes the excitation. The hydrogen bond is getting stronger after excitation of the high frequency vibration, leading to a contraction of the peptide backbone. As a result, solitons and/or polarons may be formed.

Polaron formation in protein model systems has been investigated extensively in the mid-80's. Most of the studies focused on acetanilide (C_6H_5-NHCO-CH_3, ACN), a molecular crystal consisting of quasi one-dimensional chains of hydrogen bonded peptide groups with structural properties that are similar to those of α-helices. The vibrational spectra of this crystal showed "anomalies", which have been attributed to vibrational self-trapping.[2–12] However, self-trapping was observed through an indirect effect in this works, i.e. through the temperature dependence of the linear absorption spectrum of the C=O mode. The temperature dependence is considered to be a signature of anharmonicity of the molecular potential energy surface (the linear absorption spectrum of an entirely harmonic molecular system would not change with temperature). We recently started to use an alternative, more direct approach to investigate anharmonicity, namely nonlinear vibrational spectroscopy.[13–15] The nonlinear vibrational response of a harmonic system vanishes exactly and is exclusively sensitive to the anharmonic part of the potential energy surface. Anharmonicity at the same time also gives rise to nonlinear dynamics. Hence, nonlinear spectroscopy is extremely valuable to study nonlinear phenomena such as vibrational self-trapping.

In this chapter, we discuss in detail what can be learned about vibrational self-trapping in peptide model crystals with the help of femtosecond nonlinear pump-probe spectroscopy. The chapter is organized as follows: We first introduce, in very simple words, nonlinear vibrational spectroscopy and its relation to anharmonicity (Sec. 2.1). An intriguing feature of Davidov's idea is that both coupling mechanisms, the excitonic coupling and the exciton-phonon coupling, are well established and can be observed independently by direct experiment. We will discuss both coupling mechanisms separately by primary introducing them theoretically, and then demonstrating how they manifest themselves in nonlinear vibrational spectroscopic experiments (Sec. 3 and Sec. 4). Finally, we shall combine both mechanisms to form a vibrational polaron (Sec. 5).

2. Nonlinear Spectroscopy of Vibrational Modes

2.1. *Harmonic and Anharmonic Potential Energy Surfaces*

One can always Taylor-expand the electronic potential energy surface along certain nuclear coordinates

$$V(q_1, q_2) = V_0 + \frac{\partial V}{\partial q_1} q_1 + \frac{\partial V}{\partial q_2} q_2 + \frac{1}{2}\frac{\partial^2 V}{\partial q_1^2} q_1^2 + \frac{\partial^2 V}{\partial q_1 q_2} q_1 q_2 + \frac{1}{2}\frac{\partial^2 V}{\partial q_2^2} q_2^2 + \dots . \tag{1}$$

Here, we restrict ourselves to two coordinates q_1 and q_2 for the sake of simplicity. In general, there will be, of course, $3N-6$ such coordinates, where N is the number of atoms in the molecule. The Taylor-expansion has been truncated after second order, which is called the harmonic approximation. Without loss of generality, we can get rid of the term V_0 by choosing the arbitrary energy zero point properly, we can get rid of the terms $(\partial V/\partial q_1)q_1$ and $(\partial V/\partial q_2)q_2$ by choosing the origin of the coordinate system to be a local minimum of the potential energy surface, and we can get rid of the mixed term $(\partial^2 V/\partial q_1 q_2)q_1 q_2$ by performing a normal mode transformation. Hence, without loss of generality, Eq. 1 can be reduced to:

$$V(q_1, q_2) = \frac{1}{2}\frac{\partial^2 V}{\partial q_1^2} q_1^2 + \frac{1}{2}\frac{\partial^2 V}{\partial q_2^2} q_2^2 + \dots , \tag{2}$$

or, using dimensionless coordinates:

$$V(q_1, q_2) = \frac{1}{2}\hbar\omega_1 q_1^2 + \frac{1}{2}\hbar\omega_2 q_2^2 + \dots .$$

The Hamiltonian is

$$\hat{H} = \frac{1}{2}\hbar\omega_1(\hat{p}_1^2 + \hat{q}_1^2) + \frac{1}{2}\hbar\omega_2(\hat{p}_2^2 + \hat{q}_2^2) + \dots ,$$
$$= \hbar\omega_1\left(b_1^\dagger b_1 + \frac{1}{2}\right) + \hbar\omega_2\left(b_2^\dagger b_2 + \frac{1}{2}\right) + \dots . \tag{4}$$

with $b^\dagger$ and b being the creation and annihilation operators of the harmonic oscillator:

$$b = \frac{1}{\sqrt{2}}(\hat{q} + i\hat{p})$$
$$b^\dagger = \frac{1}{\sqrt{2}}(\hat{q} - i\hat{p}) \tag{5}$$

In the harmonic approximation, the various normal modes *decouple completely.* The harmonic approximation (i.e. the normal mode picture) is the lowest level description of vibrational spectroscopy. In most cases, the harmonic approximation is extremely good, justifying its wide use in interpreting vibrational spectroscopy. As a criteria of the accuracy of the harmonic approximation one can define the parameter

$$\delta \equiv \frac{\epsilon_{01} - \epsilon_{12}}{\epsilon_{01}} , \tag{6}$$

where ϵ_{01} is the energy gap between ground and first excited state of one normal mode vibrator and ϵ_{12} that between first and second excited state. The energy levels of a harmonic oscillator are equidistant and we obtain $\delta = 0$. A typical value is $\delta = 1\%$ for e.g. a C=O stretching vibration.

Nonlinear phenomena, such as vibrational self-trapping, originate from mixed higher order terms in the expansion Eq. 3, such as:

$$V(q_1, q_2) = \frac{1}{2}\hbar\omega_1 q_1^2 + \frac{1}{2}\hbar\omega_2 q_2^2 + f_{122} q_1 q_2^2 + f_{112} q_1^2 q_2 + \dots , \tag{7}$$

with e.g.

$$f_{122} = \frac{1}{2}\frac{\partial^3 V}{\partial q_1 q_2^2} . \tag{8}$$

Unlike the bilinear term $(\partial^2 V/\partial q_1 q_2) q_1 q_2$, these higher order mixed terms remain after the normal mode transformation, and couple various normal modes to each other. These anharmonic couplings are often very weak but exist between a large number of normal modes, in which case we observe unspecific but potentially very efficient energy dissipation within the molecule. In some cases, however, these anharmonic couplings are specific and exceptionally strong within a small subset of normal modes. These are the cases we are interested in in nonlinear science.

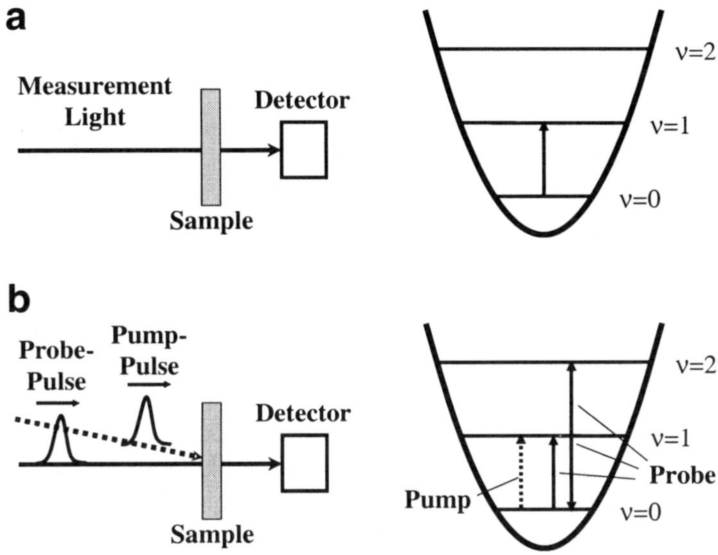

Fig. 2. (a) Linear (absorption) spectroscopy of a harmonic oscillator. (b) Nonlinear (pump-probe) spectroscopy of a harmonic oscillator.

2.2. *Linear and Nonlinear Spectroscopy*

Figure 2a depicts the *linear* absorption response of a harmonic oscillator. In simple words, the measurement light is interacting with the system, probing the possibility to absorb light with a frequency equivalent to the 0–1 transition of the oscillator (see Appendix for a more thorough discussion). The strength of the absorption band relates to the transition dipole matrix element

$$\mu_{01}^2 \equiv \langle 0|\mu|1\rangle^2 = \left(\frac{\partial\mu}{\partial x}\right)^2 \langle 0|x|1\rangle^2 \,. \tag{9}$$

The linear response of a harmonic oscillator is, of course, *non-zero*, provided that the transition dipole moment $\partial\mu/\partial x$ is non-zero.

Figure 2b depicts the *nonlinear* pump-probe response of a harmonic oscillator. In such an experiment, the sample is interacting with two short pulses: the pump and the probe pulse. The pump pulse excites the system from the ground state $\nu = 0$ to the first excited state $\nu = 1$, and the probe pulse measures the change of absorption as a result of that excitation. The pump-probe response consist of three contributions, depicted in Fig. 2b as solid arrows: (i) the bleach, i.e. the loss of 0–1 absorption due to the depletion of the ground state, (ii) the 1–0 stimulated emission, and (iii) the 1–2 excited state absorption. Bleach and stimulated emission both are negative and appear at the ω_{01} frequency with an intensity proportional to $\langle 0|x|1\rangle^2$, while excited state absorption is positive and appears at the ω_{12} frequency with an intensity proportional to $\langle 1|x|2\rangle^2$. For a harmonic oscillator, we find $\omega_{01} = \omega_{12}$ (the energy levels are equidistant) and $2\langle 0|x|1\rangle^2 = \langle 1|x|2\rangle^2$. Hence, in the case of a harmonic oscillator, the three contributions to the signal cancel exactly and the pump-probe response vanishes. This statement, which is pictured here for the simple example of $3^{\rm rd}$-order pump-probe spectroscopy on a single harmonic oscillator, can be phrased much more generally: The nonlinear optical response of a system of harmonic normal modes *vanishes exactly.* This is also true for more sophisticated forms of nonlinear spectroscopy, which might involve more than two pulses, such as photon echo experiments or even higher order experiments.

Combining that statement with the conclusion of the previous paragraph (Sec. 2.1), we find a very intriguing relationship: A system of harmonic oscillators decouples completely, and at the same time, the nonlinear optical response of such a system vanishes exactly. When we turn this argument around, it reads: Nonlinear phenomena originate from the anharmonic part of the potential energy surface (Eq. 7), and at the same time, nonlinear vibrational spectroscopy is exclusively sensitive to the anharmonic part of the potential energy surface. This is why nonlinear spectroscopy should be extremely valuable to study nonlinear phenomena such as vibrational self-trapping. The linear spectroscopic response is non-zero in any case so that indirect manifestations of nonlinearity (such as the temperature dependence of the linear absorption spectrum) have to be investigated in order to study nonlinear phenomena.

However, two remarks should be made at this point: (i) The relationship between a non-zero nonlinear response and the anharmonicity of the potential energy surface is non-trivial and has to be established in a case-to-case manner. (ii) There are, of course, terms of anharmonicity that lead to a non-zero nonlinear spectroscopic response but not necessarily to interesting nonlinear science. Most trivial example are terms of the form $f_{111}q_1^3$ in the expansion Eq. 7, which lead to intrinsic anharmonicity of the vibrator (i.e. a shift of the ω_{12} frequency with respect to ω_{01}), but which do not couple the various normal modes.

3. Proteins and Vibrational Excitons

3.1. *Theoretical Background*

Inspection of an α-helix (Figure 1) shows that the dipoles of the individual C=O groups are all aligned in the same direction and add up to a huge total dipole moment. Vibration of one of these C=O groups is coupled to that of all other C=O groups through dipole-dipole interaction. Following the terminology in electronic spectroscopy, the various C=O groups form a vibrational exciton, also called a vibron. Reducing the system to a truly one-dimensional chain, the Hamiltonian describing such a vibron reads:

$$\hat{H} = \sum_i \hbar\Omega \left(B_i^\dagger B_i + \frac{1}{2} \right) + \sum_i \beta \left(B_i^\dagger B_{i+1} + B_i B_{i+1}^\dagger \right) , \qquad (10)$$

where i runs over all sites in the chain, Ω is the vibrational frequency of the C=O modes, β the coupling (which decays very quickly with distance, so that we in general consider only nearest neighbor coupling), and $B_i^\dagger$ and B_i are the creation and annihilation operators of site excitations. Depending on distance and relative orientation, the coupling β can be of the order of about 10 cm^{-1} and hence, can lead to appreciable exciton delocalization. The exciton picture has been used extensively to investigate regular secondary structure motifs in proteins.[16,17] The characteristic shift of the amide I band (i.e. essentially the C=O band) of α-helices relative to that in β-sheets is attributed to the coupling pattern which directly reflects the geometry of the peptide backbone.

Since the coupling in the Hamiltonian (10) is bilinear, the Hamiltonian is in fact harmonic, and one could, in principle, reformulate the problem in a harmonic eigenstate basis by employing a normal mode transformation. Hence, along the lines of the discussion of Sec. 2.1, the Hamiltonian (10) is not sufficient to describe nonlinear spectroscopy. However, when we take into account the intrinsic (on-site) anharmonicity of the C=O groups, we obtain a Hamiltonian which also allows us to describe nonlinear spectroscopy:[18]

$$\hat{H} = \sum_i \left[\hbar\Omega \left(B_i^\dagger B_i + \frac{1}{2} \right) - \Delta B_i^\dagger B_i^\dagger B_i B_i \right] + \sum_i \beta \left(B_i^\dagger B_{i+1} + B_i B_{i+1}^\dagger \right) . \qquad (11)$$

A special form of nonlinear vibrational spectroscopy, 2D-IR spectroscopy, has recently been introduced, which allows one to directly observe the excitonic coupling in polypeptides.[18] Figure 3a shows a prototype

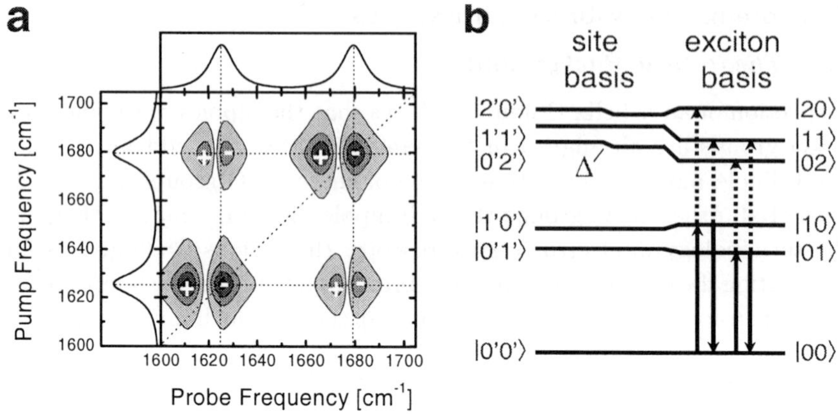

Fig. 3. (a) A prototype 2D-IR spectrum together with (b) the level scheme giving rise to the 2D response. Bleach and stimulated emission (solid arrows in the level scheme) give rise to negative peaks in the 2D spectrum, while excited state absorption (dotted arrows) give rise to positive peaks in the 2D spectrum.

2D-IR spectrum. In such an experiment, a spectrally narrow pump pulse selectively excites individual excitonic states and the broad band probe pulse probes the absorption to the two-excitonic states, the stimulated emission back to the ground states and the bleach of the ground state. The response is plotted as a function of pump and probe frequency. The key to understand the 2D-IR spectrum is the level scheme (Fig. 3b), which originates from direct diagonalization of the Hamiltonian (11). In 3rd-order spectroscopy, only states up to double excited states are reached. Since the Hamiltonian (11) conserves the number of excitations (each term contains the same number of creation and annihilation operators), it is block-diagonal, and it is sufficient to restrict the basis to the ground state $\{|0,0\rangle\}$, the two single excited states $\{|1,0\rangle, |0,1\rangle\}$ and the three possibilities of double excitations $\{|2,0\rangle, |1,1\rangle, |0,2\rangle\}$. In order to construct a 2D-IR spectrum, we calculate the population probabilities of the one-excitonic states, depending on the pump frequency, and subsequently calculate the up- and downward transitions seen by the probe-pulse according to Fig. 3. Each transition is weighted by the population probability of the state prepared by the pump pulse and by its transition strength.

The 2D-IR spectrum separates into a variety of signals, which can be classified as diagonal and cross peaks. Each peak in the 2D-IR spectrum Fig. 3 (left) is related to one arrow in the level scheme in Fig. 3 (right).[19]

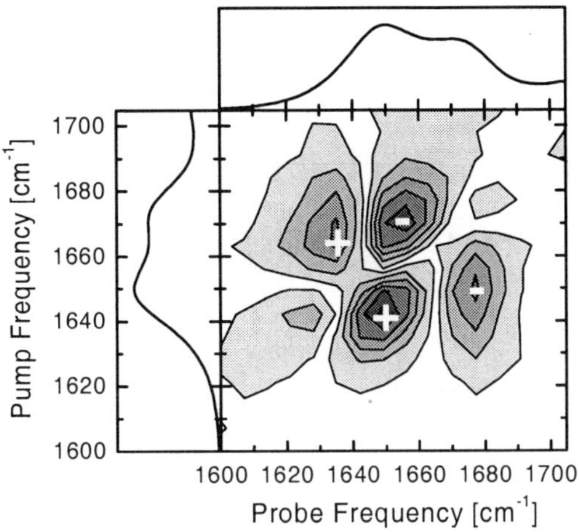

Fig. 4. 2D-IR spectrum of the amide I band of a small peptide (trialanine) dissolved in water. Adapted from Ref. 20.

The diagonal peaks, each consisting of a pair of a negative and a positive band, reflect the anharmonic response of the excitonic state, while the cross peaks are a direct manifestation of the excitonic coupling term β.[19]

3.2. *Experimental Observation*

Figure 4 shows an experimental 2D-IR spectrum of the amide I band of a small peptide fragment (trialanine) dissolved in water. The peptide contains two C=O groups, and hence is a dimer. The diagonal peaks are suppressed in that spectrum by making use of the polarization dependence of the 2D-IR response (see Ref. 20 for details). Both C=O groups are coupled to each other through excitonic interaction, giving rise to the cross peaks observed in the 2D spectrum. Since excitonic coupling depends on the geometry of the peptide backbone, we were able to deduce the structure of the molecule by analyzing its 2D-IR spectrum.[20] A similar coupling pattern has been found for a small 21 amino acid α-helix dissolved in water.[21] It has been estimated that the excitons are delocalized over almost two helix turns, proving that vibrational excitons exists even in the presence of disorder and the strong perturbation from the surrounding solvent.

2D-IR spectroscopy has been reviewed recently[19,22,23] and is used extensively to study structure and dynamics of small peptides.[20,21,24–27] 2D-IR spectroscopy is a complementary technique to 2D-NMR spectroscopy. Its

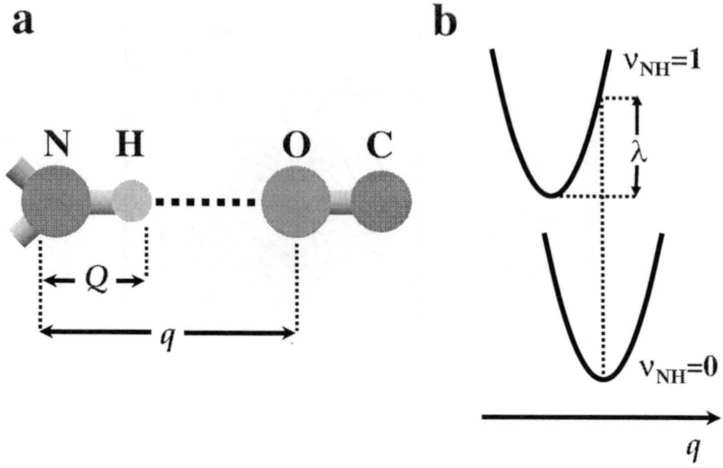

Fig. 5. (a) A prototype N–H···C=O hydrogen bond. (b) Potential energy surfaces of the $\nu_{NH} = 0$ and the $\nu_{NH} = 1$ level.

strength is the high time resolution which allows us to discriminate quickly interchanging conformational substates,[28,29] to observe ultrafast conformational fluctuations[30] and to follow fast conformational transitions of small peptides.[31]

4. Hydrogen Bonds and Anharmonicity

4.1. *Theoretical Background*

A hydrogen bond is mediated by a proton sitting in-between two negatively charged atoms, the more negative of which is called the hydrogen bond donor and the other the hydrogen bond acceptor. Figure 5a shows a prototype hydrogen bond between the N–H and the C=O group of two peptide units in an α-helix. Hydrogen bonding leads to a dramatic distortion of the potential energy surfaces that determine the nuclei positions, giving rise to strongly anharmonic or even double-well potentials. Most of our knowledge about hydrogen bond potentials stems from vibrational spectroscopy.[32–34] The absorption band of the high-frequency vibration of the donor-proton bond (i.e. the N–H bond in the example considered here) changes considerably upon formation of a hydrogen bond. The most evident effects are a strong red-shift with respect to the free group, an intensity increase and band broadening, often accompanied by a peculiar band shape with rich substructure. These effects can be reasonably explained using a two-dimensional model of the electronic ground state potential energy surface as a function of the high-frequency N–H stretching coordinate Q and the low-frequency hydrogen bond coordinate q (see Fig 5a). A Taylor expansion of that potential energy surface yields[33,34]

$$V(Q,q) = \frac{1}{2}\hbar\Omega Q^2 + \frac{1}{2}\hbar\omega q^2 + \frac{\chi}{2}Q^2 q + \ldots, \tag{12}$$

where the first two terms represent the harmonic normal modes of the high frequency N–H stretching vibration and the low frequency hydrogen bond vibration with oscillation frequencies Ω and ω, respectively. One of the striking features of hydrogen bonds is an extraordinarily large anharmonicity, described by the third term in the expansion (12). This anharmonicity leads to appreciable nonlinear behavior, a property which makes hydrogen-bonded crystals an interesting object to study in nonlinear science. The same expansion can be obtained for the C=O mode and the hydrogen bond vibration, and we will refer to either the N–H or the C=O mode as the high-frequency mode. The nonlinear coupling χ of the N–H mode is $\approx 5 - 10$ times larger than that of the C=O mode.

Under quantization, Eq. 12 translates into (see Eq. 5):

$$\hat{H} = \hbar\Omega\left(B^\dagger B + \frac{1}{2}\right) + \hbar\omega\left(b^\dagger b + \frac{1}{2}\right) + \frac{\chi}{2\sqrt{2}}B^\dagger B\left(b^\dagger + b\right) + \ldots. \tag{13}$$

where $B^\dagger$ (B) and $b^\dagger$ (b) are the creation (annihilation) operators of the high frequency mode and hydrogen bond mode, respectively. Terms, which do not retain the number of high frequency excitations (i.e. non-resonant terms), are discarded. The Hamiltonian (13) can be diagonalized analytically.[12] However, a much more intuitive picture is obtained when taking into account the more than one order of magnitude difference of the frequencies of both modes, allowing one to introduce an adiabatic separation of time scales. This is done in exactly the same way as the Born-Oppenheimer approximation separates off the fast motion of the electrons from the slow motion of the nuclei[k]. In the case of a hydrogen bond, the motion of the high frequency vibration adiabatically adapts to the position of the coordinate q of the hydrogen bond. When the coordinate q is held fixed, one can recast Eq. 12 in the form:

$$V(Q,q) = \frac{1}{2}\hbar\Omega_{eff}Q^2 + \frac{1}{2}\hbar\omega q^2 + \ldots, \tag{14}$$

[k]In fact it can be shown that with the coupling term in the Hamiltonian (13), the adiabatic approximation yields the same result as the exact, analytic solution.

with

$$\hbar\Omega_{eff} = \hbar\Omega + \chi q\,. \tag{15}$$

The vibration frequency of the fast mode mode, and hence its excitation energy, varies linearly with the hydrogen bond distance Q. Such a linear dependence is observed experimentally for a large variety of hydrogen bonded crystals,[32] which allows one to estimate the nonlinear coupling constant χ.

Starting from Eqs. 14 and 15, one obtains potential energy surfaces (see Fig. 5b):

$$E_\nu(q) = \hbar\Omega_{eff}\left(\nu + \frac{1}{2}\right) + \frac{1}{2}\hbar\omega q^2 \tag{16}$$

Each potential energy surface corresponds to a vibrational excitation level $\nu = 0, 1, \ldots$ of the high frequency mode, and describes the total energy of the system as a function of the hydrogen bond coordinate q. Figure 5b is the displaced oscillator picture which is well known from electronic Franck-Condon transitions. However, one should keep in mind that all this is happening on the *electronic ground state* potential surface! It is the nonlinear coupling term χ which gives rise to the displacements of the potential energy surfaces. The reorganization (binding) energy λ, i.e. the energy the system gains by relaxation towards the bottom of excited state potential energy surface after a vertical "Franck-Condon"-like excitation (see Fig. 5b), is given by:

$$\lambda = \frac{\chi^2}{2\hbar\omega}\,. \tag{17}$$

The minimum of the $\nu = 1$ potential energy surface is shifted towards smaller q, i.e. the hydrogen bond is getting stronger after excitation of the high frequency vibration (see Fig. 5b).

4.2. *Experimental Observation*

Figure 6 shows the pump-probe response of hydroxy-deuterated 2-(2'-hydroxyphenyl)benzothiazole (HBT-d), a molecule with an intramolecular OD$\cdots$O hydrogen bond. In that experiment, the OD stretching mode of the hydrogen bond has been resonantly excited with a femtosecond IR pulse. The pump pulse was shorter than the oscillation period of the low-frequency hydrogen bond mode coupled to the OD stretching band (a ring deformation band, see Fig. 6, inset). As a result of excitation of the OD band, the hydrogen bond gets stronger, is contracted and a coherent vibrational wave

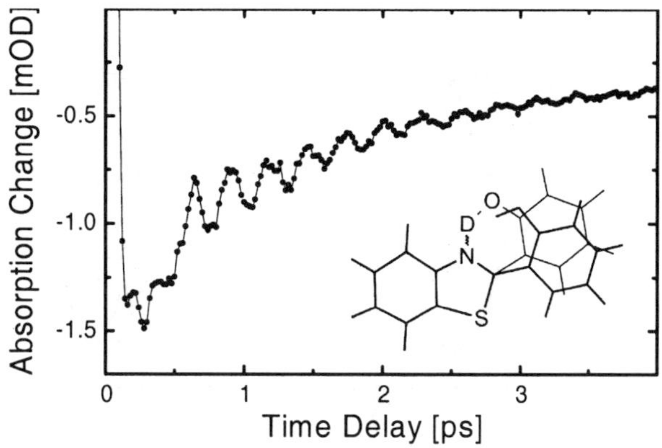

Fig. 6. Femtosecond IR pump-probe response of hydroxy-deuterated 2-(2'-hydroxyphenyl)benzothiazole (HBT-d, see inset), when impulsively exciting the OD stretching mode of the intramolecular hydrogen bond. Adapted from Ref. 35.

packet of low-frequency hydrogen bond mode is impulsively excited, which can be seen in the pump-probe signal as pronounced beatings (Fig. 6). Observation of these beatings is a direct consequence of the displacement of the $\nu_{OD} = 0$ and $\nu_{OD} = 1$ potential energy surfaces, i.e. it is a direct consequence of a nonlinear term χ coupling the high-frequency OD-stretching mode to the low-frequency hydrogen bond mode.

It is very well established in nonlinear pump-probe spectroscopy that there are two possibilities of exciting vibrational wave packets:[36] (a) a excited state wave packet and/or (b) a ground state wave packet (Fig. 7). In the first case, the pump pulse projects the ground state wave function onto the excited state, where it is non-stationary and starts to oscillate as a wave packet. In the second case, the first field interaction from the pump pulse projects the wave function up, where it propagates a little bit, and the second field interaction projects the wave packet down, where it is now displaced and continues to oscillate. One can also view the second possibility as an oscillating hole. A closer analysis of the experimental data suggested that the beatings in Fig. 6 originate from a ground state wave packet.[35] Similar coherent responses have been found also in other molecules with intramolecular hydrogen bonds.[37,38]

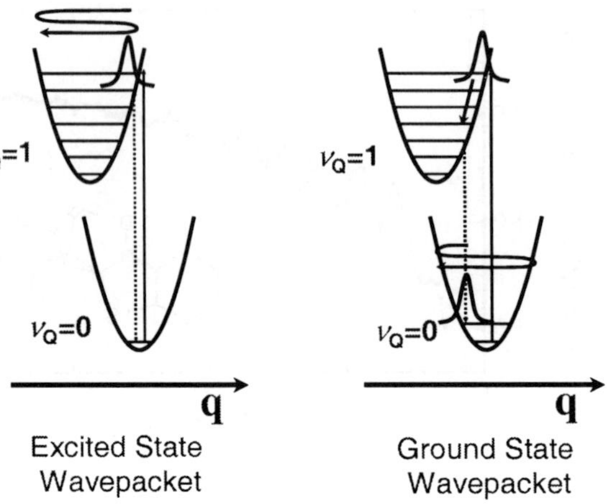

Fig. 7. The two possibilities of the pump pulse interacting with the system of displaced oscillators, generating either an excited state wave packet (left) or a ground state wave packet (right).

5. Vibrational Self-Trapping

5.1. *Theoretical Background*

In the previous two paragraphs, we have established two coupling mechanisms: excitonic coupling and nonlinear coupling to low frequency hydrogen bonds. In regular polypeptides, which are stabilized by hydrogen bonds (α-helices and β-sheets), both mechanisms coexist at the same time. This leads to the Holstein polaron Hamiltonian, which is the combination of Eq. (10) (exciton Hamiltonian, see Sec. 3) and Eq. (13) (anharmonic coupling, see Sec. 4):

$$\begin{aligned}
\hat{H} &= \hat{H}_{ex} + \hat{H}_{ph} + \hat{H}_{imt}\,, \\
\hat{H}_{ex} &= \sum_i \hbar\Omega\left(B_i^\dagger B_i + \frac{1}{2}\right) + \sum_i \beta\left(B_i^\dagger B_{i+1} + B_i B_{i+1}^\dagger\right) + \ldots\,, \\
\hat{H}_{ph} &= \sum_i \hbar\omega\left(b_i^\dagger b_i + \frac{1}{2}\right) + \ldots\,, \\
\hat{H}_{int} &= \sum_i \chi B_i^\dagger B_i \left(b_i^\dagger + b_i\right) + \ldots\,,
\end{aligned} \tag{18}$$

This Hamiltonian has been used as starting point to understand vibrational self-trapping in hydrogen bonded crystals. Most studies concentrated on

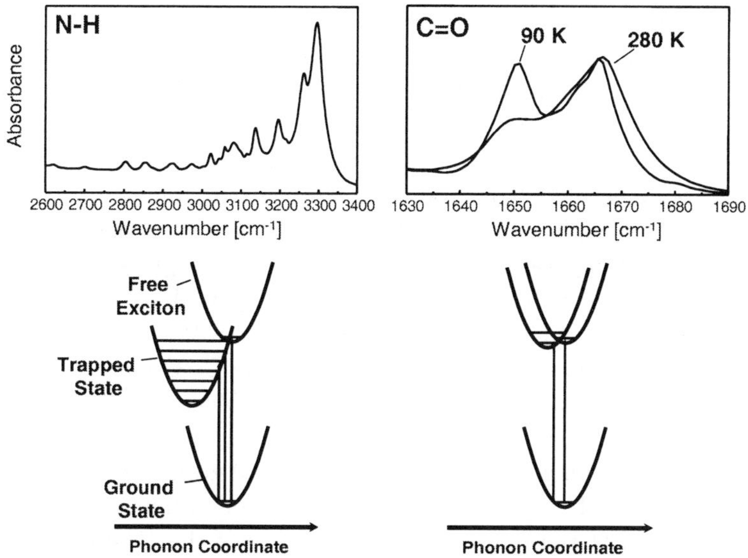

Fig. 8. Absorption spectra of the N–H and the C=O band of crystalline ACN and the potential energy surfaces giving rise to these spectra.

crystalline (ACN),[2,3,5,7,8,10,12] a hydrogen-bonded molecular crystal with a structure that resembles to some extent that of the hydrogen bonds in a α-helix. The Hamiltonian (18) can be diagonalized analytically in two limiting cases: (a) the nonlinear coupling χ vanishes or (b) the exciton coupling β vanishes. In the first case, we get back the free exciton (see Sec. 3), while in the second case the individual sites in the crystal are decoupled and the problem is the same as that of an isolated molecule (see Sec. 4). In intermediate cases (i.e. neither χ nor β can be neglected), the Holstein polaron Hamiltonian has to be diagonalized numerically, which can be done with high accuracy.[39]

5.2. *Experimental Observation*

Figure 8 shows the absorption spectra of crystalline ACN in the region of the N–H and the C=O stretching vibration. The C=O mode consists of a single peak at 1666 cm^{-1} at room temperature, which splits into two bands with a second "anomalous" peak at 1650 cm^{-1} at low temperatures.[2] The N–H stretching mode, on the other hand, is weakly temperature dependent and exhibits a main peak at 3295 cm^{-1} accompanied by an almost-regular sequence of 9 satellite peaks towards lower frequencies.[40,4] The origin of the splitting of the C=O band has been the topic of many theoretical and experimental studies, including IR absorption, Raman scattering and neutron scattering measurements.[2–12] Careri, Scott and coworkers attributed the anomalous band to vibrational self-trapping.[3,5,12] The theoretical treatments[3,7,8,12] typically neglect excitonic coupling β and describe the system as displaced oscillator (Fig. 8). In that case, one expects to observe at zero temperature a Franck-Condon like progression of absorption lines from the $\nu_Q = 0$ ground state to the displaced $\nu_Q = 1$ excited state with a spacing given by the phonon frequency ω. When the dimensionless displacement is small (which is the case for the C=O band in ACN), only the zero-phonon transition (i.e. the band at 1650 cm^{-1}) carries noticeable oscillator strength, while all other transitions are weak. At larger temperatures, the phonon ground state (in the $\nu_Q = 0$ ground state) is thermally depopulated, and the intensity of the zero-phonon line diminishes. It has been shown that the temperature dependence of the 1650 cm^{-1} band follows a $e^{-\gamma T^2}$ law, which is exactly what is expected for small nonlinear coupling χ.[3,7,8] This agreement is currently considered to be the strongest evidence for vibrational self-trapping in ACN.[12]

The N−H band of crystalline ACN has been studied much less.[4,40] Our current picture is that the coupling mechanism is essentially the same for both bands,[8] except for the amount of the binding energy λ of the self-trapped state. In the case of the C=O band, the binding energy is smaller than one phonon quantum, while it amounts to 9 phonon quanta in the case of the N−H band (see Fig. 8).

We recently started to perform nonlinear vibrational spectroscopy on crystalline ACN to better understand the nature of the peculiar vibrational states.[13–15] We have investigated both the C=O and the N−H band, which shall be discussed in the following:

C=O Band: Figure 9 shows two pump-probe spectra obtained by excitation of either of the two absorption lines of crystalline ACN at a temperature of 90 K. The pump-pulses were spectrally narrow in this experiment (spectral width ≈14 cm^{-1}) and selectively excited only one of the two bands. When resonantly pumping the "anomalous" band (1650 cm^{-1}) of ACN, the band bleaches (negative response) and a positive band emerges at 1644 cm^{-1}. When resonantly pumping the "normal" band (1666 cm^{-1}), on the other hand, hardly any bleach of the band itself is observed. Only the anomalous band responds with a signal which is similar in shape, but is slightly smaller than when pumping it directly. We have discussed the

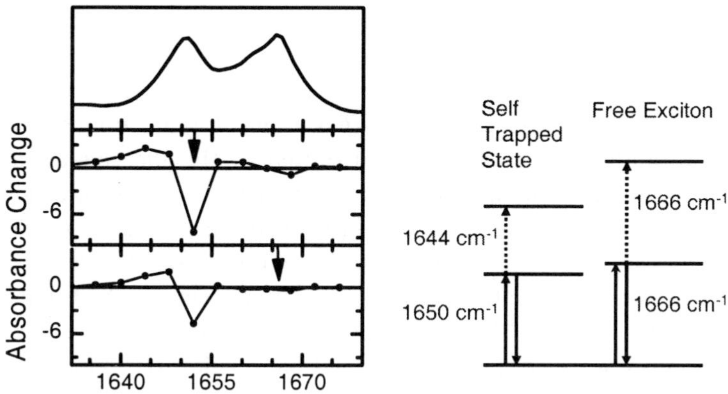

Fig. 9. Left: Linear absorption spectrum (top) and pump probe spectra of the C=O mode of crystalline ACN at 93 K for two different narrow band pump pulses chosen to be resonant with each of the absorption bands. The arrows mark the center frequency of the pump pulse. Right: Level scheme of the system, explaining the distinctively different response of both modes. Adapted from Ref. 14.

distinctly different response of both bands of ACN in detail in Ref. 14. In brief, we have shown that the effective anharmonicity $\Delta\epsilon_{kk}$ of an excitonic state in a one-dimensional chain scales like its participation ratio P_k:

$$P_k = \frac{\Delta\epsilon_{kk}}{\Delta} = \sum_i q_{ki}^4 , \tag{19}$$

where the q_{ki} are the expansion coefficients of the exciton $|k\rangle$ in a site basis $|i\rangle$:

$$|k\rangle = \sum_i q_{ki}|i\rangle , \tag{20}$$

and Δ is the intrinsic (on-site) anharmonicity of the amide-I vibrators (see Eq. 11). The participation ratio is a commonly used measure of delocalization of excitonic systems.[41–43] It is $P_k = 1$ for a completely localized state (localized for example due to disorder) and $P_k = 1/N$ for a perfectly delocalized state, where N is the size of the aggregate. Therefore, the effective anharmonicity $\Delta\epsilon_{kk}$ of a vibrational exciton is a direct measure of its degree of delocalization. This finding can be qualitatively explained in simple words: In a delocalized state, the oscillation amplitude of each individual site scales with $1/\sqrt{N}$, where N is the delocalization length. Hence, with increasing delocalization length N, each site is exploring a decreasingly smaller region of the potential energy surface, in which the harmonic approximation becomes increasingly more accurate. A completely delocalized state therefore will be effectively harmonic.

When resonantly pumping a vibrational state, one expects to observe a bleach and stimulated emission signal at its original frequency and an excited state absorption, which is red shifted by the effective anharmonicity of the state. The anomalous band shows such an anharmonic shift, and hence, is localized according to Eq. (19). The normal band, on the other hand, shows no nonlinear response which can only be the case when it is effectively a harmonic state with $\Delta\epsilon_{kk} = 0$ (it has been verified that the lifetime of the normal band is long enough to potentially observe a bleach[14]). Hence, according to Eq. (19), the normal band is largely delocalized, i.e. a free exciton described by the term H_{ex} in Eq. (18).

Figure 9, right, emphasizes the pump-probe response with the help of a level scheme. When resonantly pumping the self-trapped state, one observes a bleach and stimulated emission signal (down-going arrow) at the original frequency 1650 cm^{-1} and an excited state absorption signal (up-going arrow) that is slightly red shifted to 1644 cm^{-1} as a result of the effective anharmonicity of that state. When resonantly pumping the free exciton, on the other hand, the various signals cancel exactly since the frequency and transition strength of the excited state absorption (up-going arrow) exactly match that of stimulated emission and bleach (down-going arrow). This study clearly shows that the "anomalous" band is localized, while the "normal" band is a delocalized exciton at low temperatures. However, we have also found that the free exciton localizes due to thermally induced disorder at room temperature.[14]

N–H Band: Figure 10 shows the pump-probe response after impulsively exciting the N–H band of crystalline ACN with a spectrally broad, femtosecond laser pulse at room temperature. Just like in the case of an isolated molecule (HBT, see Sec. 4), quantum beats are observed, which are due to a coherent excitation of crystal phonon modes coupled to the N–H band. Fourier transformation of the signal in Fig. 10 reveals two frequencies: a dominant peak at 48 cm^{-1} and a weaker peak at 76 cm^{-1}. We have verified that the wave packet is a ground state wave packet.[13]

One can view such a ground state wave packet as a stimulated impulsive Raman process. Two bands can indeed be identified in the normal, non-resonant Raman spectrum of ACN,[3] whose frequencies match perfectly that of the beating frequencies observed in the pump-probe experiment. However, unlike the non-resonant Raman spectrum which is rather congested in the low frequency range, the impulsive Raman process in Fig. 10

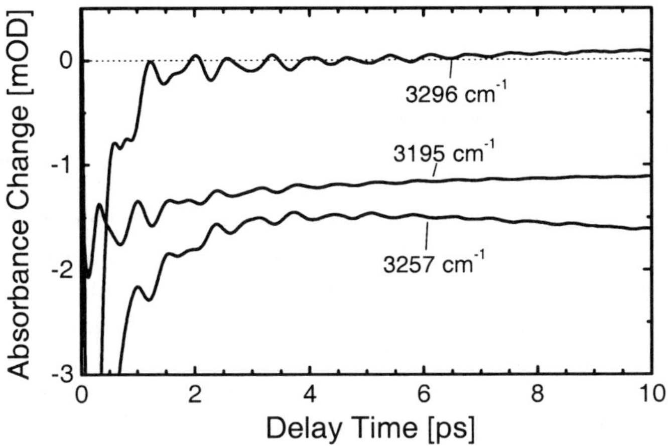

Fig. 10. Pump-probe response of the N−H band of crystalline ACN excited with an ultrashort laser pulse at various probe positions. Adapted from Ref. 13.

is a "resonant Raman effect" enhanced by the N-H absorption band as a consequence of anharmonic coupling to some of the phonons. In contrast to the common resonant Raman effect, which is resonantly enhanced due to coupling to an electronic state, we observe here resonant enhancement due to coupling to a vibrational transition. The number of "resonant Raman active" modes is considerably reduced to only two phonons at 48 cm^{-1} and 76 cm^{-1}. These are the phonons, that modulate the hydrogen bond distance and thus mediate self-trapping.

6. Conclusion and Outlook

Vibrational self-trapping in crystalline ACN has been studied extensively in the mid 80's through the temperature dependence of the C=O mode. Here, we apply a different approach, namely nonlinear spectroscopy. Nonlinear spectroscopy is specifically sensitive to the anharmonic part of the potential energy surface, which at the same time gives rise to nonlinear phenomena. The experiments unambiguously prove that both the N−H and the C=O band of crystalline acetanilide (ACN), a model system for proteins, show vibrational self-trapping. The C=O band is self-trapped only at low enough temperature, while thermally induced disorder destroys the mechanism at room temperature. The binding energy of the N-H band, on the other hand, is considerably larger and self-trapping survives thermal fluctuations even at room temperature.

However, it should also be mentioned that the theory of vibrational self-trapping of ACN is not entirely understood. For example, a numerical diagonalization of the Holstein polaron Hamiltonian (18) reveals the temperature dependence of the "anomalous" band (the self-trapped state) in the correct way, but at the same time fails to reproduce the existence of the "normal" band (the free exciton). We have introduced in Fig. 8 an additional potential energy surface in a purely phenomenological way – the free exciton – and our pump-probe experiments are in full agreement with that interpretation. However, the Holstein polaron Hamiltonian (18) does not justify that picture. It is believed that this failure is the result of the fact that crystalline ACN is not truly one-dimensional, and that inter-chain couplings between the quasi one-dimensional hydrogen bonded chains exist. It is well established in polaron theory that self-trapping is a barrier-less process in one-dimensional systems, but has to overcome a barrier in two or three dimensions.[44] In the latter case, the free exciton would be a metastable state and Fig. 8 would appear to be justified. In fact, we have directly observed the transition of the free exciton into the self-trapped state, which occurs on a 400 fs time scale.[13]

Vibrational self-trapping has been observed so far only in model crystals, but experimental evidence from polypeptides and proteins remained scarce.[45,46] The high degree of disorder in a real protein, as compared to the perfect symmetry of a crystal, and the perturbation of surrounding solvent molecules certainly tend to destroy vibrational self-trapping. We recently started to investigate stable model-α-helices using nonlinear vibrational spectroscopy. At least in the case of the N-H band, the trapping energy seems to be so high (in the order of 500 cm^{-1}, i.e. much larger than k_BT) that self-trapping should also be observable in the non-perfect environment of a real protein. In any case, our experiments make evident that the large nonlinearity of the hydrogen bonds, which stabilize regular protein structures, gives rise to a rich variety of collective phenomena that have yet to be explored.

Acknowledgments

Part of the work summarized in this chapter has been performed by Dorte Madsen, Jens Stenger (Fig. 6) and Sander Woutersen (Fig. 4) in the group of Thomas Elsaesser at the Max Born Institute, Berlin. Financial support from the 'Deutsche Forschungsgemeinschaft' and the 'Schweizer Nationalfond' is highly acknowledged.

Appendix: Feynman Diagram Description of Linear and Nonlinear Spectroscopy.

Linear and nonlinear spectroscopy is in general described in terms of a power expansion of the electric light field interacting with the molecular system,[36] leading to a very intuitive picture, which often is depicted in terms of so called double sided Feynman diagrams (see Fig. 11). In such diagrams, the vertical lines represent the time evolution of the *ket* and the *bra* of the molecular density matrix. Interaction of the laser light fields with the density matrix are depicted by arrows pointing towards or away from the system. At thermal equilibrium, one starts with the system in the ground state (density matrix element $|0\rangle\langle 0|$). In linear spectroscopy (Fig. 11a), the system interacts with the laser once, generating a $|1\rangle\langle 0|$ coherence matrix element of the density matrix, which then irradiates a first order polarization $P^{(1)}$ (depicted by the dotted arrow). In 3$^{\rm rd}$-order spectroscopy (Fig. 11b), the laser pulses interact with the system three times, again generating a coherence (either $|1\rangle\langle 0|$ or $|2\rangle\langle 1|$), which is irradiating the 3$^{\rm rd}$-order polarization $P^{(3)}$. In the case of pump-probe spectroscopy, the first two field interactions come from the pump pulse, while the third field interaction comes from the probe pulse (which is a result of the particular phase matching condition of the pump-probe-experiment, see Ref. 36 for details). In contrast to linear spectroscopy, however, there are many possibilities of the three light pulses interacting with the system. Under certain circumstances (validity of the rotating wave approximation, etc.) these can be reduced to three contributions depicted in Fig. 11b: the stimulated emission, bleach and excited state absorption.

The $n^{\rm th}$-order polarization $P^{(n)}$ depends on the $n^{\rm th}$-order response function $S^{(n)}$ and the electric field E interacting with the system:

$$\begin{aligned} P^{(1)}(t) &= \int_0^\infty dt_1 E(t-t_1) S^{(1)}(t_1) \\ P^{(3)}(t) &= \int_0^\infty dt_3 \int_0^\infty dt_3 \int_0^\infty dt_1 E(t-t_3)E(t-t_3-t_2)\cdot \\ &\qquad E(t-t_3-t_2-t_1)S^{(3)}(t_3,t_2,t_1) \end{aligned} \tag{21}$$

When using quasi-delta laser pulses, the integrals disappear and the $n^{\rm th}$-order polarization is the same as the $n^{\rm th}$-order response function. The response functions have simple forms if we assume a purely homogeneously

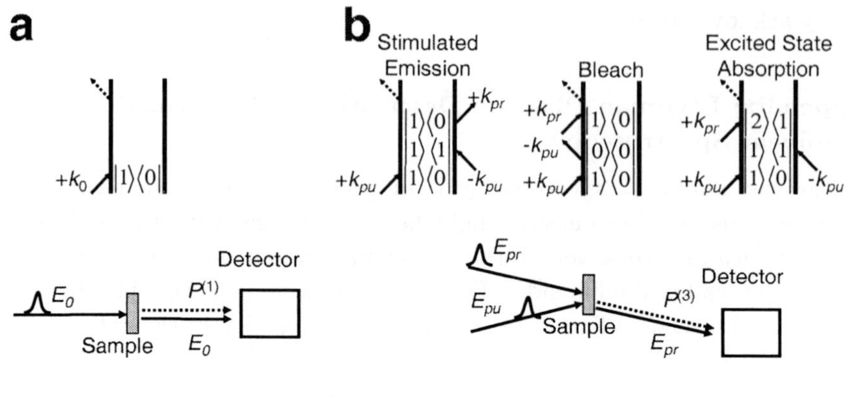

Fig. 11. (a) Feynman diagram (top) and principle of the experimental setup (bottom) for linear absorption spectroscopy. (b) Feynman diagram (top) and principle of the experimental setup (bottom) for nonlinear pump-probe spectroscopy.

broadened line (i.e. the so-called Bloch-limit). Then, we obtain for the linear polarization:

$$P^{(1)}(t) = -\frac{i}{\hbar} E_0 \mu_{01}^2 e^{-i\omega_{01}t} e^{-\gamma t} \tag{22}$$

where γ is the homogeneous dephasing rate. Under the same conditions, the 3rd-order polarization consists of three terms describing stimulated emission, bleach and excited state absorption:

$$P^{(3)}(t) = P_{SE}^{(3)}(t) + P_{Bl}^{(3)}(t) + P_{Ex}^{(3)}(t)\,, \tag{23}$$

with

$$\begin{aligned} P_{SE}^{(3)}(t) &= \frac{i}{\hbar^3} |E_{pu}|^2 E_{pr} \mu_{01}^4 e^{-i\omega_{01}t} e^{-\gamma t} e^{-T/T_1}\,, \\ P_{Bl}^{(3)}(t) &= \frac{i}{\hbar^3} |E_{pu}|^2 E_{pr} \mu_{01}^4 e^{-i\omega_{01}t} e^{-\gamma t} e^{-T/T_1}\,, \\ P_{Ex}^{(3)}(t) &= -\frac{i}{\hbar^3} |E_{pu}|^2 E_{pr} \mu_{01}^2 \mu_{12}^2 e^{-i\omega_{12}t} e^{-\gamma t} e^{-T/T_1}\,, \end{aligned} \tag{24}$$

where T_1 the vibrational relaxation time and T the pump-probe delay time. For a harmonic oscillator, we find $\omega_{01} = \omega_{12}$ and $2\mu_{01} = \mu_{12}$ and the three contributions to the 3rd-order polarization cancel exactly. Both in linear and in nonlinear pump-probe-spectroscopy, the detector measures the nth-order polarization heterodyne detected by the probe laser field:

$$\begin{aligned} I_{lin} &\propto |E_0 + iP^{(1)}|^2 = |E_0|^2 + 2\Im(E_0 \cdot P^{(1)}) + |P^{(1)}|^2\,, \\ I_{PumpProbe} &\propto |E_{pr} + iP^{(3)}|^2 = |E_{pr}|^2 + 2\Im(E_{pr} \cdot P^{(3)}) + |P^{(3)}|^2\,. \end{aligned} \tag{25}$$

where the third term is in general neglected.

Eq. (24) describes the 3rd-order response function of a three-level system. In 3rd-order spectroscopy, one is reaching only the first and the second excited state, and a three-level system is sufficient to describe the response of a single (harmonic or anharmonic) oscillator. Of course, in most of the cases studied in this paper, we deal with coupled oscillators, resulting in much more than three spectroscopic states (see e.g. Fig. 3). In that case, more than just three terms contribute to Eq. (24) (one for each arrow in the level scheme), but the principal idea remains the same. A more general discussion of nonlinear spectroscopy is given in Ref. 36.

References

1. A. S. Davidov, *Phys. Scr.* **20**, 387 (1979).
2. G. Careri, U. Buontempo, F. Carta, E. Gratton and A. C. Scott, *Phys. Rev. Lett.* **83**, 304 (1983).
3. G. Careri, U. Buontempo, F. Galluzzi, A. C. Scott, E. Gratton and E. Shyamsunder *Phys. Rev.* **B30**, 4689 (1984).
4. G. B. Blanchet and C. R. Fincher, *Phys. Rev. Lett.* **54**, 1310 (1984).
5. J. C. Eilbeck, P. S. Lomdahl and A. Scott, *Phys. Rev.* **B30**, 4703 (1984).
6. C. T. Johnston and B. I. Swanson, *Chem. Phys. Lett.* **114**, 547 (1985).
7. D. M. Alexander, *Phys. Rev. Lett.* **54**, 138 (1985).
8. D. M. Alexander and J. A. Krumhansl, *Phys. Rev.* **B33**, 7172 (1986).
9. M. Barthes, *J. Mol. Liq.* **41**, 143 (1989).
10. A. C. Scott, I. J. Bigio and C. T. Johnston, *Phys. Rev.* **B39**, 12883 (1989).
11. W. Fann, L. Rothberg, M. Roberson, S. Benson, J. Madey and S. Etemad, *Phys. Rev. Lett.* **64**, 607 (1990).
12. A. C. Scott, *Phys. Reports* **217**, 1 (1992).
13. J. Edler, P. Hamm and A. C. Scott, *Phys. Rev. Lett.* **88**, 067403 (2002).
14. J. Edler and P. Hamm, *J. Chem. Phys.* **117**, 2415 (2002).
15. J. Edler and P. Hamm, *J. Chem. Phys.* (in press, 2003).
16. S. Krimm and J. Bandekar, *Adv. Protein Chem.* **38**, 181 (1986).
17. H. Torii and M. Tasumi, *J. Chem. Phys.* **96**, 3379 (1992).
18. P. Hamm, M. Lim and R. M. Hochstrasser, *J. Phys. Chem.* **B102**, 6123 (1998).
19. S. Woutersen and P. Hamm, *J. Phys.: Condens. Matter* **14**, R1035 (2002).
20. S. Woutersen and P. Hamm, *J. Phys. Chem.* **B104**, 11316 (2000).
21. S. Woutersen and P. Hamm, *J. Chem. Phys.* **115**, 7737 (2001).
22. P. Hamm and R. M. Hochstrasser, in *Ultrafast Infrared and Raman Spectroscopy*, edited by M. D. Fayer (Marcel Dekker, New York, 2001), pp. 273–347.
23. M. T. Zanni and R. M. Hochstrasser, *Curr. Opin. Struc. Biol.* **11**, 516 (2001).
24. P. Hamm, M. Lim, W. F. DeGrado and R. M. Hochstrasser, *Proc. Natl. Acad. Sci. USA* **96**, 2036 (1999).

25. M. T. Zanni, N. H. Ge, Y. S. Kim and R. M. Hochstrasser, *Proc. Natl. Acad. Sci. USA* **98**, 11265 (2001).
26. C. Scheurer and S. Mukamel, *J. Am. Chem. Soc.* **123**, 3114 (2001).
27. A. M. Moran, S. M. Park, J. Dreyer and S. Mukamel, *J. Chem. Phys.* **118**, 3651 (2003).
28. S. Woutersen and P. Hamm, *Chem. Phys.* **266**, 137 (2001).
29. S. Woutersen, R. Pfister, P. Hamm, Y. Mu D. S. Kosov and G. Stock, *J. Chem. Phys.* **117**, 6833 (2002).
30. S. Woutersen, Y. Mu, G. Stock and P. Hamm, *Proc. Natl. Acad. Sci. USA* **98**, 11254 (2001).
31. J. Bredenbeck, J. Helbing, R. Behrendt C. Renner, L. Moroder J. Wachtveitl and P. Hamm, *J. Phys. Chem. B* (in press), (2003).
32. G. C. Pimentel and A. McClellan, *The Hydrogen Bond* (W. H. Freemann, San Francisco, 1960).
33. D. Hadži and S. Bratos, in *The Hydrogen Bond*, edited by P. Schuster, G. Zundel and C. Sandorfy (Elsevier, Amsterdam, 1976), Vol. II, Chap. 12.
34. O. Henri-Rousseau and P. Blaise, *Adv. Chem. Phys.* **103**, 1 (1998).
35. D. Madsen, J. Stenger, J. Dreyer, E. J. Nibering, P. Hamm and T. Elsaesser *Chem. Phys. Lett.* **341**, 56 (2001).
36. S. Mukamel, *Principles of Nonlinear Optical Spectroscopy* (Oxford University Press, Oxford, 1995).
37. J. Stenger, D. Madsen, J. Dreyer, E. J. Nibering, P. Hamm and T. Elsaesser, *J. Chem. Phys.* **A105**, 2929 (2001).
38. K. Heyne, N. Huse, E. T. J. Nibbering and T. Elsaesser, *Chem. Phys. Lett.* **369**, 591 (2003).
39. J. Bonča, S. A. Trugman and I. Bastistić, *Phys. Rev.* **B60**, 1633 (1999).
40. N. B. Abbot and A. Elliott, *Proc. Roy. Soc. (London)* **A243**, 247 (1956).
41. D. Thouless, *Phys. Rep.* **13**, 93 (1974).
42. M. Schreiber and Y. Toyozawa, *J. Phys. Soc. Jpn.* **51**, 1537 (1992).
43. T. Meier, Y. Zhao, V. Chernyak and S. Mukamel, *J. Chem. Phys.* **107**, 3876 (1997).
44. K. S. Song and R. T. Williams, *Self-Trapped Excitons* (Springer, Berlin Heidelberg New York, 1996).
45. A. Xie, L. van der Meer, W. Hoff and R. H. Austin, *Phys. Rev. Lett.* **84**, 5435 (2000).
46. V. Helenius, J. Korppi-Tommola, S. Kotila, J. Nieminen and R. Lohikoski, *Chem. Phys. Lett.* **280**, 325 (1997).

CHAPTER 8

BREATHERS IN BIOMOLECULES ?

Michel Peyrard* and Yannick Sire**

* *Laboratoire de Physique, Ecole Normale Supérieure de Lyon*
46 allée d'Italie, 69364 Lyon Cedex 07, France
E-mail: Michel.Peyrard@ens-lyon.fr

** *Laboratoire Mathématiques pour l'Industrie et la Physique,*
INSA, dép. GMM, 135 avenue de Rangueil, 31077 Toulouse Cedex 4, France
E-mail: sire@gmm.insa-tlse.fr

This report, initially prepared to introduce a discussion on "breathers in biomolecules", discusses some experimental results and their connection with theory. Classical vibrations are considered first and quantum effects are discussed in a second part. The paper does not intend to be a complete review of nonlinear localisation in molecules, but instead it introduces and discusses a few examples with the aim of stimulating thoughts and hopefully further experimental studies or theoretical investigations paying particular attention to the experimental aspects.

1. Introduction

In the last decade, numerous studies have been devoted to localised o-scillatory modes in nonlinear lattices, the so-called "discrete breathers". Proofs of existence of these exact solutions have been given for different lattices, approximate analytical solutions and exact numerical ones have been obtained. However, there are still few experimental observations of discrete breathers at the molecular scale. One major difficulty is to clearly distinguish these localised modes from the standard vibrational modes of molecules and crystals.

Biomolecules would be good candidates for these observations because they are highly deformable objects, undergoing large conformational changes that probe the nonlinear parts of the interaction potentials between their components, which are often weakly coupled to each other. Theoretical studies have speculated that discrete breathers could exist in DNA and proteins and perhaps play some role in energy storage and transfer in biological molecules.

In this report, initially prepared to introduce a discussion on "breathers in biomolecules", we discuss some experimental results and their connection with theory. Classical vibrations are considered first and quantum effects are presented in a second part. The last part contains a discussion. This paper does not intend to be a complete review of nonlinear localisation in molecules, but instead it introduces and discusses a few particular examples, emphasising experimental investigations, with the aim of stimulating thoughts and hopefully further experiments or theoretical investigations paying particular attention to the experimental aspects.

2. Classical vibrations

2.1. *Local modes in small molecules*

Many experiments have been done to test the possible existence of local modes in small molecules. Vibrational modes are observed through Raman and infrared spectroscopy. Anharmonicity can be detected by the presence of overtones but observing localisation is difficult because spectroscopic experiments probe the dynamics in Fourier space and not in real space. Moreover, as the wavelength of light is much larger than distances between the atoms of a molecule, spectroscopic experiments do not provide spatial information on the vibrational modes that it probes. One could think that it is therefore impossible to use them to determine if some vibrations are localised. This is not true but one has to rely on indirect informations.

One possible method has been used by Halonen *et al.*[1] who recorded high resolution Fourier transform infrared spectra to measure the stretching fundamental ($1000A_1/F_2$), the first ($2000A_1/F_2$) and the second ($3000A_1/F_2$) stretching vibrational overtones of mono-isotopic deuterated stannane $^{120}SnD_4$. Earlier high resolution studies had observed the fundamental, first, second, fifth and seventh $Sn-H$ stretching vibrational overtones of stannane $^{120}SnH_4$. The trick to observe localisation is to take advantage of the dynamical symmetry breaking that it introduces. Molecules such as SnH_4 or SnD_4 have the symmetry of a spherical top. If only one of the $Sn-H$ or $Sn-D$ bonds is excited to a high vibrational level, the average length of this bond becomes larger than the length of the others due to the anharmonicity on the potential connecting the atoms. The symmetry of the molecule changes to that of a prolate symmetric top as shown schematically in Fig. 1. A detailed analysis of high resolution spectra can detect the symmetry change. It is interesting to notice that localisation can already be

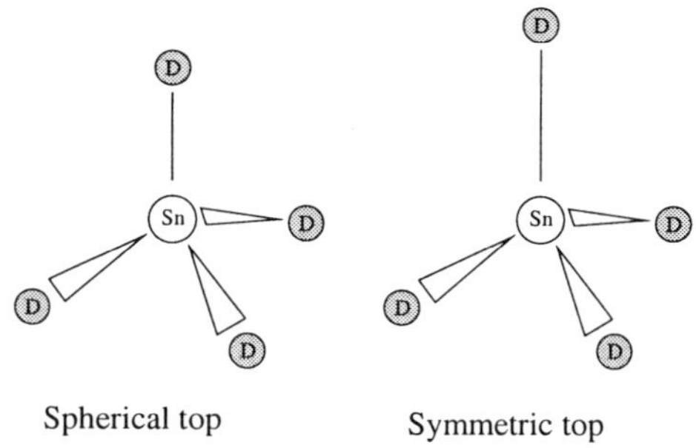

Fig. 1. Schematic picture of the dynamical symmetry change created by the excitation of one particular bond in SnD_4 to a high vibrational level.

detected in the analysis of the second stretching overtone of SnD_4. A similar observation has also been made for SnH_4. On the contrary, localisation has not been observed in SiD_4. This result appears rather natural and is in perfect agreement with the criterion of existence of discrete breathers which shows that, above a given threshold for the coupling between oscillators, localisation cannot be sustained. Silicium has an atomic mass of 28, while Sn has an atomic mass of 118.7. For a given level of excitation, the motion of a heavy central atom has a smaller amplitude than the motion of a light central atom. Therefore, it excites the other bonds less efficiently, leading to a weaker coupling between oscillators (in the limit of a central atom of infinite mass, that stays immobile, the coupling vanishes). Other spherical top molecules, such as GeH_4, have been studied with similar techniques.[2–8] In the same way, local modes have been observed in benzene.[9] For older experiments on local modes in small molecules, we refer the reader to Refs. 10 and 11.

2.2. *Local modes in large molecules*

Most large dynamical systems are thought to have ergodic dynamics, whereas small systems may not have free interchange of energy between degrees of freedom. Such an assumption is made in many areas of physics and chemistry. In Ref. 12, the authors examine the transition to facile vibrational energy flow in a large set of organic molecules as molecular size is increased. The problem is to determine an easily computable parameter that traces the transition in the energy flow with high fidelity. It appears that the relevant parameter T that determines the transition to ergodic mixing is the local density of states coupled by anharmonic resonances. The authors of this study noticed that the vibrational potential surface of a molecule can be exactly factorised when the number of vibrational degrees of freedom approaches infinity as long as the normal modes are sufficiently random combinations of cartesian coordinates. This leads to the following expression for the localisation parameter T:

$$T(E) = \frac{2\pi}{3}\left(\sum_{Q}\langle|V_Q|\rangle\rho_Q(E)\right)^2 \tag{1}$$

where Q is a distance between states in the state space, $\rho_Q(E)$ is the local energy of states resonantly coupled to states which are at "distance" Q away. Assuming that the coupling between two normal modes is random, $\langle|V_Q|\rangle$ represents the mean field value at a distance Q. If $T > 1$, there is an easy energy flow. On the contrary, for $T < 1$, energy flow is confined to a subset of the energetically allowed states. The value $T = 1$ is then the transition criterion.

The computations have been done for all C – H stretching fundamentals of a set of different molecules and they show that T is a good indicator of the transition to ergodicity. The transition from weak anharmonic resonances and restricted energy flow to free energy flow takes place over a narrow range of the number of locally coupled states. However, surprisingly large molecules can show non-ergodic behaviour because the crucial parameter is the *local* density of states. As the number of atoms increases, the density of states and the transfer rate increases and one could expect that it becomes harder and harder to detect localised modes. But if only a local density of states has to be considered, non-ergodicity is still possible for large molecules.

This analysis suggests therefore that localised modes can be expected in large molecules, but that a transition to ergodicity must nevertheless be expected when molecules become large enough. This would mean that discrete breathers should not exist in very big molecules such as biological molecules. But this analysis has been performed in organic molecules which are still fairly small compared to biological molecules. These intermediate size molecules behave as a whole, in the sense that all their atoms are strongly correlated to each other. Even in the cases of the most branched molecules, every atom can be reached from another one by a path that includes a small number of bonds (less than 10). Biological molecules are much larger than that, and an independent behaviour of some parts appears to be possible. The assumption that normal coordinates are sufficiently random combinations of the cartesian coordinates, made in the analysis of Bigwood *et al.*[12], may not be valid for a large molecule such as DNA or for big proteins. In Sec. 2.5 we discuss the case of fluctuational opening of DNA, which appears to contradict the analysis of Bigwood *et al.*

2.3. *Local modes in crystals*

Moving to an even larger system, one can consider a crystal. Due to its regular structure, with translational symmetry, one could think that localised vibrations cannot exist. There are, however, some experimental data that show the contrary.

At low temperature, energy deposited on a crystal surface by a laser pulse or an ohmic heater can flow rapidly away from the excitation point by a ballistic phonon propagation. However, at high excitation levels, a portion of the absorbed energy is trapped near the excitation region. The authors of Ref. 13 have developed an experimental technique (spatial filtering) which investigates both the time and the spatial dependence of a laser-induced heat pulse. They used in their experiments both Ge and Si, which can be produced as high quality crystals, and they showed that a major portion of the delayed flux in Ge is spatially very diffusive but a component emanates directly from the excitation region. It is short-lived at our scale (0.1 to 1.5 μs depending on the excitation level), but very long-lived in terms of typical phonon vibration period. Consequently, in Ge, a laser pulse produces a local storage of energy, more efficiently at higher excitation levels, which is consistent with a phenomenon of nonlinear localisation. Such storage is also observed in metal-coated GaAs and strongly doped Ge, which proves that the phenomenon is not electronic. The analysis of the results shows that they are compatible with the following scenario: within the heated region, phonon-phonon scattering produces a phonon distribution roughly described by a local temperature well above the bulk crystal temperature. This region radiates phonons as a Planck source.

As in molecules, the observation of localisation in crystals requires special techniques. In Ge and Si its detection has been achieved because the authors introduced a fruitful spatial filtering technique for phonons that allows direct observation of the ballistic ones that are initially stored at the excitation region.

We refer the reader to Refs. 14 and 15, for an observation of a rotational mode in 4-methyl-pyridine. In this paper, the authors study the torsional dynamics of the methyl groups in solid 4-methyl-pyridine (C_6H_7N) at low temperatures. The potential barrier for internal rotation of the methyl group is very low in the isolated molecule and remains small in the crystalline state. Measurements with totally hydrogenated and totally deuterated 4-methyl-pyridine and the observation of a continuous frequency shift in the hydrogenated 4-methyl-pyridine spectrum with increasing concentration of deuterated molecules lead the authors to conclude that they were observing collective motions for the methyl group. The results have been described in terms of the response of a breather of the quantum sine-Gordon equation and a quite good agreement between model and experiments has been obtained. However, in this case the evidence of the localisation relies on the model and not on a direct experimental observation.

2.4. *Localisation of vibrations and chemical reaction rates*

Localisation could affect chemical reaction rates because the energy in some bonds may be temporarily higher than its equilibrium value. But chemistry can also be used as a tool to study the energy flow in a molecule, providing another indirect approach to localisation. This method has been implemented by Uzer *et al.*[16] who studied hydrogen peroxide HO_2H and its isotopic variant HO_2D. The idea is to highly excite, through its overtones, one particular $O-H$ bond. The energy which is injected at one end of the molecule (see Fig. 2) is then transfered to the other bonds and it can cause their dissociation. The analysis of the products gives a picture of the energy distribution between bonds at the time of dissociation. The authors observed that, within the lifetime of the energised molecules, the internal energy redistribution in not complete. Indeed, the second OH stretch in HO_2H (or the OD stretch in HO_2D) receives very little energy.

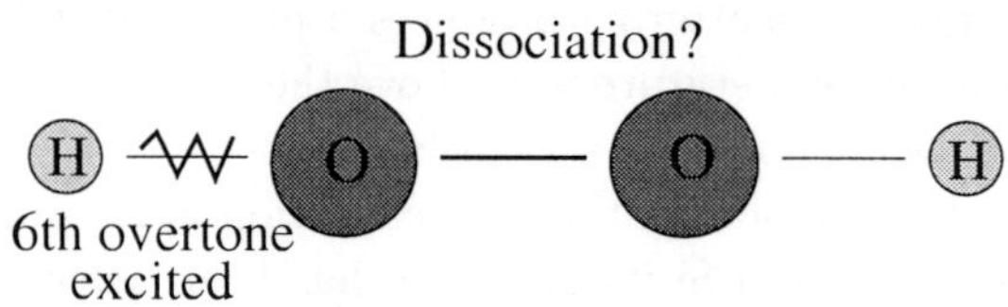

Fig. 2. Schematic picture of the method used by Uzer *et al.*[16] to study the energy distribution in a molecule.

From a nonlinear point of view, their results lead to the conclusion that the flow of energy from an excited O – H stretch into the dissociative O – O bond involves multiple resonances and anharmonic effects. To model the anharmonic stretching forces, the authors used a Morse potential for the O – H bond and the O – O bond. The parameters of the potential are adjusted according to realistic values. The conclusion of the analysis is that, although energy redistribution is not complete at dissociation time, this effect is small. The remark that one can make is similar to the one that we made above for the theoretical study of Bigwood *et al.* (Sec. 2.2): the molecules which have been investigated are far smaller than biological molecules. The rather fast energy flow from one part to another in such small molecules is not surprising and may not mean that localisation cannot be observed in much bigger systems such as biological molecules as it has been observed in a crystal (Sec. 2.3). It is nevertheless interesting to notice that an incomplete energy redistribution can already be observed on the time scale of a dissociation experiment in a molecule as small as HO_2H. Moreover, the authors suggest that such anharmonic effects should occur in other molecules as small as HO_2H and HO_2D .

2.5. *Fluctuational opening in DNA*

The "breathing" of DNA, consisting in the temporary opening of the base pairs, has been known from biologists for decades. This is attested by proton-deuterium exchange experiments. DNA is put in solution in deuterated water, and one observes that the imino-protons, which are the protons forming hydrogen bonds between two bases in a base pair, are exchanged with deuterium coming from the solvent. As these protons are deeply buried in the DNA structure, the exchange indicates that bases can open, at least temporarily, to expose the imino protons to the solvent.[17] The determination of the lifetime of a base pair, i.e. the time during which it stays closed, has been the subject of some controversy[18] because the rate limiting step in the exchange may be either the rate at which base pairs open, or the time necessary for the exchange. Accurate experiments, using NMR to detect the exchange, showed that the lifetime of a base pair is of the order of 10 ms. These experiments also show that the protons of one particular base pair can be exchanged while those of a base pair next to it are not exchanged. This indicates that the large conformational changes that lead to base pair opening in DNA are highly localised and moreover it shows that the coupling between successive bases along the DNA helix is weak. It is tempting from these results to conclude that the "breathing of DNA" observed by biologists (and used by them for instance to introduce dyes in DNA) is due to a discrete breather in the dynamics of the bases. This is also what comes out from simple models of DNA dynamics, but the present experiments cannot demonstrate this because they do not give any information on the nature of the open state. It could be an oscillatory state (i.e. a breather) but the experiments do not rule out a single flipping of the bases out of the stack of the double helix without any oscillation.

There are however some data which show that an oscillatory mode is associated to the opening of the base pairs. They concern the opening which occurs when the two strands are separated by heating. In Ref. 19, the authors studied thermal denaturation of the B form of double-stranded DNA by differential scanning calorimetry and Raman spectroscopy using 160-pair fragments of calf thymus DNA. The changes observed in specific Raman band frequencies and intensities as a function of temperature reveal that thermal denaturation is accompanied by disruption of base pairs, unstacking of the bases and disordering of the B form backbone. Furthermore the intensity of some Raman bands at 1240 cm^{-1} and 1668 cm^{-1} exhibits the same increase with temperature in the $340 - 360$ K range as the variation of enthalpy ΔH of the denaturation transition measured by differential scanning calorimetry. This shows that there is a link between an increase of the amplitude of the vibration of the bases and DNA melting. Moreover, as melting occurs first through local opening of DNA, the experiments appear to be compatible with the following pathway to melting: one vibrational mode associated to base-pair opening has an amplitude that grows locally until it becomes so large that it leads to a local opening, forming a precursor of melting. In such a picture the fluctuational openings observed at room temperature would be very similar, and they would not evolve toward local melting simply because temperature is not high enough.

As discussed in Sec. 2.1, the wavelength of the light used in spectroscopic experiments is so large compared to interatomic distances that standard spectroscopy cannot conclude about the localisation of the motions. The great interest of DNA is that one now knows how to perform local reactions at sites which can be precisely determined. This opens the possibility to observe the local dynamics of the molecule through the response of a chromophore attached to the region of interest. In Ref. 20, the authors studied the local dynamics of calf thymus double-helical DNA by measuring the visible absorption band of the cationic dye ethidium bromide, both free in solution and bound to DNA, for a temperature range $360 - 30$ K in two different solvent conditions. The comparison of the thermal behaviour of the absorption band of free and DNA-bound ethidium bromide gives information on the local dynamics of the double helix in the proximity of the chromophore. The experiments show that above 280 K, the anharmonicity of the motion of the dye increases much faster when it is bound to DNA than when it is free in solution. This is attributed to an onset of wobbling of the dye in its intercalation site, which is likely to be connected with the onset of local opening/unwinding of the double helix. The experiments clearly show that large amplitude "premelting" motions of DNA occur well below the denaturation temperature. The use of a dye provides a method to probe these motions locally. However, although the experiments of Ref. 20 can probe the "micro-environement" of DNA vibrations, they have not determined whether the motions, which are observed in the regions of DNA which are close to the dye, are really localised in space or if the same motions exist everywhere along the helix. If the dynamics of DNA are inhomogeneous in space, the chromophores situated in a region of DNA which is only weakly vibrating should show a response different from the response of chromophores situated in regions of DNA which are moving with a large amplitude. A detailed analysis of the shape of the chromophore bands versus temperature could perhaps detect the existence of localised motions, but it is again an indirect observation, subjected to the accuracy of the analysis. Combining such an analysis with the use of biomolecular engineering to insert probes in specific points of DNA, would provide useful data on DNA dynamics in selected regions.

The experiments on DNA clearly show the existence of large amplitude, highly localised conformational changes. Although there are indications that a vibrational mode is associated to opening, experiments do not prove unambiguously that local conformational changes are due to discrete breathers, but they nevertheless show that the studies which conclude that a large molecule should have a nice ergodic behaviour (Sec. 2.2), precluding large amplitude localised motion, do not apply to DNA.

3. Quantum self-trapping

Acetanilide ($CH_3 - CONH - C_6H_5$) is commonly regarded to be a simple model system to study the nature of vibrational excitations in protein secondary structures because it consists of quasi-one-dimensional chains of hydrogen bonded peptide units with structural properties that are very similar to those in a protein α-helix. In crystalline acetanilide two types of coupling mechanisms have been considered to lead to self trapping: the normal modes of the peptide units – CO – NH –, coupled by dipole-dipole interactions are delocalised states also called the vibrational excitons. These vibrational excitons are coupled to lattice phonons by nonlinear interaction. Excitation of an exciton then leads to a deformation of the lattice, thereby localising the state.

In Ref. 21, the authors present a femto-second pump-probe study of the peculiar amide I band of acetanilide. They observe a perfect harmonicity of the 1666 cm^{-1} subpeak. This is associated to the delocalisation of this state at low enough temperatures (93K): if the energy injected by the excitation is distributed over many sites, the vibrational amplitude stays low, explaining the harmonic behaviour which is observed. But low temperature experiments also detect an anomalous peak at 1650 cm^{-1} which is strongly anharmonic and hence assigned to a self-trapped state. With increasing temperature, thermal disorder localises the 1666 cm^{-1} band by Anderson localisation and destroys self-trapping. Theory explains the experimentally observed anharmonicity of the self-trapped state as well as the harmonicity of the free exciton. The existence of self trapping in acetanilide has been the subject of some debate and alternative theories involving Fermi resonance or structural defects,[22–25] including for instance the coupling of the electronic defect of a color center with a vibrational mode,[26] have been proposed to explain the behaviour of the anomalous 1650 cm^{-1} band. But, according to these theories, the 1666 cm^{-1} and 1650 cm^{-1} bands should have similar behaviours, which is in contradiction with the experiments. Ref. 21 provides a convincing proof for self-trapping of the amide I band in crystalline acetanilide because it uses a pump-probe method that is specifically sensible to anharmonic effects since it can eliminate all harmonic contributions. There are nevertheless some open questions in the analysis of the results. In Ref. 27, the authors, comparing temperature dependence, assigned a band at 3250 cm^{-1} to the first overtone of the self-trapped state at 1650 cm^{-1}. According to this assumption, the authors of Ref. 21 should expect to see an excited state absorption band at 1600 cm^{-1} (i.e. 3250 – 1650 cm^{-1}) in their pump-probe experiment after selectively exciting the 1650 cm^{-1} mode. In fact, they observed a small signal at this frequency but the major part of the signal emerges at 1644 cm^{-1}. This discrepancy is still unexplained but it could lie in the dynamical aspects of the formation of the self-trapped state. The phonon system needs time to respond to amide I excitation. This point has been neglected in Ref. 27. The experimental results point out that the theory is still incomplete.

Moreover theoretical investigations have concentrated their attention on the C – O stretching mode of the peptide unit – CO – NH – , which has been studied in Ref. 21 but other experiments show that the N – H stretching mode may also play some role in the self-trapping of energy. Femtosecond IR spectroscopy[28] of delocalised N – H excitations of crystalline acetanilide confirms self-trapping in hydrogen-bonded peptide units. After selective excitation of the free exciton, self-trapping occurs within a few hundred fs. An important point is that the parameters involved in the theory can be determined by independent experiments. The investigations of the C – O and N – H bands of crystalline acetanilide show that the large nonlinearity of the hydrogen bonds, which stabilise protein structures, can lead to self trapping. The lifetime of these localised excitations is about 20 ps which would be sufficient to transport energy over significant distances of a few tens of angströms. The latest experiments[21] suggest however that nonlinear self trapping is not relevant for biology because increasing temperature to biological temperatures, which induces increased thermal disorder, destroys the self-trapping.

Another phenomenon of self-trapping has been observed by Xie *et al.*[29] As in Ref. 28, they use pump-probe experiments in the IR to measure vibrational relaxation rates. They study myoglobin which is an important biological protein. Their results must be interpreted within a parallel with the previous ones [28] because, as for crystalline acetanilide, myoglobin has an almost α-helix structure. They show experimentally that myoglobin has an unusually long, nonexponential excited state relaxation. On the contrary, other biological molecules such as β-sheet protein, or the simple amino acid alanine do not have such a long-lived state. This shows that α-helix in proteins is essential to support nonlinear states. The authors of this work suggest that the long-lived ground state population depletion, produced by nonlinear generation of a long-lived state in the α-helix of a protein, indicates that nonlinear excitations may play an important and significant role in energy transfer in biomolecules, but this is still speculative.

While a chain of hydrogen bonded groups appears to be necessary to lead to self trapping, it does not necessarily have to be a chain of peptidic bonds. Some spectroscopic studies of $L-$alanine *crystals* find a temperature dependence of an "anomalous" torsional mode of NH_3^+ which splits into two components at low temperature. The results suggest that this mode could be a self-localised vibrational mode due to the coupling to hydrogen bond stretching modes.[30] The presence of an overtone attests of the high anharmonicity of the NH_3^+ torsion mode.

Some experiments[31] suggest a possible necessary condition of self-trapping. One can observe that self trapping disappears by deuterating the CH_3 group into CD_3, which changes its vibration frequencies. In the case of acetanilide, a CH_3 torsion has the same frequency as one of the hydrogen bond stretching modes. This result suggests that this coupling might be necessary to have self-trapping. Studies in progress[32] on small peptides in water suggests that the existence of breathers in such molecules requires a very high excitation level.

Up to now we discussed localisation which only involved *vibrational* degrees of freedom. The nonlinearity necessary for self-trapping was either due to the intrinsic nonlinearity of the interaction potentials, as in SnH_4 or DNA, or due to a coupling between two vibrational degrees of freedom as in crystalline acetanilide where the excitons couple to the motion of the peptide groups. Nonlinearity of vibrational modes can also be achieved through the coupling with another degree of freedom, which is not vibrational, such as an *electronic* degree of freedom. This phenomenon has been observed in charge transfer compounds such as the halide-bridged transition metal complex $[Pt(en)_2][Pt(en)_2Cl_2](ClO_4)_4$, where (en) = ethylene-diamine, also denoted by PtCl.[33] The studies, performed by Raman spectroscopy, again had to rely on a clever trick to demonstrate localisation. In this case, they take advantage of the presence of two isotopes of Cl, ^{35}Cl and ^{37}Cl. If the observed vibrational modes had been extended over many sites, they would have shown the response of a system with an average mass for Cl. Instead, for localised modes, one can distinguish a vibrational mode involving ^{35}Cl from another one involving ^{37}Cl. In Ref. 34, the authors developed a nonlinear model to account for the Raman redshifts of the overtones in isotopically pure $Pt^{35}Cl$ and $Pt^{37}Cl$. Then the theory has been extended to naturally abundant, isotopically disordered PtCl.[35] Using parameters for the model which are consistent with independent calculations of some properties of the system, the phonon dispersion curve and the coupling of electronic with lattice degrees of freedom, allows the authors to reproduce the features of the position of the Raman peaks as well as the first-overtone spectrum without introducing any additional parameter. This provides a fairly convincing evidence that the proposed multiquanta breather model is appropriate for PtCl.

4. Discussion

The results that we have presented here show that, although there are still few evidences of breathers at the molecular scale, there are some indications that nonlinear localisation does exist at such a scale. The clearer evidence is probably the one provided by pump probe experiments on crystalline acetanilide because the measurements can selectively detect nonlinear effects. Since acetanilide has a chain of peptide bonds similar to α-helix proteins, this is an indication that nonlinear self trapping could also exist in proteins, although it has not yet been observed. The experiments are much more difficult for proteins because it is hard to get good quality samples and because the $\alpha-$helices are only one part of the molecule. In DNA, the existence of highly localised conformational changes is established, but whether they are breathers, or at least initiated by breathers, is still an open question. Experiments show that a vibrational mode increases in amplitude when the base pairs start to have large openings near the temperature of thermal denaturation, but the possible localisation of this mode has not been established. This would be difficult by spectroscopy, as discussed above, unless some specific methods could be found.

The small number of experimental observations raise several questions for experimentalists

(1) what are the optimal experimental methods that one should use to observe localised modes?
(2) are there new or recent experimental methods, that were not available a few years ago, that one should use? In this respect nonlinear spectroscopy using pump probe methods have shown a way to explore.
(3) what are the optimal samples?

Another open question is the relevance of nonlinear localisation for biology. The experiments on acetanilide introduce a doubt that it could be relevant because self trapping seems to vanish at biological temperatures. But this may not be true for other systems with a stronger nonlinearity. For DNA the local openings are of biological importance because they allow intercalation in the stack of base pairs.

The link between experiments and theory should be developed. Theoreticians can perhaps make suggestions to experimentalists, but their role is also important to analyse the experimental results. The case of PtCl provides a characteristic example. While it is hard for spectroscopy alone to conclude on localisation, if the experimental results can be analysed in a convincing way by a model in which nonlinear localisation plays a crucial role, it is an indication that this phenomenon is present in the material [38]. However, such a synergy between theory and experiment is only conclusive if the theory is based on some parameters which are independently determined and moreover if the same theory can describe a full set of experimental facts. It is usually not hard to find a theory that explains one particular value (such as the frequency of a mode or the denaturation temperature of DNA). For PtCl, beyond the frequencies of the modes, the theory also obtains the frequency of the overtones and their broadening due to the abundance of the different Cl isotopes.

Nonlinearity can be intrinsic to the interaction potentials or it can arise through the coupling to other degrees of freedom, which can be vibrational or electronic. The case of the electron phonon coupling raises an interesting question: the connection between nonlinear effects, discrete breathers and phase transitions.[36] A weak coupling can lift a local degeneracy and lead to a splitting of some vibrational modes.[37] A stronger coupling leads to a lattice distortion, it is the Jahn-Teller effect.[39] When an ion of a crystal sits in a site of high symmetry, its ground state may be degenerate. A distortion of the lattice can lift the degeneracy and lead to a decrease of the energy of the ground state proportional to the amplitude ξ of the distortion. But the distortion itself costs an energy that grows as ξ^2. This leads to an equilibrium for a finite value of ξ. There is a formal analogy between this situation and the case of a breather which is stabilised by a local distortion of the crystal lattice. This suggests that discrete breathers could be related to lattice instabilities similar to those that lead to phase transitions. In acetanilide, there are indications that the material is in a pre-transitional state. Another example is found in Raman scattering of L-alanine:[30] some vibrational modes at low frequency have a discontinuity at 220 K which suggests a structural phase transition. Birefringence depolarisation detects a continuous and microscopic distortion of the lattice below 220 K although crystallographic studies do not find a change in the space group. Measurements of the lattice parameter from room temperature to liquid helium temperature show that, instead of decreasing, it increases and has some steps.[36] These experiments do not show a real phase transition but indicate pre-transitional effects. There are also some indications of a lattice instability in acetanilide, indicated by a change of the sound speed in the direction of the hydrogen bonds at 150 K.[36] As strong anharmonic effects have to be expected near a phase transition, studies of nonlinear localisation near phase transitions or structural instabilities might deserve more attention.

Acknowledgments

Part of this work has been supported by the EU Contract HPRN-CT-1999-00163 (LOCNET network). We would like to thank M. Barthès (Montpellier) for helpful comments.

References

1. M. Halonen, L. Halonen, H. Bürger and W. Jerzembeck, *J. Chem. Phys.* **108**, 9285 (1998).
2. D. V. Willets, W. J. Jones and A. G. Robiette, *J. Mol. Spectrosc.* **55**, 200 (1975).
3. H. Qian, Q. Zhu, H. Ma and B. A. Thrush, *Chem. Phys. Lett.* **192**, 338 (1992).
4. Q. Zhu, B. Zhang, Y. Ma and H. Qian, *Spectrochim. Acta* **A46**, 1217 (1990).
5. Q. Zhu, H. Ma, B. Zhang, Y. Ma and H. Qian, *Spectrochim. Acta* **A46**, 1323 (1990).
6. Q. Zhu and B. A. Thrush, *J. Chem. Phys.* **92**, 2691 (1990).
7. F. Sun, X. Wang, J. Liao and Q. Zhu, *J. Mol. Spectr. Spec.* **184**, 12 (1997).
8. A. Campargue, J. Vetterhöffer and M. Chenevier, *Chem. Phys. Lett.* **192**, 4353 (1992).
9. L. Halonen, *Chem. Phys. Lett.* **87**, 221 (1982).
10. R. T. Birge, H. Spomer, *Phys. Rev.* **28**, 259 (1926).
11. J. W. Ellis, *Phys. Rev.* **33**, 27 (1929).
12. R. Bigwood, M. Gruebele, D.M. Leitner and P.G. Wolynes, Proc. Natl. Acad. Sci. USA **95**, 5960 (1998).
13. M. Greenstein, M. A. Tamor and J.P. Wolfe, *Phys. Rev.* **B26**, 5604 (1982).
14. F. Fillaux, chapter 2, *Energy localization and transfer*, T. Dauxois, A. Litvak-Hinenzon, R. S. Mackay, A. Spanoudaki Eds., Advanced Series on Nonlinear Dynamics, World Scientific (2003).
15. F. Fillaux, C. J. Carlile, *Phys.Rev.* **B42**, 5990 (1990).
16. T. Uzer, J. T. Hynes and W. P. Reinhardt, *J. Chem. Phys.* **85**, 5791 (1986).
17. M. Gueron, M. Kochoyan and J. L. Leroy, *Nature* **328**, 89 (1987).
18. M. Frank-Kamennetskii, *Nature* **328**, 17 (1987).
19. J. G. Duguid, V. A. Bloomfield, J. M. Benevides and G. J. Thomas, *Biophys. J.* **71**, 3350 (1996).
20. A. Cupane, C. Bologna, O. Rizzo, E. Vitrano, L. Cordone, *Biophys. J.* **73**, 959 (1997).
21. J. Edler and P. Hamm,*J. Chem. Phys.* **117**, 2415 (2002).
22. G. C. Blanchet and C. R. Fincsher, *Phys. Rev. Lett.* **54**, 1310 (1985).
23. C. T. Johnston, B. I. Swanson, *Chem. Phys. Lett.* **114**, 547 (1985).
24. A. Tenenbaum, A. Campa and A. Giasanti, *Phys. Lett.* **A121**, 126 (1987).
25. W. Fann, L. Rothberg, M. Roberson, S. Benson, J. Madey, S. Eternad and R. Austin, *Phys. Rev. Lett.* **64**, 607 (1990).

26. D. B. Fitchen, *Physics of Color Centers,* Ed. W. B. Fowler (Acad. Press, New York, 1968), p. 293.
27. A. Scott, E. Gratton, E. Shyamsunder and G. Careri, *Phys. Rev.* **B32**, 5551 (1985).
28. J. Edler and P. Hamm, *Phys. Rev. Lett.* **88**, 067403 (2002).
29. A. Xie, L. Van der Meer, W. Hoff and R. A. Austin, *Phys. Rev. Lett.* **84**, 5435 (2000).
30. M. Barthès, A. Fahre Vik, A. Spire, H. N. Bordallo and J. Eckert, *J. Phys. Chem.* **A106**, 5230 (2002).
31. M. Barthès, H. Kellouai, G. Page, J. Moret, S. W. Johnson and J. Eckert, *Physica D: Nonlinear Phenomena* **68**, 45 (1993).
32. P. Hamm, work in progess
33. B. I. Swanson *et al.*, *Phys. Rev. Lett.* **82**, 3288 (1999).
34. N. K. Voulgaris, G. Kalosakas, A. R. Bishop and G. P. Tsironis, *Phys. Rev.* **B64**, 020301 (2001).
35. G. Kalosakas, A. R. Bishop and A. P. Shreve, *Phys. Rev.* **B66**, 094303 (2002).
36. M. Barthès, F. Denoyer, H. N. Bordallo, E. J. Lorenzo, A. Robert and F. Zontone (to be published in *Eur. Phys. J. B*).
37. F. S. Ham, *Phys. Rev.* **A138**, 1727 (1965).
38. Reference 14 explains that vibrational spectroscopy only probes extended states and cannot excite or probe individually equivalent atoms in molecules or equivalent sites in crystals. This appears to be in contradiction with our reports of the experimental observation of localisation by vibrational spectroscopy. While this statement is indeed correct for standard spectroscopy techniques, special methods can be designed to overcome this difficulty, such as the analysis of the dynamical symmetry breaking due to a localised vibration in SnD_4.
39. J. Callaway, *Quantum theory of the solid state* (Academic Press, San Diego, 1991).
40. S. A. Trugman *et al.*, *Phys. Rev.* **B60**, 1633 (1999).
41. Q. Zhu, H. Qian, B. A. Thrush, *Chem. Phys. Lett.* **186**, 436 (1991).

CHAPTER 9

STATISTICAL PHYSICS OF LOCALIZED VIBRATIONS

Nikos Theodorakopoulos

Theoretical and Physical Chemistry Institute,
National Hellenic Research Foundation,
Vasileos Constantinou 48, 116 35 Athens, Greece
E-mail: nth@eie.gr

The paper discusses the role of nonlinear excitations at finite temperatures. The motivating principle is the description of one-dimensional thermodynamic instabilities (e.g. the Peyrard-Bishop (PB) model of DNA denaturation) as driven by the thermal statistical properties of domain-walls of macroscopic energy. The possible relevance of independent local modes to the finite-temperature spectra of the PB model is discussed. Finally, a minimal model of the helix-coil transition in polypeptides is presented; the model exhibits cooperativity comparable to that observed experimentally in polyalanine and, as such, it offers an interesting perspective for identifying finite-energy nonlinear structures as "agents" of the helix-coil transition.

1. Introduction/Outlook

This talk was prepared as a contribution to a discussion session. It is therefore by necessity a mixture of a "highlighted" progress report on an ongoing project and a somewhat speculative outlook for the near future.

One of the principal driving forces behind the introduction of solitons to the condensed matter/statistical physics community during the 70's was the need to describe structural phase transitions[1,2]. As it happened, the standard "paradigm" of the time, the harmonically bound one-dimensional chain with a bistable on-site potential, the ϕ^4 model, did not even exhibit a thermodynamic phase transition. Nonetheless, a substantial theoretical apparatus was developed over the years. Theorists, and to some extent experimentalists, learned how to deal with a robust non-linear object, the soliton, and the "phonon cloud" which characterizes its thermal properties. In some exceptional cases, those of integrable and near-integrable continuum[3] and discrete[4] systems, the "soliton gas" turned out to be a useful conceptual and computational tool for a wide variety of static and dynamic phenomena. Soliton statistical mechanics had matured - even in the absense of the phase transition it had originally been meant to describe.

A quarter-century later it was realized that the Peyrard-Bishop (PB)[6] model of thermal DNA denaturation provided the ideal case of a simple, yet non-trivial, one-dimensional model with a genuine thermodynamic phase transition driven by a domain-wall (DW) of infinite energy[7]. The statistical properties of that soliton-like excitation determine the transition temperature. Although some important details remain to be clarified - in particular regarding what happens in the extremely discrete limit - this is a key result obtained within LOCNET, which explicitly relates phase transitions to exact nonlinear structures. I will briefly review the theory behind it in Section 2.

Less progress has been made with the identification of generic dynamical signatures of discrete breathers (DB) at finite temperatures away from the limit of complete integrability[l]. Note however that self-focusing models (variants of the discrete nonlinear Schrödinger equation) have been reported[5] to exhibit signatures of long-lived nonlinear coherent structures at finite temperatures. The analysis of temperature-dependent dynamical structure factors of the PB model[8] reveals a strong central peak (CP) feature which becomes dominant at very long wavelengths near the critical point. We do not currently understand the exact microscopic mechanism for the emergence of the CP. A preliminary mode-coupling analysis however suggests that the memory kernel has features which are similar to those of force autocorrelations of DB's. A summary of pertinent results will be given in Section 3.

The description of DNA denaturation by the PB model and its refinements, which account for the nonlinear stacking interaction[9] and the helicoidal structure[10] opened a new perspective for microscopic, Hamiltonian models of cooperativity in biomolecules; such models, which take into account some of the salient energetic and dynamic features of biomolecules

---

[l]This should be contrasted with the case of Ref. [4c], which describes the thermodynamics of an integrable variant of the discrete, isotropic one-dimensional Heisenberg model in terms of an ideal gas of DB's (cf. above) and therefore clearly identifies populations of thermally excited DB's.

are likely to displace, in due time, the probabilistic description of all related cooperative phenomena. One of the areas which has attracted attention recently is the spontaneous helix formation in polypeptides; this is mainly justified by the uninterrupted interest in the secondary structure of proteins[m]; in addition, the recent availability of powerful all-atom computational schemes[12] has enhanced theoretical interest in the helix-coil transition which occurs in polypeptide chains.

Section 4 presents a preliminary report of a minimal microscopic model of helix formation and melting, based on the geometry of semiflexible chains and the anisotropy of the hydrogen bond. The model exhibits cooperativity comparable to that observed in poly-alanine. Further work is necessary in order to establish its static nonlinear structures and its dynamics; however, in view of its successful description of the "main cooperative event", this seems to be a promising research direction.

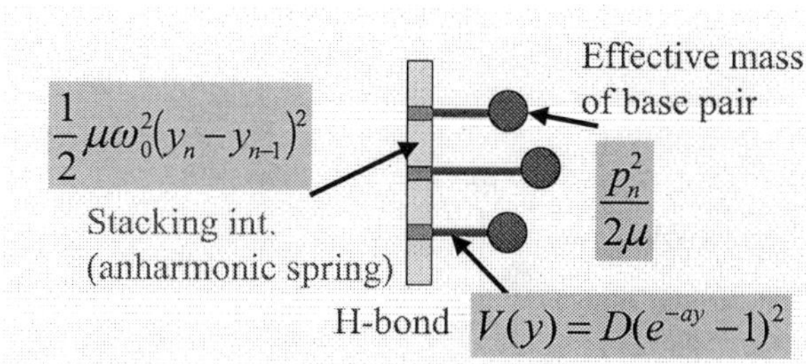

Fig. 1. A schematic illustration of the Peyrard-Bishop model of DNA denaturation.

2. Thermal DNA denaturation: A domain-wall driven transition?

The PB model of thermal denaturation of DNA is described[6] by the classical Hamiltonian

$$H = \sum_n \left[\frac{1}{2\mu} p_n^2 + \frac{1}{2} \mu \omega_0^2 (y_n - y_{n-1})^2 + V(y_n) \right] \quad , \tag{1}$$

[m] "The year 2001 marked the 50th anniversary of the α-helix, the first proposed protein secondary structure... However, despite numerous efforts ..., the detailed thermodynamic basis for helix formation and for the helix propensities of the amino acids is not yet well understood. Accurate values for such basic thermodynamic parameters as the changes in enthalpy and heat capacity on helix formation are still under debate..."[11]

where y_n and p_n are the relative transverse displacement and momentum, respectively, of the n-th base pair, μ its reduced mass, $V(y) = D(e^{-ay} - 1)^2$ is an on-site Morse potential which models the hydrogen bonding, and ω_0, a, and D are constants. The dimensionless parameter $R = Da^2/\mu\omega_0^2$ expresses the relative strength of on-site and elastic interactions. The model has been proposed in other physical contexts, such as the wetting of interfaces[13].

In general, the thermodynamics of (1) can be obtained by numerical solution of the singular transfer integral (TI) eigenvalue problem[14]; in the continuum approximation ($R \ll 1$), it is possible to map the TI problem to a Schrödinger-like equation[7]; the phase transition corresponds to the disappearance of the last bound state of that equation, and the concomitant divergence of the average transverse displacement $< y >$. This happens at a temperature

$$T_c = 2\left(\frac{2}{R}\right)^{1/2}\frac{D}{k_B} \quad . \tag{2}$$

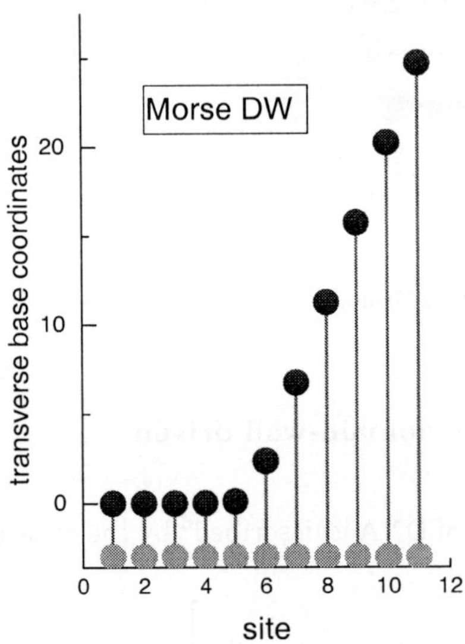

Fig. 2. The upper set of points describes the shape of the domain wall (3); the figure can be interpreted as a schematic picture of DNA unzipping if the two sets of points are taken to represent the two strands.

A domain wall (DW) is a static solution of the equations of motion which interpolates between the stable, global energy minimum $\{y_n = 0, \forall n\}$ and the metastable state where the harmonically bound particles are oscillating on the "flat-top" of the Morse potential; its form, in the limit $R \ll 1$, is[7]

$$y_n \approx \frac{1}{a} \ln \left[1 + e^{\pm\sqrt{2R}(n-n_0)}\right] \tag{3}$$

(cf. Fig. 2). The total energy of the DW is infinite - more precisely, it is proportional to the system's size. This must be so, since otherwise there would be no phase transition[15]. However, it is possible to calculate the amount of energy needed to shift the position of the DW by one lattice site. This amount is equal (for the upper sign of (3)) to

$$\Delta E_{DW} = E_{DW}(n_0 + 1) - E(n_0) = -2D \tag{4}$$

i.e. the wall moves spontaneously to the right and "zips" the system back to its global equilibrium position; conversely, there is a energy cost of $2D$ to shift the DW by one lattice site to the left, and thus "unzip", or "melt", one site from its global equlibrium position.

At finite temperatures the energetic cost of melting can be offset by an entropic gain; the entropy difference can be calculated e.g. by taking into account the different phonon frequencies, optical and acoustical, respectively, which characterize small oscillations in the left and right portions of the chain of Fig. 2; the result is[7]

$$\frac{\Delta S_{DW}}{k_B} = -\frac{1}{\pi} \int_0^{\infty} dq \; \ln \left[\frac{\omega_{ac}(q)}{\omega_{opt}(q)}\right] = \left(\frac{R}{2}\right)^{1/2} \tag{5}$$

and has a simple interpetation: the free energy cost for shifting the DW by one lattice site, $\Delta F_{DW} = \Delta E_{DW} - T\Delta S_{DW}$ vanishes exactly at the temperature (2). In other words, this analysis strongly suggests that the phase transition is driven by the balance of energetic cost vs. entropic gain involved in stabilizing a DW, so that it can move without resistance along the chain and "unzip" it.

3. ILMs in DNA dynamics?

We have done[8] extensive MD simulations of the model (1) for a value $R = 10.1$. Fig. 3 shows the spectra of dynamical correlations

$$S_{AA}(q, \omega) = \frac{1}{N} \int_{-\infty}^{+\infty} \frac{dt}{2\pi} \, e^{-i\omega t} \sum_{m,n} e^{iqa(n-m)} < A_n(t) A_m(0) > \quad , \tag{6}$$

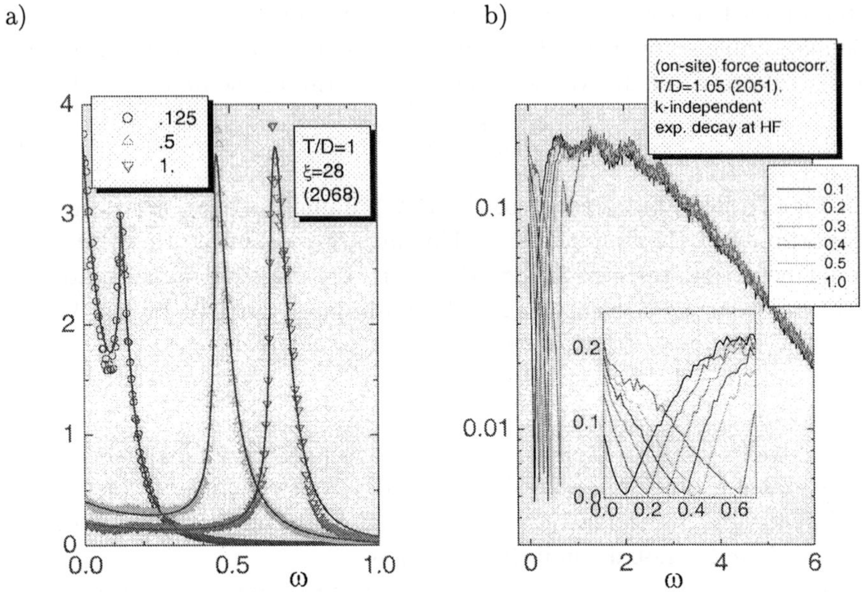

Fig. 3. Panel a: the normalized spectra S_{yy} at $k_B T/D = 1.0$. Fits are obtained with the Mori-Zwanzig Ansatz (Eqs. 7 -9) for three different values of q. The CP develops at long wavelengths, $q\xi/(2\pi) < 1$. Panel b: the force-force autocorrelations, S_{ff}; note the constant, q-independent slope at high frequencies, and the dips at low frequencies; the onset shows details of the dips, which occur at exactly the acoustic frequencies of the harmonic chain (vibrations on the flat top of the Morse potential).

where $A_n = y_n$, slightly below the critical point. The analysis of peak positions and linewidths reveals that these "subcritical" spectra depend mainly on whether the wavelength $2\pi/q$ is smaller or larger than the (static) critical correlation length ξ. In the first case, the spectra reflect the dynamics of the "islands" of the high-temperature phase; in the second, they are dominated by a strong central peak (CP) feature, with a - roughly - non-critical width. The emergence of the CP can be described, at a phenomenological level, in terms of a Mori-Zwanzig formalism, where the (normalized) spectrum is given by

$$\hat{S}_{yy}(q,\omega) = \frac{1}{\pi\omega} Im \frac{\omega_{0q}^2}{\omega_{0q}^2 - \omega^2 - i\omega\Gamma} \quad , \tag{7}$$

with $\omega_{0q}^2 = T/S(q)$, $S(q)$ the static structure factor, and Γ is the memory kernel. We have obtained satisfactory preliminary spectral fits (cf. Fig. 3a) with an Ansatz

$$Re\Gamma = \Gamma_0 \exp(-\omega/\omega_c) + \Gamma_1 \tag{8}$$

where

$$\Gamma_1 = \begin{cases} a(1-\omega/\hat{\omega}_q)^{3/2} & \text{if } \omega < \hat{\omega}_q \\ b(\omega-\hat{\omega}_q)^{1/2}\exp(-\omega/\omega_c) & \text{if } \omega > \hat{\omega}_q \end{cases}, \tag{9}$$

where $\hat{\omega}_q$ are the bare acoustic frequencies of (1) (i.e. the frequencies corresponding to the flat top of the Morse potential), a, b, Γ_0, ω_c are constants; $Im\Gamma$ has also been approximated by its peak value.

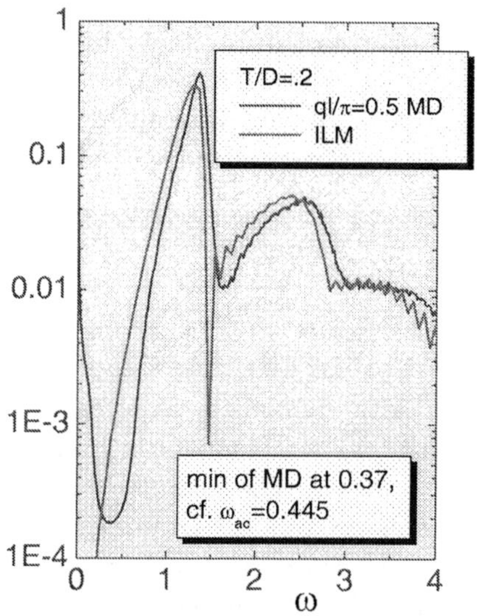

Fig. 4. Local force autocorrelations at low temperatures; note the dip near, but not quite at the bare acoustic frequency. The wavy curve is a theoretical estimate based on an independent localized mode picture, i.e. the normalized spectrum of (11).

Certain distinctive features of the phenomenological memory kernel (8) are evident in the spectra of the on-site force autocorrelations $f_n = -V'(y_n)$ (Fig. 3b). In particular (i) the dip occurs exactly at $\omega = \hat{\omega}_q$, just as in Eq. 9, and (ii) the constant slope at high frequencies is equal to ω_c.

A more clear-cut signature of independent local modes (ILM) occurs at low temperatures. Since $R = 10.1 \gg 1$, it is a reasonable first approximation to assume that, at low temperatures, the relevant ILM' s will be mostly excited at a single-site[16], i.e.

$$f_{\lambda,\delta}(t) = \lambda^2 \epsilon_\lambda^{1/2} \frac{\cos(\lambda t + \delta) - \epsilon_\lambda^{1/2}}{\left[1 - \epsilon_\lambda^{1/2} \cos(\lambda t + \delta)\right]^2} \tag{10}$$

where $0 < \lambda < \lambda_{max} = \sqrt{2}$, $0 < \delta < 2\pi$ depend on the initial conditions and $\epsilon_\lambda = 1 - \lambda^2/2$ is the energy of the ILM. The canonical average of autocorrelations of (10) is

$$< f(t)f(0) >= \int_0^{2\pi} \frac{d\delta}{2\pi} \int_0^{\lambda_{max}} d\lambda \; Q(\lambda) \; f_{\lambda,\delta}(t) f_{\lambda,\delta}(0) \tag{11}$$

where $Q(\lambda) = Z^{-1} \exp(-\epsilon_\lambda/T)$ and Z is determined from the normalization condition $\int d\lambda Q(\lambda) = 1$. The MD force spectra at $T = 0.2$ are shown in Fig. 4 along with a numerical Fourier transform of (11). Overall agreement is satisfactory, except for the very low frequency part of the spectrum.

4. Helix formation and melting in polypeptides

4.1. *Definitions, Notation*

I introduce a minimal Hamiltonian model of helix-forming polypeptides. The model consists of a polymer chain with monomers at sites $\{\vec{R}_n\}$, bonds $\{\vec{\Delta}_n \equiv \vec{R}_{n+1} - \vec{R}_n\}$ of fixed length $|\vec{\Delta}_n| = \ell$ and fixed bond angles $\{\cos\theta_n \equiv \vec{\Delta}_{n-1} \cdot \vec{\Delta}_n/\ell^2\}$ with $\theta_n = \theta \; , \forall n$. The chain's configuration can be completely described by the sequence of local azimuthal angles $\{\phi_n\}$("freely rotating chain", cf. Fig. 5a). The hydrogen bonding between nth and $(n+3)$rd monomers is modeled by a Morse potential whose overall strength is direction-dependent; thus, if $\vec{\Delta}_n^{(3)} \equiv \vec{R}_{n+3} - \vec{R}_n$ is the vector connecting nth and $(n+3)$rd monomers, the interaction U is strongest if $\vec{\Delta}_n^{(3)}$ is perpendicular to the plane defined by the monomers $n-1, n, n+1$, i.e. if the direction cosine

$$\chi(\phi_{n+1}, \phi_{n+2}) = \frac{\vec{\Delta}_n^{(3)} \cdot \vec{P}_n}{|\vec{\Delta}_n^{(3)}||\vec{P}_n|} \quad , \tag{12}$$

where $\vec{P}_n = \vec{\Delta}_n \times \vec{\Delta}_{n-1}$, is a maximum. This is quantified by the Ansatz

$$U_{n+1,n+2} = \{\chi(\phi_{n+1}, \phi_{n+2})\}^\sigma \; V_M \left[\rho(\phi_{n+2}) - \rho^*\right] \tag{13}$$

where $\rho(\phi_{n+2}) = |\vec{\Delta}_n^{(3)}|$, $\rho^* = \rho(\phi^*)$, $V_M(x) = D[\exp(-2ax) - 2\exp(-ax)]$, σ is an even integer, D and a are constants, and ϕ^* is determined from the condition

$$\frac{\partial}{\partial \phi_{n+1}} \chi(\phi_{n+1}, \phi_{n+2}) = 0 \quad , \{\phi_n = \phi^* \quad \forall n\} \quad . \tag{14}$$

In effect, the model assumes that the equilibrium length of the H-bond is "tuned" to the value of the azimuthal angle which also achieves maximum orientational ordering, i.e. an H-bond which starts at a given monomer is locally perpendicular to the plane defined by that monomer and its nearest neighbors.

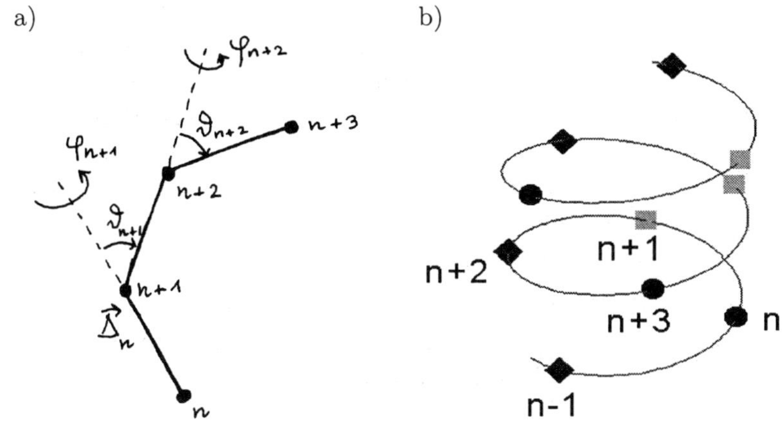

Fig. 5. Panel (a) illustrates the geometry of the freely rotating polymer. Bond lengths $|\vec{\Delta}_n|$ and bond angles $\pi - \theta_n$ remain fixed. The complete configuration can be described by the set of azimuthal angles of rotation; the convention is that the angle ϕ_{n+1} refers to rotation of the vector $\vec{\Delta}_{n+1}$ around the axis defined by $\vec{\Delta}_n$. Panel (b) shows the polymer on a helix, at given constant $|\vec{\Delta}_n| = l, \theta_n = \pi/2, \phi_n = \phi^* = 51.8°, \forall n$ (cf. Eq. 14). The vectors $\{\vec{\Delta}_n^{(3)} \equiv \vec{R}_{n+3} - \vec{R}_n\}$ are all at an angle $\gamma^* = 24.8°$ with the helical axis. Different symbols are used to emphasize the distinct coupling along the three "strands".

The total potential energy has the form

$$V(\{\phi_n\}) = \sum_n U_{n,n+1} \quad , \tag{15}$$

where end-of-the-chain effects have been neglected.

The following parameters were used: (i) $\theta = 90°$, which results in $\phi^* = 51.8°$, a rotation angle between successive monomers $\beta = 101°$, a tilt angle of the H-bonds with respect to the helical axis given by $\gamma^* = 24.8°$, and a ratio $\rho^*/\ell = 1.328$; (ii) $\ell = 4.066$ Å (corresponding to $\rho^* = 5.4$ Å); (iii) $D = 0.9 \times 10^{-20}$ *Joule* from the calorimetrically determined enthalpy $\Delta H = 1.3 kcal/mole$[17], and (iv) $a = 3.3 \text{Å}^{-1}$, a value which follows from the above D and the estimate of 19.5 N/m used for the H-bond spring constant of the α-helix[18]; finally, the H-bond anisotropy parameter has been chosen as $\sigma = 6$.

4.2. *Thermodynamics*

The configurational partition function of (15) can be calculated using standard TI techniques. The results are summarized in Fig. 6. The model predicts a smooth helix-coil transition with an effective melting point $k_B T_m/D = 0.32$ and a width of the specific heat peak (full width at half-maximum) $W/D = 0.15$. These values are of comparable magnitude with the ones deduced experimentally in the case of polyalanine[17], $k_B T_m^{expt}/D = 0.44$, and $W^{expt}/D = 0.08$.

A few comments are in order here

(1) The cooperativity comes in large part from the anisotropy of the H-bond. As shown in Fig. 6a, an isotropic($\sigma = 0$) interaction also results in a peak in the specific heat, but the peak is much broader and occurs at a lower temperature.

(2) In effect, the interaction favors the formation of a certain range of azimuthal angles around ϕ^* *and* a certain degree of parallel ordering (due to the anisotropy prefactor). This is illustrated schematically in Fig. 7a, which shows that, in order for the energy to fall within 1/4 of the Morse well, the angle ϕ must lie within 1/10th of its total available range of approximately $0 < \phi < \pi$ (for a particular helicity). This tendency can be emulated by a variant of the Potts model, which allows for Q internal states but has an attractive interaction only when nearest neightbors are in a *particular* state[19]. Fig. 7b shows that the $Q = 9$ variant of that Potts-like model reproduces quite closely the broad features of specific heat (peak position and width).

(3) These model considerations can be further improved if one takes account of the H-bond anisotropy in a properly symmetric function; it is then seen that one must demand that the vector $\vec{\Delta}_n^{(3)}$ should also be perpendicular to the plane defined by the monomers $n+2, n+3, n+4$ which

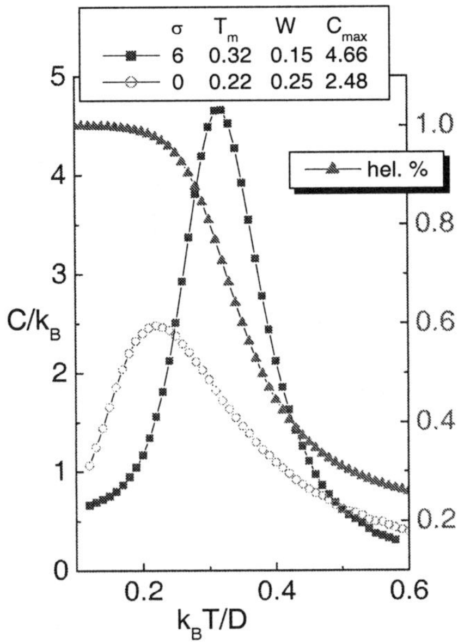

Fig. 6. A summary of the thermodynamics of the helix-coil transition of the model (cf. text). The specific heat (■, left y-axis) exhibits a peak of finite width; at roughly the same temperature range, the helix fraction (▲, right y-axis) drops drastically. For comparison, the specific heat in the case $\sigma = 0$ (no anisotropy) is also shown (○); cooperativity, as measured by e.g. the dimensionless width of the peak, is substantially weaker.

lie "above" it in the chain. Although the partition function which corresponds to such an interaction cannot be treated within the TI scheme, that of its Potts-like emulation (attractive interaction between *three* successive sites being at one particular of Q states) can be computed[20] and produces excellent agreement with the polyalanine data.

In summary, the model presented seems to provide a reasonable, semi-quantitative description of the thermodynamics of the helix-coil transition in polypeptides. Investigation of its dynamics is therefore likely to provide useful insight into the possible role of localized excitations in this fundamental biophysical process.

Acknowledgments

This work has been supported in part by EU contract HPRN-CT-1999-00163 (LOCNET network).

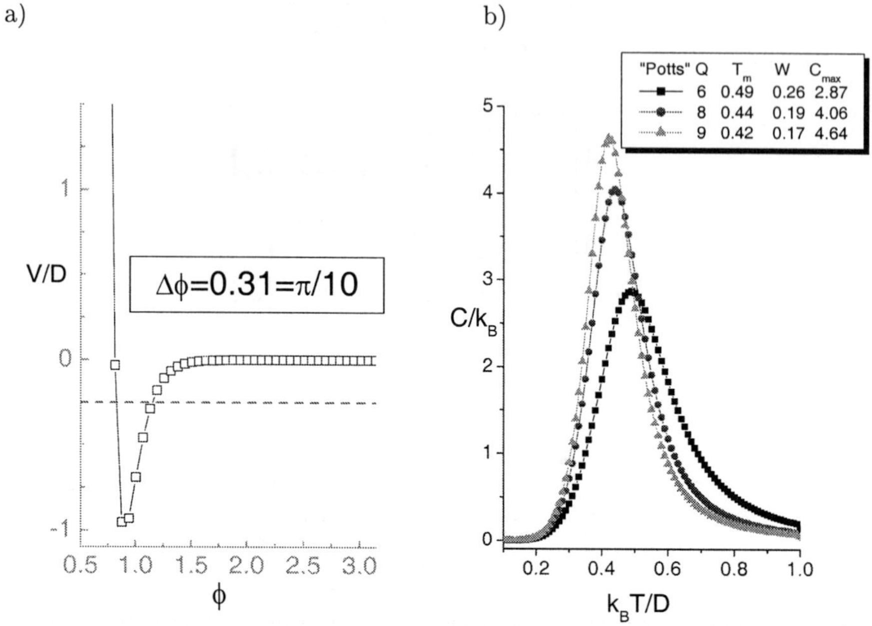

Fig. 7. Panel a: With the parameters used (cf. text) the Morse potential can bring about serious "focusing" of the azimuthal angle ϕ. The figure shows that an energy of 1/4 of the depth of the well corresponds to a spread $\Delta\phi = \pi/10$; therefore, the combined effect of the anisotropy factor and the Morse potential can be emulated by a Potts-like interaction (cf.text). Panel (b) shows the specific heat obtained for various "Potts-like" parameters.

References

1. S. Aubry, *J. Chem. Phys.* **62**, 3217 (1975); *ibid* **64**, 3392 (1976).
2. J. A. Krumhansl and J. R. Schrieffer, *Phys. Rev. B* **11**, 3535 (1975).
3. (a) J.F. Currie, J. A. Krumhansl, A.R. Bishop and S. E. Trullinger, *Phys. Rev. B* **22**, 480 (1980); (b) N. Theodorakopoulos *Phys. Rev. B* **30**, 4071 (1984); (c) J. Timonen, M. Stirland, D.J. Pilling, Yi Cheng and R.K. Bullough, *Phys. Rev. Lett.* **56**, 2233 (1986); (d) N-N Chen, M.D. Johnson and M. Fowler, *ibid* **56**, 904 (1986).
4. (a) N. Theodorakopoulos and N.C. Bacalis, *Phys. Rev. B* **46**, 10706 (1992); (b) N. Theodorakopoulos and M. Peyrard, *Phys. Rev. Lett.* **83**, 2293 (1999); (c) N. Theodorakopoulos, *Phys. Rev. B* **52**, 9507 (1995).
5. (a) B. Rumpf and A.C. Newell, *Phys. Rev. Lett.* **87** 054102 (2001); (b) K.O. Rasmussen, T. Cretegny, P.G. Kevrekidis and N. Gronbech-Jensen, *ibid* **84**, 3740 (2000); (c) R. Jordan and C. Josserand, *Phys. Rev. E* **61**, 1527 (2000).
6. M. Peyrard and A.R. Bishop, *Phys. Rev. Lett.* **62**, 2755 (1989).

7. T. Dauxois, N. Theodorakopoulos and M. Peyrard, *J. Stat. Phys.* **107**, 869 (2002).
8. N. Theodorakopoulos, M. Peyrard and T. Dauxois, *Proceedings, International Conference on "Localization and energy transfer in nonlinear systems", Escorial, Spain, 2002*, World Scientific (in press); http://arxiv.org/cond-mat/0211287 , and to be published.
9. T. Dauxois, M. Peyrard and A. R. Bishop, *Phys. Rev. E* **47**, R4 (1993).
10. M. Barbi, S. Lepri, M. Peyrard and N. Theodorakopoulos, to be published.
11. M.M. Lopez, D-H. Chin, R.L. Baldwin, and G.I. Makhatadze, *Proc. Natl. Acad. Sci. USA* **99**, 1298 (2002).
12. U.H.E. Hansmann and Y. Okamoto, *J. Chem. Phys.* **110**, 1267 (1999); *ibid* **111**, 1339 (1999).
13. D. M. Kroll and R. Lipowski, *Phys. Rev. B* **28**, 5273 (1983).
14. N. Theodorakopoulos, *Phys. Rev. E* **68**, 026109 (2003).
15. L. D. Landau and E. M. Lifshitz, *Statistical Physics*, Pergamon Press (1980).
16. J. L Marin and S. Aubry, *Nonlinearity* **9**, 1501 (1996).
17. J.M. Scholtz, S. Marqusee, R.L. Baldwin, E.J. York, J.M. Stewart, M. Santoro and D.W. Bolen, *Proc. Natl. Acad. Sci. USA* **88**, 2854 (1991).
18. A.C. Scott, *Phys. Rev. A* **26**, 578 (1981).
19. Sh. A. Hairyan, E. Sh. Mamasakhlisov and V.F. Morosov, *Biopolymers* **35**, 75 (1994).
20. N. Theodorakopoulos, to be published.

CHAPTER 10

LOCALIZATION AND TARGETED TRANSFER OF ATOMIC-SCALE NONLINEAR EXCITATIONS: PERSPECTIVES FOR APPLICATIONS

G. Kopidakis* and S. Aubry**

** Department of Materials Science and Technology, University of Crete*
P.O. Box 2208, 71003 Heraklion, Crete, Greece
E-mail: kopidaki@materials.uoc.gr
*** Laboratoire Léon Brillouin (CEA-CNRS), CEA Saclay*
91191-Gif-sur-Yvette Cedex, France
E-mail: aubry@llb.saclay.cea.fr

We review nonlinearity-induced localization in discrete systems with emphasis on theory and numerical calculations on models used in materials physics to describe interatomic interactions. We discuss how the concept of discrete breather or intrinsic localized mode could become an important tool for understanding nanoscale phenomena in molecules and solids. A particularly attractive field of application for nonlinear localized excitations is the controlled and directed transport of energy. We discuss the recently proposed targeted transfer of localized excitations based on the concept of nonlinear resonance and its potential applications. As an area for such applications, we present directional ultrafast electron transfer at low temperatures using this selective transfer mechanism and we give examples from biological electron transfer. We finally discuss possible applications in nanotechnology and biomolecules.

1. Introduction

Many physical processes related to wave propagation are described by nonlinear differential equations that have spatially localized solutions. Localization of excitations due to nonlinearity, often referred to as self-focusing or self-trapping, has been studied extensively theoretically, numerically and experimentally. Solitons and breathers, which are exact solutions of continuous integrable nonlinear models that describe macroscopic, mesoscopic and even microscopic phenomena, have been observed in diverse fields such as fluid mechanics, optics, atomic and condensed matter physics, *etc.* The advances in theory, computational resources and experimental techniques have led to an unprecedented interest in the field of nonlinear dynamics over the last few decades. A recent spectacular example of enormous activity related to nonlinear excitations is the field of Bose-Einstein condensates. The relatively recent beautiful and neat experiments, rapidly became the experimental testbed and favorite toy not only for a lot of theorists in atomic physics but also in interdisciplinary nonlinear science, were the nonlinear Schrödinger equation had already been exhaustively studied and results could apply directly. The current interest in fields where macroscopic and mesoscopic nonlinear excitations are directly observed in experiments somewhat overshadows research on nonlinear excitations in real materials and biological matter at the atomic level. At this scale, the manifestation of localized excitations is indirect and a linear theory, quantum mechanics, applies. However, as we try to show in this short discussion, nonlinear excitations are important in modeling and simulations of real materials. We discuss nonlinear localized modes in relation to vibrational and electronic properties, but many results are universal and apply to spin and magnetic properties, as well as to the forementioned, and many more, mesoscopic and macroscopic phenomena.

In condensed matter physics one very often reduces the complexity of the problems by using appropriate elementary excitations that describe phenomena in the simplest manner. In periodic crystals, most ground state properties are obtained by taking advantage of lattice translational invariance and harmonic approximations for interatomic potentials. Then, one deals with propagating oscillations of atoms, the phonons, of spins, the magnons, of metallic surface free electrons, the plasmons, *etc.* We refer to these as linear excitations. But there are cases where nonlinear equations appear even in the atomic scale. The concept of polaron, an elementary excitation consisting of an electron and the lattice distortion it induces and in which it is localized, is well established since first proposed by Landau[1] and is used to explain electronic transport properties of many materials. In this case, the coupling of a quantum electron with a deformable medium generates nonlinearity in the effective Schrödinger equation.[2] Another well studied example (theoretically) of atomic scale nonlinear excitation is the Davydov soliton.[3,4] High frequency vibrations coupled to low frequency modes give rise to a self-trapped state which was proposed for energy transport in proteins. Some difficulties with the mobility and robustness at biologically relevant temperatures of the discrete Davydov soliton have reduced the initial excitement, but recent experiments have proved that such self-localized exciton states do exist.[5] Many chemists are familiar with the concept of localized nonlinear excitation. Local modes were succesfully used in the early days of spectroscopy to interprete vibrational spectra of molecules.[6] then forgotten and rediscovered.[7,4,8] These localized vibrations in molecules fit into the general framework of the relatively recent Discrete Breather theory. Originally developed for infinite periodic lattices, this theory predicts the existence of spatially localized and temporally periodic solutions as a consequence of discreteness and nonlinearity. This type of nonlinear excitation was termed Discrete Breather (DB) or Intrinsic Localized Mode (ILM). In the following, we briefly review some basic theoretical results on DBs but there are complete and comprehensive presentations on the topic.[9,10] We present some of the properties of these nonlinear localized excitations in detail because we believe that work on DBs offers new conceptual, theoretical and computational tools to go beyond integrable models in all research fields.

Why DBs should be present and play an important role in condensed matter and materials physics? The very existence of discrete breathers (DBs) requires discreteness and nonlinearity. Both of these conditions are met in real matter: it consists of atoms that interact through nonlinear interatomic potentials. It is very interesting then to explore the possibility of energy localization through DBs at the atomic level and its manifestation and consequencies. Moreover, once this dynamic localization is understood, novel mechanisms for energy transfer at the nanoscale may be envisioned. This brief presentation of some of our results in the field of nonlinear localized modes and their selective transfer is restricted to a pedagogical discussion and to relatively simple examples. For the complete discussion and technical details, the reader is referred to the original publications.

Within the context of classical dynamical systems, DB theory is by now mathematically well founded and a large number of analytical and numerical works have established their domain of existence, the mechanisms for their creation and their manifestation in a variety of physical systems. A number of experiments in macroscopic and mesoscopic systems have confirmed theory and numerics beyond any doubt. These include arrays of nonlinear optical waveguides,[11] arrays of Josephson junctions,[12] micromechanical oscillator arrays,[13] *etc.* However, when atomic vibrations or electronic properties are involved, the situation is not completely clarified, even though experimental evidence is present in the vibrational spectra of nonlinear materials,[14] local modes in molecules,[4] spin excitations,[15] myoglobin,[16] *etc.* This is mainly because DBs in atomic scale can only be observed indirectly. In addition, quantum effects have to be taken into account. While phenomena at the atomic scale are governed by quantum mechanics, classical mechanics continues to be of enormous importance for their study. Beside the fact that our physical intuition is based on classical concepts, realistic computer simulations involving atoms in materials physics, such as molecular dynamics, remain classical. Even when electrons are treated quantum mechanically, atoms obey Newton's equations of motion. It is a very challenging task to understand atomic localized anharmonic vibrations that come out of classical theory and simulations, their relevance to and their manifestation in actual experiments. For instance, a lot of effort has been devoted in chemical physics in order to exploit the concepts and methods of semiclassical mechanics in the theory of highly excited molecules.[8]

Understanding these phenomena at the atomic scale is not only important from the fundamental point of view, it is also very crucial at a time when nanotechnology and molecular electronics take center stage in materials research and applications. Nanodevices, molecular chips and memories have already been fabricated and appear to be much more promising in terms of performance and capacity than conventional semiconductor technology products. Moreover, a lot of puzzling phenomena of energy storage and selective transfer in biomolecules could be explained in terms of the ideas and concepts developed in our field. It is clear that many conventional condensed matter physics approaches, which were very successful up to now, become obsolete and even irrelevant in biology and electronics at the molecular level. We try to give a simple discussion of our work on localization and targeted transfer of nonlinear excitations that may be relevant to nanoscale phenomena. As an example where our theories could apply, we present the ultrafast electron transfer in reaction centers of photosynthetic units, an example where dynamics takes over from thermodynamics. Conventional electron transfer theories are based on thermodynamics and explain thermally activated processes. We show that in a lot of very interesting cases, such as photosynthetic reaction centers, nonlinear electron dynamics could account for all experimentally observed features.

This discussion is organized as follows: In Sec. 2 we introduce DBs in periodic lattices. Theoretical and numerical studies until recently were exclusively devoted to these systems and toy models were used for interparticle interactions. We review in addition in some detail results we obtained for realistic models.[17] More specifically, carbon-hydrogen stretch vibrations are shown to become DBs for large oscillation amplitudes in a variety of structures (molecules, solid surfaces, *etc*) and nonlinear localized vibrations in the diamond structure are discussed. We then briefly review DBs in random systems, where the lack of translational invariance may provide a linear spectrum with localized states. Starting from small systems, that could represent molecules with different atomic elements, and extrapolating to infinite systems, that could represent amorphous materials, we explain the interplay of disorder and nonlinearity.

In Sec. 3 we review targeted energy transfer, *i.e.*, the nonlinear resonant mechanism for energy transfer between nonlinear oscillators or DBs. The case of the integrable dimer is presented as an example, as well as coupled random chains sustaining DBs.

In Sec. 4 we discuss how electron transfer becomes a nonlinear dynamical problem at low temperatures and introduce our model. The principle of targeted transfer is then used to describe electron transfer between two molecules which is mediated by a third appropriately chosen "catalyst".

We finally conclude in Sec. 5, with a summary and future perspectives.

2. Discrete Breathers

The discrete breather or intrinsic localized mode is a localized nonlinear excitation which we think is highly relevant for atomistic calculations and simulations. Although the term is often used in a broader sense, in this discussion we call Discrete Breathers (DBs) the spatially localized time-periodic solutions of discrete classical nonlinear Hamiltonian systems. Discreteness provides gaps and bounds to the linear oscillations spectrum and nonlinearity makes the amplitude of oscillation frequency dependent. Thus, DBs may be found with frequency and harmonics outside the linear spectrum. Ever since these generic solutions were proposed,[18] rigorous proofs of existence have been given. A lot of them work close to the so-called anti-continuous limit, which may be appropriately defined for different models.[9,19,20,21,22,23] One example is the first rigorous proof of existence of DBs.[19] In infinite chains of weakly coupled anharmonic oscillators, the system can be reduced to an array of uncoupled anharmonic oscillators, where the existence of a local mode is trivial (a single oscillating element). Then, the implicit function theorem is used to prove that this exact solution persists when the oscillators are coupled. Other methods are purely variational and do not use any anti-continuous limit[24] or are based on the center manifold reduction.[25] DBs existence proofs hold for systems with optical or/and acoustic phonons (the term phonon is normally reserved for atomic vibrations but we often use it to describe any linear excitation). Moreover, they show that DBs are universal and may be found in finite or infinite systems, at any dimension, periodic or non-periodic and disordered, of arbitrary complexity, *etc.* In integrable models, localized solutions are not generic and do not persist under (weak) model perturbations. The key mathematical property that renders DBs highly relevant in physics is that they are universal, exact, robust, and, very often, linearly stable one parameter family (parametrized by their frequency-action) solutions of discrete nonlinear models independent of their complexity. Real materials of course are more complex than the models used to describe them and temperature always perturbs exact solutions, but stability properties imply that DBs may survive for a very long time on thermalized background and interacting with phonons. Besides giving a solid mathematical basis for DBs, existence theorems lead to numerical methods for their exact calculation.[26] This is very important since DB solutions can almost never be expressed in analytic form, not even for toy models. We have used extensively self-consistent methods of numerical continuation based on Newton-Raphson type schemes (appendices in Refs. 27, 28).

2.1. *DBs in periodic lattices*

From the point of view of the condensed matter physicist, the existence of spatially localized states in periodic crystals is rather counter-intuitive. In these systems, one usually works with propagating states and localized states are associated with impurities, disorder, *etc.* DB theory predicts that localized states may appear in perfectly periodic systems if the amplitude of a local oscillation becomes large and the system is driven to frequencies (and with harmonics) outside the linear spectrum, *i.e.*, the "band" of frequencies of linear excitations. We call these states extraband DBs (EDBs). The existence of intraband DBs (IDBs) in periodic crystals, *i.e.*, localized states with frequencies that belong to the linear spectrum, is mathematically forbidden. From physical considerations alone, IDBs cannot survive in this case, since any resonance of the DB with the extended phonons (when the DB frequency or harmonics belong to the linear spectrum) will radiate the DB away. Studies on DBs are largely focused on periodic lattices, where the effect of localization is entirely due to nonlinearity. Theoretical and numerical calculations were until restricted to toy models. We consider now as examples of such simple models one dimensional chains of harmonically coupled identical nonlinear oscillators. These so-called Klein-Gordon chains are described by Hamiltonians of the form

$$H_{KG} = \sum_i \frac{1}{2}\dot{u}_i^2 + V_i(u_i) + \frac{C}{2}(u_{i+1} - u_i)^2 \,, \tag{1}$$

where the atoms i with displacement u_i and unit mass are submitted to an on-site nonlinear potential $V(u_i)$ with minimum at $u_i = 0$ and are elastically coupled to their nearest neighbors with coupling constant C. Expanding $V(u_i)$ close to equilibrium, the harmonic part of the potential

$$V_i(u_i) = \frac{1}{2}u_i^2 + \dots \tag{2}$$

gives the dispersion relation, i.e., the linear phonon frequency at wave vector q

$$\omega(q) = \sqrt{1 + 4C\sin^2\frac{q}{2}} \,. \tag{3}$$

The phonon spectrum is the interval $[1, \sqrt{1+4C}]$ and exhibits a gap $[0, 1]$. An anti-continuous limit is obtained for this model at $C = 0$, when the anharmonic oscillators are uncoupled. Then, the motion of each oscillator is periodic and its frequency $\omega(I)$ depends on its amplitude or equivalently on its action I (which is the area of the closed loop in the phase space $(u, \dot{u})$). Thus, we generally have $d\omega(I)/dI \neq 0$. There are trivial DB solutions at the anti-continuous limit corresponding, for example, to a single oscillator oscillating at frequency ω_b while the other oscillators are immobile. If $n\omega_b \neq 1$ for any integer n, *i.e.*, the breather frequency and its harmonics are not equal to the degenerate phonon frequency $\omega(q)$ at $C = 0$, the implicit function theorem can be used for proving that this solution persists up to some non-zero coupling C as a DB solution at frequency ω_b.[19] The equations of motion corresponding to Hamiltonian (1) are:

$$\ddot{u}_i + V_i'(u_i) - C(u_{i+1} + u_{i-1} - 2u_i) = 0 \,. \tag{4}$$

The linearized equations around solution $u_i(t)$ with period t_b are used for the linear stability analysis.

$$\ddot{\epsilon}_i + V_i''(u_i)\epsilon_i - C(\epsilon_{i+1} + \epsilon_{i-1} - 2\epsilon_i) = 0 \,. \tag{5}$$

DB solutions may be found for Eqs. (4) by continuation as described in Appendix A of Ref. 27.

In Fig. 1 a DB solution is continued from the uncoupled limit ($C = 0$) for fixed ω_b with on-site potential

$$V_i(u_i) = \frac{1}{2}u_i^2 + \frac{1}{4}u_i^4 \,. \tag{6}$$

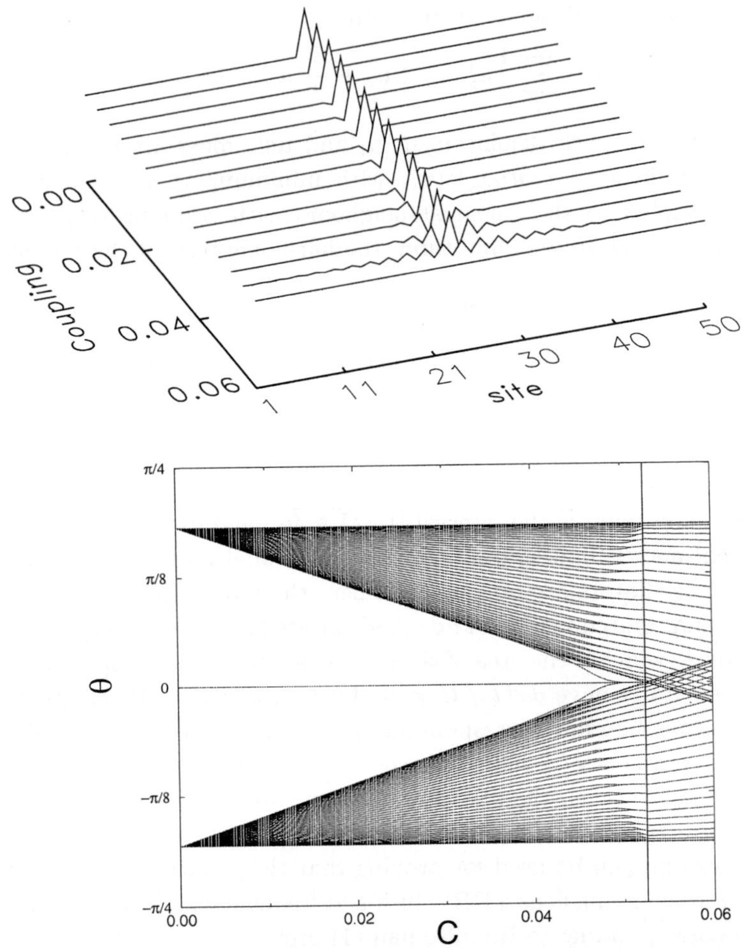

Fig. 1. Continuation from the anti-continuous limit of a DB solution for a Klein-Gordon chain with on-site quartic potential, Eq. (6) in the text, for $\omega_b = 1.1$. Top: the displacements of the sites at the beginning of the period versus the coupling constant C. Bottom: the arguments θ of the Floquet matrix are plotted. Continuation stops at resonance with upper edge of phonon band at $C = 0.0525$.

This is a "hard" potential, *i.e.*, the amplitude of oscillation increases when the frequency ω_b increases. Extraband DBs appear for $\omega_b^2 > 1 + 4C$. The DB solution disappears when it becomes resonant with the phonons $\omega_b^2 =$

$1 + 4C$, *i.e.*, at $C = 0.0525$. In Fig. 2(top) a DB solution is continued from

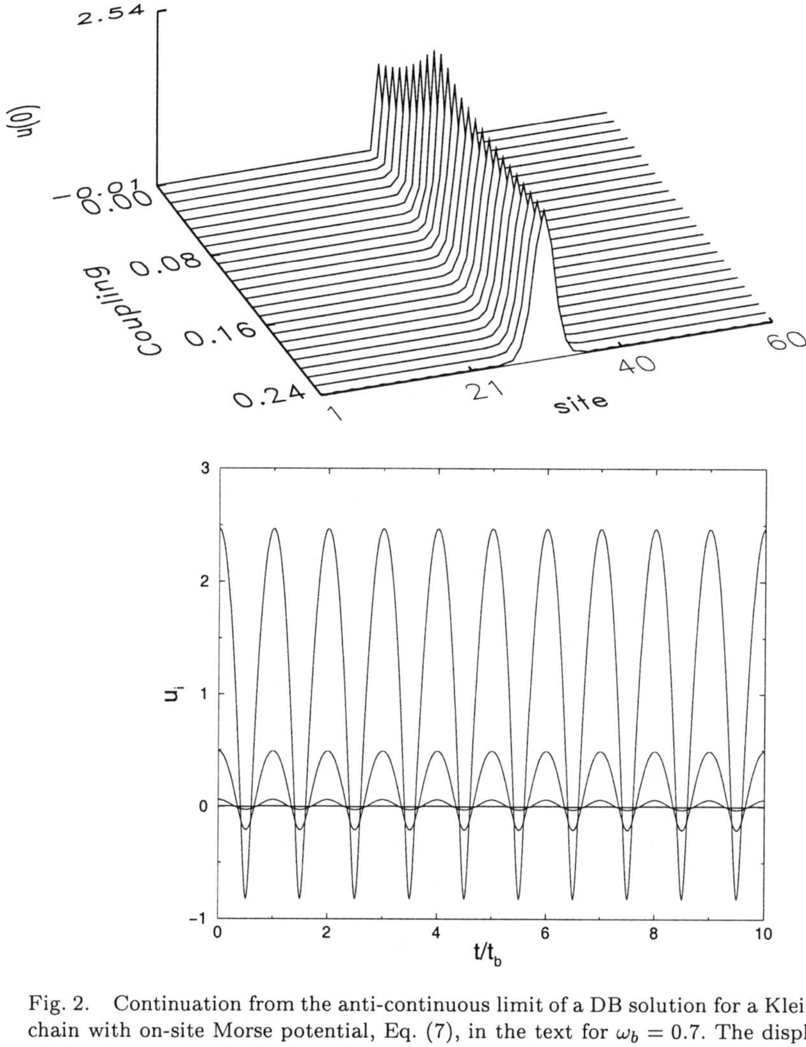

Fig. 2. Continuation from the anti-continuous limit of a DB solution for a Klein-Gordon chain with on-site Morse potential, Eq. (7), in the text for $\omega_b = 0.7$. The displacements of the sites at the beginning of the period versus the coupling constant C (top) and the time evolution of the displacements for $\omega_b = 0.7, C = 0.08$ (bottom) are plotted. Continuation stops at the resonance of the first harmonic with the lower edge of the phonon band at $C = 0.24$.

the uncoupled limit for fixed ω_b with the Morse on-site potential at $\omega_b = 0.7$

$$V_i(u_i) = \frac{1}{2}(1 - e^{-u_i})^2 \tag{7}$$

and the time evolution of the DB solution for $C = 0.08$ is shown (bottom). Morse is a "soft" potential, *i.e.*, the amplitude of the solution decreases when ω_b increases, and is the basis for many realistic potentials. Extraband DBs appear for $\omega_b < 1$ (as long as the harmonics are not resonant with the phonon band).

Similar, but appropriately modified, methods are used to find DBs in the so-called FPU chains, where interaction potentials between sites are nonlinear.[24]

The self-consistent methods for the numerical calculation of DBs become computationally expensive for large systems when realistic models are used. A large number of real time numerical simulations in these simple models show that DBs may appear spontaneously under conditions out of thermodynamic equilibrium. For instance, extremely rapid quenching of thermalized systems creates "hot spots" that are nothing but DBs.[29] In many cases we find approximate DBs may be found in simulations simply by initially exciting one site (*e.g.*, with a relatively large displacement from equilibrium) and perhaps its nearest neighbors (to a smaller degree) and follow the time evolution of the system. In Fig. 3 we show the example of a DB in the Morse potential created by local excitation. One can see from the time evolution of the displacements in Fig. 3 (top) that the solution is very close to the exact DB of Fig. 2(bottom). The frequency spectrum (obtained by Fourier transforming the velocities) in Fig. 3(bottom) clearly shows the fundamental DB frequency and its harmonics. The weakly excited phonons are also present in this figure, since this is an approximate DB solution (see insert in Fig. 3, bottom). It is interesting that the DB persists with this background of phonons for very long times. This is expected from stability analysis and for weak perturbations but the DB lifetime when interacting with phonons, in general, depends on model, temperature, *etc.*

DBs are usually studied in toy models, in order to understand their complex universal properties. It is however crucial to demonstrate their existence in more complex realistic models used in materials physics. Realistic models usually contain a large number of parameters which are obtained by fitting to structural, vibrational, elastic, *etc.*, properties of materials that come out of experiments or *ab initio* calculations. A lot of empirical models for interatomic interactions have been developed with classical potentials and, although electronic properties are not included, some of them are

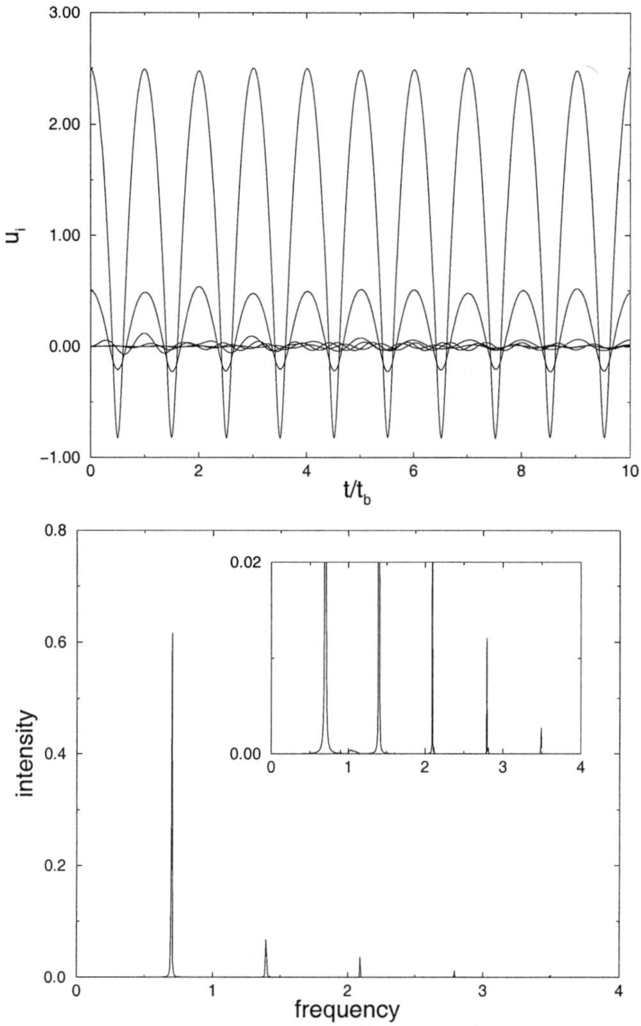

Fig. 3. Time evolution for the displacements (top) and the corresponding spectrum (bottom) for a DB obtained with a local excitation for the model of Fig. 2.

very successful in predicting structural and vibrational properties of specific materials. One such potential developed for silicon[30] and for carbon[31] is based on the Morse potential, where DBs are well studied (briefly discussed earlier), appropriately modified so that the strength of interaction between atoms varies according to the number of neighbors (weaker bonds for higher number of neighbors) and with fine tuned parameters. Since DBs existence is independent of dimension and model complexity, DBs may be found in this case. The rigid diamond structure may give localized stretch-Bond!Stretching vibrations with frequencies above the phonon spectrum when the interatomic potential is convex, as predicted in Ref. 24. As an initial condition of a molecular dynamics simulation in the microcanonical ensemble (NVE, *i.e.*, constant number of atoms, volume, energy) we have locally excited a bond by stretching it a significant length away from the equilibrium distance and followed the time evolution of the system. For certain values of the initial displacement the excitation stays localized almost exclusively on the excited bond instead of dispersing throughout the crystal for both silicon and carbon. In silicon, the displacement profiles fit to the standard picture of DBs, while in carbon, the equilibrium positions of the two oscillating atoms are shifted away from each other (*i.e.*, local defect is created). A complete discussion will be given elsewhere, as well as results with more sophisticated models for these materials.

A clear picture, which is consistent with experiments and ideas developed in chemical physics, emerges from the study of the carbon-hydrogen stretch vibration in several structures containing these two basic elements. We have demonstrated the existence of DBs in such systems within semi-empirical models through Tight-Binding Molecular Dynamics (TBMD) simulations.[17] TBMD is a compromise between empirical classical molecular dynamics simulations, that can handle a large number of atoms and for relatively long times but do not describe electronic properties, and first principles molecular dynamics, which are the most accurate but are limited to small systems and short times. TBMD preserves the quantum mechanical treatment of the electronic part, which is included in a Hamiltonian with matrix elements that are adjustable parameters determined by fitting to *ab initio* calculations. Having a reliable and transferable model for carbon-carbon interactions,[32] with all different bonding environments and coordination numbers, is very important. The same applies for carbon-hydrogen interactions.[33] The models we use have been successful in predicting structural, electronic, vibrational and elastic properties in diverse systems that range from carbon microclusters, fullerenes and hydrocarbon molecules to liquid and amorphous carbon.[34,35] The fact that purely anharmonic effects such as the temperature dependence of vibrational frequency shifts and linewidths in diamond and graphite are described in perfect agreement with experiment[36] is particularly important in the present context. We have described some of our results in detail in Ref. 17. Here we only give a few

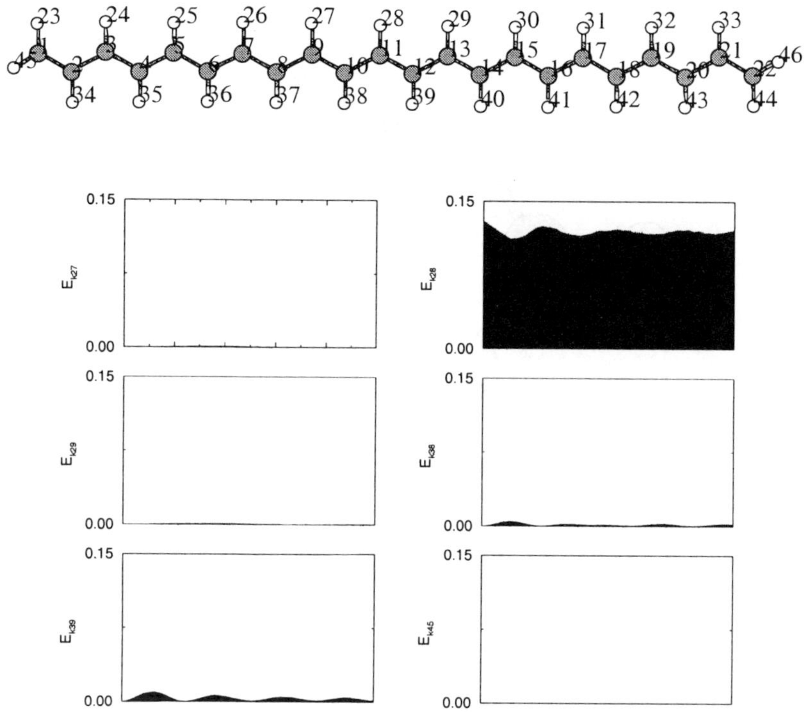

Fig. 4. Top: equilibrium configuration of $C_{22}H_{24}$, $r_{ch} = 1.1$ Å, $r_{cc} = 1.42$Å (single), $r_{cc} = 1.38$ Å (double). Bottom: kinetic energy of some H atoms, E_{ki} (in eV) versus time during the first 7 ps of the relaxation. H-atom 28 initially displaced by 0.17 Å along the C-H stretch direction. The energy remains localized on H-atom 28.

examples.

In Fig. 4, an example of a DB in the $C_{22}H_{24}$ molecule (its equilibrium geometry is shown on top) is presented. With the molecule in the ground state, a hydrogen atom is pulled from equilibrium in the carbon-hydrogen stretchBond!Stretching direction by 15%, well within the anharmonic part of the C-H potential. The time evolution is followed during many picoseconds and the excitation stays localized on the initially excited bond. Displacements and spectra have the same features as Fig. 3, the DB of the toy model with Morse potential. On the contrary, small amplitude initial excitations quickly disperse throughout the chain, as expected for linear propagating modes.

In Fig. 5, a DB in benzene is presented. A carbon-hydrogen bond is

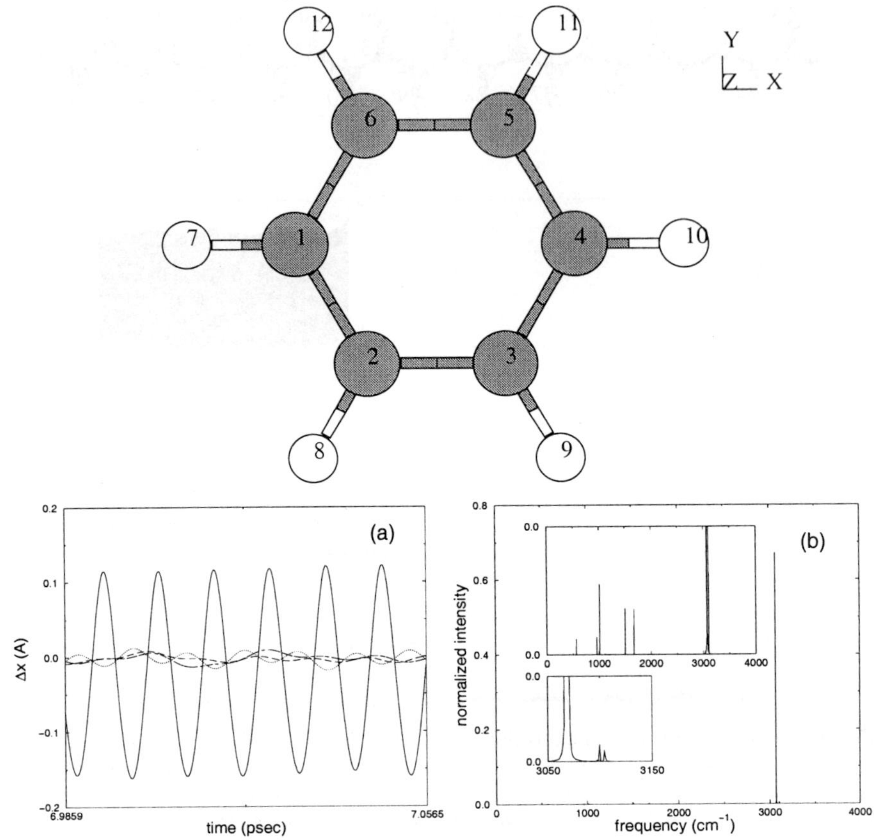

Fig. 5. Equilibrium configuration of C_6H_6, $r_{ch} = 1.100$ Å, $r_{cc} = 1.396$ Å (top). At the bottom, in (a), the displacements in the x-direction of the H-atoms from the equilibrium position when atom 7 is initially displaced by $\Delta x = 0.1743$ Å ($\Delta E = 0.38$ eV) versus time for a short time interval. The large amplitude oscillation is for the H-atom 7. The rest of the atoms oscillate with much smaller amplitudes. In (b) the vibrational spectrum from the initial condition of (a). The two figures in conjunction bear all the hallmarks of DB.

initially excited in a way similar to the previous example of Fig. 4 and the oscillation remains localized where it was initiated for many picoseconds. In the lower left part of Fig. 5 we show the displacements versus time of all hydrogen atoms during a small portion of the simulation. The initially excited hydrogen oscillates with a much larger amplitude than its neighbors. The corresponding vibrational spectrum is shown to the right (from data collected over several picoseconds) where a single peak dominates at a frequency that does not correspond to any vibrational normal mode of C_6H_6.

It takes a very large magnification (insets in the main figure (b)) to be able to see the very weakly excited normal modes. We thus have an anharmonic oscillation localized practically on a single bond and having a frequency which is smaller than the frequency of the normal C-H stretch modes, and also harmonics (not shown in the figure) that are not resonant with any linear modes. This is a DB on a background of weakly excited phonons, like the toy model of Fig. 3. These results are consistent with classical DB theory. For large displacements, the excited C-H stretching vibration is a nonlinear oscillator weakly coupled to the other C-H oscillators, which are at rest (this is numerically verified), *i.e.*, close to the anti-continuous limit. Its frequency decreases with increasing oscillation amplitude (the C-H interaction potential is i"soft").[17] When this vibration is not resonant with any linear normal modes, something plausible for C-H stretch modes that have linear frequencies well above and separated from the rest of the normal mode frequencies of the molecule, it persists as an almost exact DB solution. The early spectroscopic assignments in C_6H_6 and other molecules used the concept of local mode.[6] Highly excited vibrational states of molecules can be better explained on the basis of local modes than normal modes. What we did then here is to rediscover local modes as DBs. Something still needed though is to understand the effect that the classical treatment of atoms has on the results (especially for the low energy states).

We have also presented TBMD results for hydrogen on diamomd surfaces and on the localized carbon-carbon stretch in a diamond cluster, Fig. 6. While for the C-H stretch results are clearly consistent with DBs (with even stronger localization due to the surface geometry), Fig. 6(bottom left), the question of the C-C stretching localized vibration needs further investigation Fig. 6(bottom right). The TBMD results on this and other diamond clusters (similar but larger to the one shown here) and supercells show that there is localization but it is not clear whether it is a stable DB (in the strict sense of the term that we use here) or a relatively short lived localized vibration (contrary to the C-H DBs that persist for long simulations of the order of ns).

All examples presented so far are single site DBs. We have performed detailed studies of multibreathers, *i.e.*, DBs where more than one sites are excited in the system, in periodic models. We just mention here that these investigations shed light into nonlinear standing waves stability, ultimately related to DB formation and system thermalization.[37] Multibreather bifurcations in finite periodic systems are also related to the intraband DBs, *i.e.*, DBs with frequencies inside the discrete linear spectrum (strictly pro-

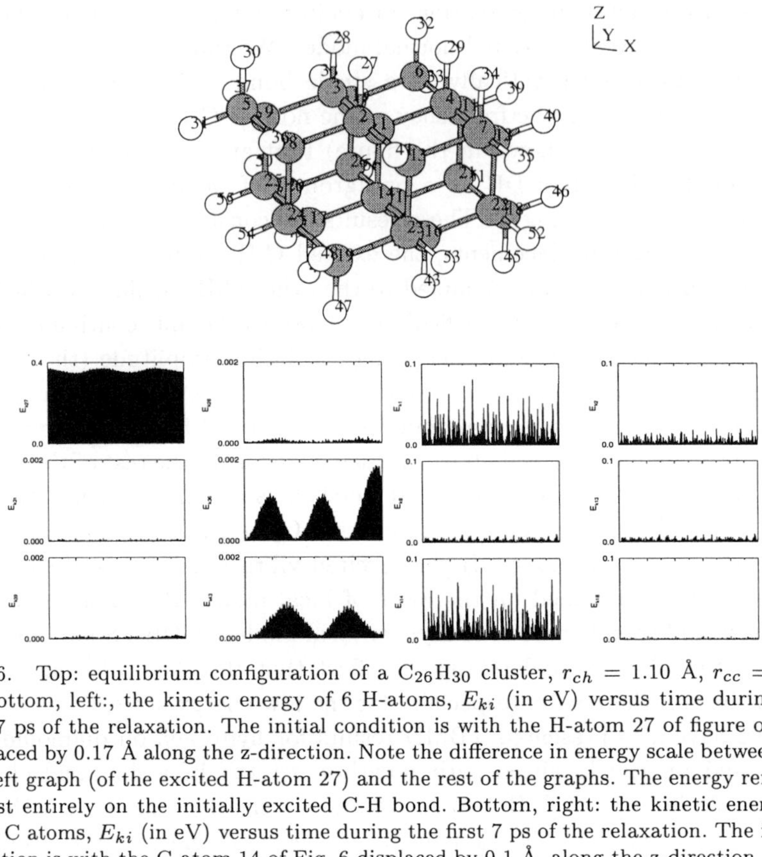

Fig. 6. Top: equilibrium configuration of a $C_{26}H_{30}$ cluster, $r_{ch} = 1.10$ Å, $r_{cc} = 1.54$ Å. Bottom, left:, the kinetic energy of 6 H-atoms, E_{ki} (in eV) versus time during the first 7 ps of the relaxation. The initial condition is with the H-atom 27 of figure on top displaced by 0.17 Å along the z-direction. Note the difference in energy scale between the top left graph (of the excited H-atom 27) and the rest of the graphs. The energy remains almost entirely on the initially excited C-H bond. Bottom, right: the kinetic energy of some C atoms, E_{ki} (in eV) versus time during the first 7 ps of the relaxation. The initial condition is with the C-atom 14 of Fig. 6 displaced by 0.1 Å along the z-direction. Most of the energy stays localized on the C-C stretch vibration (atoms 1 and 14).

hibited in infinite systems with continuous linear band), which we termed "phantom breathers".[38]

A very interesting property of DBs in some models and under some particular conditions is mobility.Breather!Moving Although not universal, the phenomenon of DB motion is attracting a lot of interest since it offers a possibility for energy transport in nonlinear discrete chains. In Fig. 7, a DB in a Klein-Gordon chain with the Morse potential, Eq. (7), is set to move. Note that moving DBs have an infinite tail of very small amplitude and in Fig. 7 the traveling excitation is not a mathematically exact solution. Although it appears that when present, moving DBs may be robust to perturbations and they could be relevant for energy transport in biomolecules

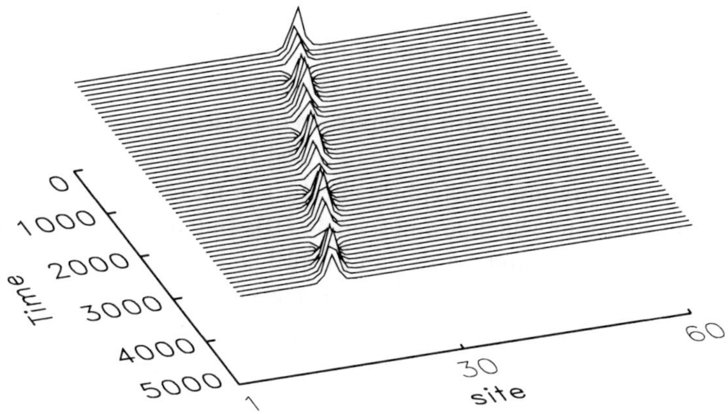

Fig. 7. Under special conditions DBs can be set to motion, but with an infinite tail. In this case, a Klein-Gordon chain with the Morse potential, Eq. (7), and $\omega_b = 0.7125, C = 0.1$. The time evolution of the displacements of the sites is plotted.

and other systems at the atomic level in some cases, it is clear that due to the complexity of biosystems, in most cases they cannot be the agents for energy targeting and focusing. In Sec. 3 we propose a different robust mechanism based on nonlinear resonances that gives selectivity and specificity in energy transfer through localized nonlinear excitations.

2.2. *DBs in random systems*

While the existence of DBs in periodic lattices, which are localized excitations due to nonlinearity without any impurities or disorder being involved, is an interesting novel paradigm, real materials are usually not periodic. We have studied in detail DBs in lattices that are not periodic, starting from small, going to large and extrapolating to infinite systems.[27,39,28] The results, which clarify the interplay of disorder and nonlinearity, are summarized here. When the Hamiltonian lattice is random (for example linearly coupled random anharmonic oscillators), the existence proof for extraband DBs (EDBs), *i.e.*, when the frequency of the DB and its harmonics do not belong to the linear phonon spectrum, is identical to the one for periodic lattices. When the linear modes are localized due to disorder (Anderson modes),[40] it was demonstrated[27,39,28] that disorder and nonlinearity in the same system play a double game. For some time-periodic solutions, they cooperate for maintaining mode localization thus generating intraband DBs (IDBs). Other time-periodic solutions where found to be spatially extended and transport energy, thus resulting in delocalization. In this case, resonances between oscillators with different frequencies are restored by the nonlinearities, since their frequencies are tuned by their amplitude.

In order to understand this subtle effect, it is necessary to investigate time-periodic solutions which are not single DBs but multiDBs. These solutions can be easily found at the anti-continuous limit by choosing an arbitrary subset of the uncoupled nonlinear oscillators (finite or infinite) oscillating at the same frequency. It can be proven in the same way as for the single DBs that when their frequency and harmonics do not belong to the linear phonon spectrum (for periodic systems as well as for random systems) these multiDB solutions persist up to some non-vanishing coupling as exact solutions. Among them, there is an infinite number of linearly stable multiDB solutions.[9] We investigated on the basis of a combination of numerical and analytical arguments how the frequency of these exact solutions could penetrate the phonon spectrum when it is discrete (for example when the model randomness is large enough).

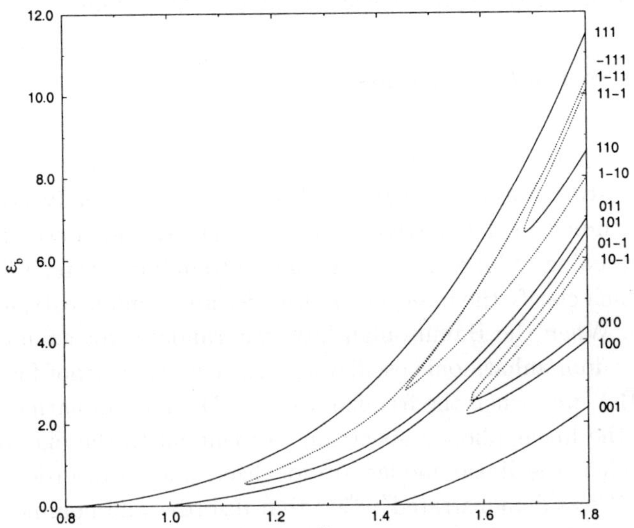

Fig. 8. Energy $\mathcal{E}_b$ versus ω_b of the DB and multiDB solutions of a system with three sites at coupling $C = 0.1$. The local frequencies in Eq. (8) are $\omega_{1,0} = 0.86953452$, $\omega_{2,0} = 0.71504480$, and $\omega_{3,0} = 1.34578317$. The frequencies of the linearized modes are $\omega_1 = 0.99414762$, $\omega_2 = 0.81450423$, and $\omega_3 = 1.42364498$.

In Fig. 8 we present a bifurcation diagram for a small Klein-Gordon oscillators system, a model similar to Eq. (6), but with randomness included in the local potential $V_i(x)$ of each oscillator in the linear frequency

$$V_i(x) = \frac{1}{2}\omega_{i,0}^2 x^2 + \frac{1}{4}x^4 \,, \tag{8}$$

where $\omega_{i,0}^2$ are local parameters chosen randomly with a uniform probability distribution in a given interval $\left[\omega_{min,0}^2, \omega_{max,0}^2\right]$. Considering the stable equilibrium solutions of the corresponding equations of motion (4) $\{u_i(t) \equiv 0\}$, we obtain from the linearized Eqs. (5) the eigenequations that give the phonon spectrum associated with the rest solution. Each phonon mode is well peaked around its maximum on a single site and thus, every eigenfrequency of the linearized system ω_i can be labeled with the specific site i. All possible multiDBs, calculated by continuation, are considered. The energy of each one as a function of its frequency is plotted on Fig. 8. The multiDBs are labeled by a coding sequence (1 for site oscillating at the anti-continuous limit with frequency ω_b, -1 for site oscillating with opposite phase, 0 for site at rest). Many such bifurcation diagrams were considered in order to understand the penetration of multiDBs in the phonon band.

As the system size increases, careful continuation of IDBs from the anti-continuous limit (or by varying some other model parameter) becomes impractical. In order to find IDBs we developed an *in situ* continuation method, the Nonlinear Response Manifold method (NLRM).[39,28]. A DB solution with frequency ω_b on a site i is found by applying a time-periodic force on i with fixed ω_b. We continue the trivial vanishing solution as a function of the amplitude of the force and follow the 1D manifold of solutions that allows this amplitude to return to zero, where we obtain a non-trivial time-periodic and time-reversible solution which is necessarily a localized IDB. This IDB is located on i or, when its existence on i is forbidden, on a nearby resonant site. Further continuation of the manifold yields new vanishings of the external force that correspond to more IDBs localized at several sites. This NLRM method gives systematically and *in situ* spatially localized IDBs at the computer accuracy, independent of model, parameter values, and without using any uncoupled, or other limit. In other words, instead of solving the dynamical equations (4) we search for time periodic solutions of the set of equations

$$\ddot{u}_j + V_j'(u_j) - C(u_{j+1} + u_{j-1} - 2u_j) = \lambda\delta_{i,j}\cos\omega_b t \,. \tag{9}$$

The sine force with amplitude λ is applied on the site i where we wish to create the single DB with frequency ω_b or at the sites where we wish

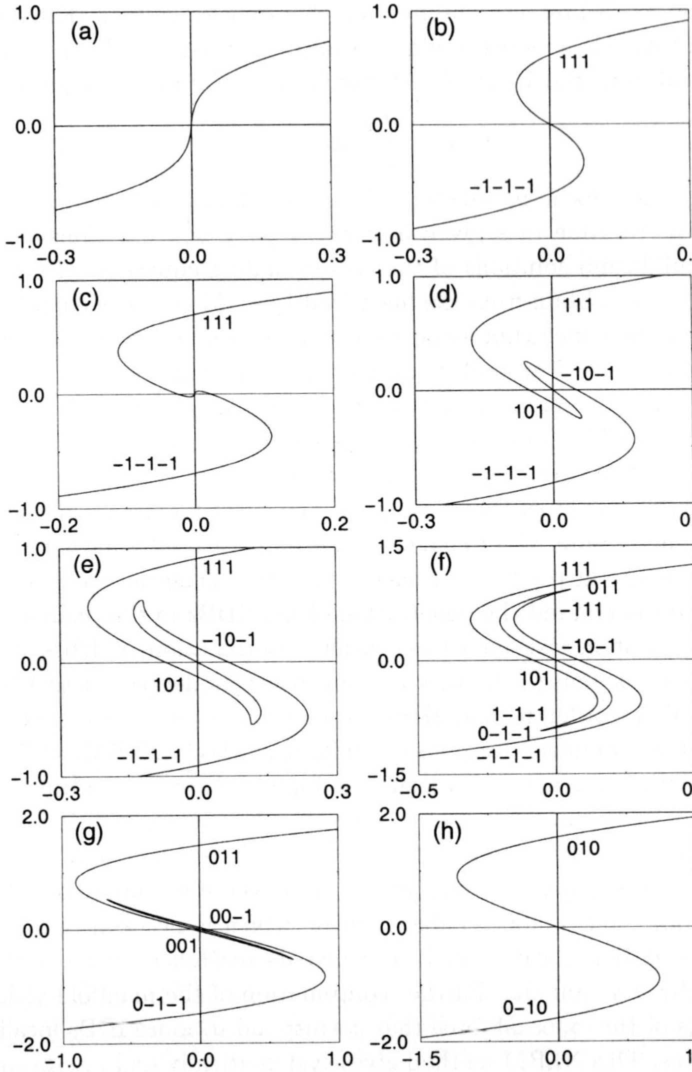

Fig. 9. Projection on site 2, $u_2(0)$ vs λ, of the nonlinear response manifold for $\omega_b = 0.8$ (a), 0.95 (b),0.99 (c), 1.05 (d), 1.10 (e), 1.15 (f), 1.50 (g), 1.60 (h) for the 3-site system of Fig. 8, the force applied at site 2.

to create a multiDB. We start from a known solution of these equations, like the trivial solution at rest for λ=0, and continue as a function of λ with a Newton method (Appendix A of Ref. 28). If the model were linear the response would be a linear function of λ. In our nonlinear problem the manifold is not a straight line in the phase space, but a folded 1D manifold. For a system with size N, this 1D NLRM is embedded in the space $\mathbf{R}^{N+1}$. It is simpler and sufficient to represent only its projection on the 2D plane corresponding to the initial position $u_i(0)$ of oscillator i and λ. In Fig. 9 we show the evolution of the NLRM for various ω_b for the tree-site system of Fig. 8. The same DBs solutions are obtained.

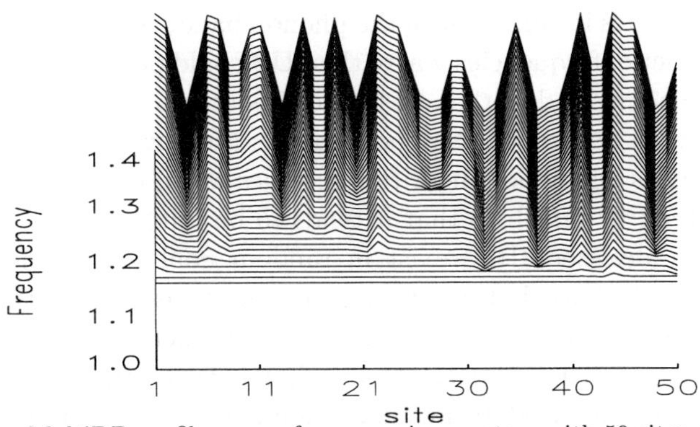

Fig. 10. MultiDB profile versus frequency in a system with 50 sites and $C = 0.05$. The linear localized mode $\omega_4 4$ becomes extended when frequency increases (for the "hard" potential, Eq. (8) in text).

Using diagrams, such as the ones presented in Figs. 8, 9 and other numerical data for several systems of different sizes, as well as analytical arguments, we obtained the following generic conlusions.

Any single EDB or any spatially localized multiDB systematically disappears by bifurcating with another multiDB before its frequency penetrates the linear phonon band edge. This bifurcation is caused by the resonance between the DB frequency and the linear Anderson modes with frequencies close to the band edge (see Fig. 8).

There are spatially extended time-periodic solutions obtained by continuation of certain multiDBs while their frequency enters the phonon spectrum till a certain intraband frequency At this point, the multiDBs disappear by bifurcating with another multiDB. For these multiDBs, the sites associated with the linear modes that become resonant as frequency varies are oscillating with appropriate phases. As a result, the resonance with the linear mode becomes ineffective (Fig. 8).

Some of these multiDBs can be continued to the linear limit at zero amplitude but remain extended modes. Conversely, the strict continuation of localized linear Anderson modes at nonzero amplitude is possible but it yields immediately extended (but sparse) multiDBs as soon as the amplitude is non-zero (Fig. 8 and Fig. 10).

Each localized linear Anderson mode can be continued, but in an approximate sense only, as a spatially localized IDB. There is a generally small discontinuity, each time IDB frequency crosses a linear mode frequency. The resonance with the linear mode transfers the DB energy to this mode, thus defocusing its energy and destroying the DB. Since for infinite systems this set of linear frequencies is dense inside the phonon band, there are infinitely many gaps in the set of frequencies where this IDB could persist. It has been proven though in some models that this set of frequencies is non-void. It is a fat Cantor set with non-zero Lebesgue measure. Moreover, in the limit of small DB amplitudes where nonlinearities becomes weak, the gap widths of this Cantor set tend to vanish so that it tends to be full.[41]

The fact that localized IDBs exist in random nonlinear models, but become extended solutions that can propagate energy as soon as their frequency is continuously varied, could lead to a non-phenomenological understanding of disordered systems. On one hand, even small nonlinearities delocalize purely localized linear modes, implying that a non-vanishing residual thermal conductivity persists at low temperature and drops to zero at 0 K. On the other hand, a substantial part of energy may be spontaneously trapped as localized IDBs for very long times, that could become macroscopic at low temperature. These effects (easily checked by molecular dynamics) provide an alternative interpretation for slow relaxation processes in glasses, such as persistent spectral hole burning, usually described by two-level system models.

3. Targeted energy transfer

When two coupled harmonic oscillators are resonant, any amount of energy initially injected on one of them oscillates between these oscillators with a frequency proportional to their coupling. In a harmonic periodic system, where many identical linear oscillators are coupled, there is energy propagation. In order for the energy to remain localized, resonance has to be broken somehow (with impurities, quasiperiodic modulation, disorder, *etc.*).

For anharmonic oscillators, the situation is more interesting, as their frequencies depend on the amplitude of oscillation. Therefore, in a dimer consisting of nonlinear oscillators, even when an appropriate amount of energy is initially on a "donor" oscillator, such that its frequency is equal to the linear frequency of the other "acceptor" oscillator at rest, resonance is broken after some energy transfer, since the frequencies change, and the transfer stops. However, we found the conditions for which the resonance between two weakly coupled nonlinear oscillators persists all along a complete energy transfer. We termed this exceptional fine tuned situation Targeted Energy Transfer (TET) because it is highly selective. We showed that the theory applies to systems consisting of many nonlinear oscillators, resulting in a selective, directed, and robust transfer of vibrational, electronic, Davydov solitons, and other excitations (depending on the physical system). Here, we briefly discuss these ideas on specific examples. An extended discussion is presented in the relevant publications.[42,43] Since we are interested in atomic-scale phenomena of the real world in biological and materials physics and chemistry, we do not discuss potential applications in Bose-Einstein condensates, Josephson junctions, nonlinear photonic crystals *etc.* Nevertheless, it would be very interesting to apply the concepts we introduce in some of these "clean" systems where theory is similar and the very well controlled experiments could demonstrate their validity.

The work reviewed here focuses more on potential applications in biological physics. Some of the current most puzzling problems in the field involve energy self-focusing and transport in biopolymers. In a photosynthetic unit, for example, consisting of an aggregate of many different chlorophyll molecules, light harvesting occurs through photon capture by the antenna-like function of the unit. Subsequently, energy self-focusing takes place and light is transferred in the form of an exciton coherently through a complex cascade of transfer within and in-between pigment proteins and upon reaching a photosynthetic unit the energy is released. This process lasts less than a picosecond and it is very efficient (about 95%) (see *e.g.* Ref. 44). Bioenergetics also requires that localized energy deposition through, *e.g.*, ATP hydrolysis is transferred almost without losses over relatively large distance enabling conformational biomolecular changes and conversion into mechanical energy (biological motors) (see *e.g.* Ref. 45). These and other molecular transfer features led several years ago Davydov to propose that solitons might be the coherent agents of energy transfer in biological molecules.[3,4] While this transfer mechanism is quite appealing, subsequent work by several groups has shown several theoretical as well as practical weaknesses to the soliton idea.[4,46]

Targeted transfer is an alternative, more general principle that could be applicable to a wide range of physical, chemical, and biological phenomena. In the subsections that follow, we describe the general theory of TET between anharmonic oscillators or nonlinear systems sustaining DBs in general, illustrate it on an integrable discrete nonlinear Schrödinger dimer, and apply it for weakly DBs in weakly coupled chains. We expect that the general methodology for detecting the possibility for TET applies to more complex models and, possibly, to *ab initio* calculations. It can be applied to vibrational, electronic, polaronic, excitonic, *etc.*, transfers, and as a result it could help resolve questions that range from bioenergetics to dynamics of chemical reactions, synthetic and biological catalysts, *etc.* Future work should connect the present approach to standard exciton transport theories. TET can also be used as a guiding principle for the design of nanodevices with specific energy transfer features. In the next section we apply TET in low temperature ultrafast electron transfer.

3.1. *Nonlinear resonance*

Consider an anharmonic system, the donor D, described by a set of conjugate variables and a Hamiltonian H_D. This could be an anharmonic oscillator or a DB (which comes as continuous family parametrized through its action) and its energy is a function of the action I_D, *i.e.*, $H_D(I_D)$, with frequency $\omega_D = \frac{dH_D}{dI_D}$. Similarly, consider an acceptor A with $\omega_A = \frac{dH_A}{dI_A}$ weakly coupled to D, so that the Hamiltonian of the system is:

$$H = H_D(I_D) + H_A(I_A) + H_V(I_D, \theta_D, I_A, \theta_A) . \tag{10}$$

The small coupling term H_V depends on all conjugate variables in general. Systems A and D could physically represent two independent molecules, two linked parts of the same macromolecule, two weakly coupled layers or atomic clusters, *etc.*, interacting through hydrogen bonds, van der Waals forces, screened Coulomb interactions, could be mediated by a solvent, *etc.* When the coupling is weak enough, it can be shown that the total action is practically conserved over very long time. Thus, during the adiabatic transfer:

$$I_T = I_D + I_A . \tag{11}$$

To prove the conditions for complete energy transfer in the generalized dimer from D to A, we changed to the conjugate variables $((I_D+I_A)/2, \theta_D+\theta_A)$ and $((I_D - I_A)/2, \theta_D - \theta_A)$ so that the topology of the space phase was conveniently represented on the surface of a sphere. The results were obtained by considering the orbits on this surface.[42] For this presentation it is more convenient to give these results in a different form.

Let I be the action transferred to A, *i.e.*, $I = I_A$, so that the action on D is $I_D = I_T - I$ (from Eq. (11)). With the assumptions of weak coupling and (almost) resonance, the total angle $\theta_0 = \theta_D + \theta_A$ rotates fast (at the order of the oscillators frequency), while the phase difference $\theta = \theta_D - \theta_A$ is slowly varying. The adiabatic approximation consists of averaging over the fast variable θ_0, so that the coupling energy H_V becomes a function of the total action, the action transferred and the phase difference only (the new variables are introduced as well):

$$< H_V(I_T, I, \theta_0, \theta) >_{\theta_0} = V_0(I_T, I) + V(I_T, I, \theta)\,. \tag{12}$$

In Eq. (12) the averaged coupling energy is written in two parts, $V_0(I_T, I)$, which is the average of $< H_V(I_T, I, \theta_0, I_\theta) >_{\theta_0}$ with respect to θ at fixed (I_T, I), and $V(I_T, I, \theta)$, which is the oscillating part with zero average.

Thus, Hamiltonian (Eq. (10)) becomes:

$$H = H_0(I_T, I) + V(I_T, I, \theta)\,, \tag{13}$$

where $V(I_T, I, \theta)$ is a small perturbation to the part of the Hamiltonian which is independent of θ:

$$H_0 = H_D(I_T - I) + H_A(I) + V_0(I_T, I)\,. \tag{14}$$

We remind that I_T is the total action and I is the action transferred to the acceptor. Since the coupling V is very small in Eq. (13), energy conservation can be written as

$$E_T = H_0(I_T, I_T) = H_0(I_T, 0) \simeq H_0(I_T, I)\,. \tag{15}$$

Eq. (15) expresses the fact that the energy E_T should be the same at the end of the transfer, when all energy is on A (first equality), at the beginning of the transfer, when all energy is on D (second equality), and any moment during the transfer (third equality). It is natural then to define the detuning function

$$\epsilon_T(I) = H_0(I_T, I) - H_0(I_T, 0)\,, \tag{16}$$

which is the variation of energy as a function of the action transferred to A. Note that, according to Eqs. (15), in the case of complete transfer, $\epsilon_T(I)$ is zero at the beginning (when everything is on D, no action transferred, $I = 0$) and at the end (when everything is on A, all action transferred, $I = I_T$). If it is zero for all values of I (*i.e.*, if the semi-equality becomes equality in Eq. (15), the transfer is perfect and, from its derivative with respect to I, the frequencies of the oscillators remain equal during the transfer, *i.e.*, in resonance. In fact, $\epsilon_T(I)$ does not have to be zero. We proved[42] that for a physically reasonable coupling with one maximum and one minimum per period with respect to θ, the necessary and sufficient condition for complete energy transfer is

$$\min_{\theta} V(I_T, I, \theta) < \epsilon_T(I) < \max_{\theta} V(I_T, I, \theta), \quad (17)$$

for all $0 \le I \le I_T$. The fact that $\epsilon_T(0) = \epsilon_T(I) = 0$, or equivalently Eq. (15) $E_T = H_0(I_T, I_T) = H_0(I_T, 0)$ defines the energy E_T and the action I_T for which complete transfer occurs gives the selectivity of the nonlinear resonant mechanism we introduce. The degree of selectivity is determined by the strength of the coupling. The weaker the coupling, the more selective the transfer becomes. The systems $D - A$ have to be tuned. The detuning function ϵ_T is, in some sense, a dynamical barrier that the system has to overcome in order to transfer, but it can be positive or negative.

3.2. *Targeted energy transfer in a nonlinear dimer*

As the simplest, lowest order approximation to a nonlinear $D - A$ system, we consider the following Hamiltonian

$$H = \mu_D|\psi_D|^2 + \frac{1}{2}\chi_D|\psi_D|^4 + \mu_A|\psi_A|^2 + \frac{1}{2}\chi_A|\psi_A|^4 - \lambda(\psi_D\psi_A^* + \psi_A\psi_D^*), \quad (18)$$

which describes two weakly coupled anharmonic oscillators, the donor (D) and the acceptor (A), with linear frequencies μ_X and nonlinear parameters χ_X ($X = D, A$) coupled through a linear term with constant λ (positive here). $(\psi_X^*, i\psi_X))$ are pairs of complex conjugate variables. The corresponding equations of motion are discrete nonlinear Schrödinger (DNLS) equations:

$$i\dot{\psi}_D = (\mu_D + \chi_D|\psi_D|^2)\psi_D - \lambda\psi_A, \quad (19)$$

$$i\dot{\psi}_A = (\mu_A + \chi_A|\psi_A|^2)\psi_A - \lambda\psi_D. \quad (20)$$

Besides their use in a variety of problems in nonlinear optics, polarons, excitons, Davydov solitons, discrete solitons, Bose-Einstein condensates, just to name a few, DNLS Hamiltonians are good approximations of Klein-Gordon systems in the limit of small amplitude where the anharmonic terms can be treated at the lowest significant order. However, they are special since they are norm-conserving. Although such DNLS integrable dimers have been extensively studied in all of the forementioned research fields, we are not aware of any previous work that gives the analytic TET solution.

Here we write down some of the equations that are necessary to explain the basic principles. The complete analytical solutions and discussion can be found in Ref. 42.

Switching to action-angle variables,

$$\psi_D = \sqrt{I_D} e^{-i\theta_D}, \ \psi_A = \sqrt{I_A} e^{-i\theta_A}, \tag{21}$$

Hamiltonian (18) becomes

$$H = \mu_D I_D + \frac{\chi_D}{2} I_D^2 + \mu_A I_A + \frac{\chi_A}{2} I_A^2 - 2\lambda\sqrt{I_D I_A} \cos(\theta_D - \theta_A) . \tag{22}$$

with the corresponding equations of motion

$$\dot{I}_D = -\dot{I}_A = -2\lambda\sqrt{I_D I_A} \sin(\theta_D - \theta_A), \tag{23}$$

$$\dot{\theta}_D = \mu_D + \chi_D I_D - \lambda\sqrt{I_A/I_D} \cos(\theta_D - \theta_A), \tag{24}$$

$$\dot{\theta}_A = \mu_A + \chi_A I_A - \lambda\sqrt{I_D/I_A} \cos(\theta_D - \theta_A) . \tag{25}$$

Using the definitions and results of the previous subsection 3.1, let the total action be I_T, the action transferred to the acceptor A be I, *i.e.*,

$$I_D = I_T - I, \qquad I_A = I, \qquad \theta = \theta_D - \theta_A \tag{26}$$

and let the total energy be E_T. Hamiltonian (22) is now written in the form of Eqs. (13,14) $H = H_0 + V(I_T, I, \theta)$, where $H_0 = H_D(I_T - I) + H_A(I) + V_0(I_T, I)$ with terms:

$$H_D(I_T - I) = \mu_D(I_T - I) + \frac{\chi_D}{2}(I_T - I)^2 , \tag{27}$$

$$H_A(I) = \mu_A I + \frac{\chi_A}{2} I^2 , \tag{28}$$

$$V_0(I_T, I) = 0 , \tag{29}$$

$$V(I_T, I, \theta) = -2\lambda\sqrt{(I_T - I)I} \cos\theta . \tag{30}$$

Complete transfer occurs when all of I_T initially on D is finally on A, Eq. (15), so that the total action (and energy) of the system are specified:

$$E_T = \mu_D I_T + \frac{\chi_D}{2} I_T^2 = \mu_A I_T + \frac{\chi_A}{2} I_T^2 \Rightarrow I_T = \frac{-2(\mu_D - \mu_A)}{\chi_D - \chi_A} . \tag{31}$$

The detuning function (16) is found in this case (using (28),(29),(31)):

$$\epsilon_T(I) = \frac{1}{2}(\chi_D + \chi_A)(I - I_T)I . \tag{32}$$

From Eq. (32) it is inferred that, in the DNLS dimer, complete transfer is produced when

$$\chi_D = -\chi_A , \tag{33}$$

for any coupling, since $\epsilon_T(I)$ is identically zero. Note that from Eq. (31) and since I_T has to be positive, the difference $\mu_D - \mu_A$ and χ_D must have opposite signs when $\chi_D = -\chi_A$. Then, the DNLS dimer is perfectly tuned when the nonlinear coefficients are opposite. In this case the Hamiltonian (22), (18) becomes $H = E_T - 2\lambda\sqrt{I_D \cdot I_A}\cos(\theta_D - \theta_A) = E_T - \lambda(\psi_D^*\psi_A + \psi_A^*\psi_D)$ and the solutions of the dynamical equations are simply $I_D = I_T\cos^2(\lambda t)$ and $I_A = I_T\sin^2(\lambda t)$, *i.e.*, the energy oscillates back and forth between D and A with frequency $\omega_T = 2\lambda$.

In addition to the ideal fine tuned situation just described, targeted transfer can be obtained according to the general condition Eq. (17), which for this dimer gives

$$\lambda > \left| \frac{(\chi_D + \chi_A)(\mu_D - \mu_A)}{4(\chi_D - \chi_A)} \right| . \tag{34}$$

When $\chi_D = -\chi_A$ complete transfer is always obtained for non-zero coupling. It is never obtained for $\chi_D = \chi_A$, unless $\mu_D = \mu_A$. Then, for this symmetric dimer, I_T is undetermined. Energy transfer always occurs (not selective) for $\lambda > \chi_D I_T/2$.

Up until now, we only discussed Hamiltonian oscillators. We will present later examples with damping. This is highly relevant for physical applications, since real $D - A$ systems are subjected to several forms of dissipation. Then, the complete periodic exchange and oscillation of energy during TET becomes very often an irreversible transfer.

While the integrable DNLS dimer presented here and studied in detail in Ref. 42 appears to be special, we have shown that the TET theory applies, as expected, to other nonlinear oscillator dimers. More specifically, TET was investigated in Klein-Gordon oscillators.[47] For a given donor D with a given initial energy, the potential of the acceptor A is numerically tuned so that the detuning function (16) was zero. In fact, it is possible to obtain a very small detuning function by fitting the nonlinear coefficient of A to the lowest order. A small coupling is enough for complete energy transfer.

A very interesting question, especially for applications at the atomic level, is whether or not the phenomenon of TET persists in quantum mechanics. In order to give an answer, we investigated quantum TET both in the DNLS and the Klein-Gordon dimer.[47] We found that a quantum wave packet initially on D, with the classical energy for TET, is almost completely transferred to A (except for the zero point energy of D for Klein-Gordon dimer, as expected). As in the classical case, the excitation oscillates between D and A, but now the wave packet loses its coherence after a number of oscillations, which becomes large in the classical limit. So, although quantum dispersion is present, TET persists even in the pure quantum regime, for a number of oscillations. This is good enough for real systems, where dissipation makes TET irreversible, even after the first transfer in some cases.

3.3. *Targeted energy transfer through discrete breathers*

TET theory applies to DBs. In this case, instead of nonlinear oscillators, the donor D and the acceptor A are more complex systems consisting of many sites and sustaining DBs. We considered the case of two weakly interacting random chains and showed that TET indeed persists.[43] A perfect and complete transfer and subsequent oscillation of energy is obtained, if the general theory (Subsec. 3.1) is applied. In the example we present, TET is achieved with the interchain coupling being two orders of magnitude smaller that the intrachain coupling. TET is a selective, efficient and robust mechanism.

The two chains are described by DNLS Hamiltonians:

$$H_X = \sum_i \{-E_{X,i}|\psi_{X,i}|^2 - \frac{1}{2}\sigma_X|\psi_{X,i}|^4 \ - C(\psi_{X,i}\psi^*_{X,i+1} + \psi^*_{X,i}\psi_{X,i+1})\}\,, \tag{35}$$

where $E_{X,i}$ are randomly chosen on-site energies but such that there is a gap between the "band" of D and A, $\psi_{X,i}$ the amplitudes on sites i, $X = D$ or A, C the "intramolecular" coupling constant (between the "atoms" of the same D or A "molecule"), and in our examples here, $\sigma_D = 1$ while $\sigma_A = -1$. The opposite sign of the nonlinear terms in Eq. (35) makes D a system of hard anharmonic oscillators (the amplitude of oscillation increases with frequency) and A of soft oscillators and the "dimer" $D - A$ is tuned according to eq.(33). A simple "intermolecular" coupling is

$$H_V = -\lambda \sum_i (\psi_{A,i}\psi^*_{D,i} + \psi^*_{A,i}\psi_{D,i})\,, \tag{36}$$

so that the equations of motion are

$$i\dot{\psi}_{X,i} + \sigma_X|\psi_{X,i}|^2\psi_{X,i} + C(\psi_{X,i+1} + \psi_{X,i-1}) + E_{X,i}\psi_{X,i} + \lambda\psi_{Y,i} = 0\,, \tag{37}$$

where $(X, Y) = (D, A)$ or (A, D). In DNLS models, the action is the norm, *i.e.*, $I_X = \sum_i \psi^*_{X,i}\psi_{X,i}$. For each isolated molecule, we calculate the family of discrete breathers that in the limit of zero amplitude becomes the mode of the gap edge. With usual continuation techniques we calculate the energy of these DBs when their frequency changes in the gap. In Fig. 11(a) we plot the energy E_D, E_A given by eq.(35) of DBs on D, A as a function of the

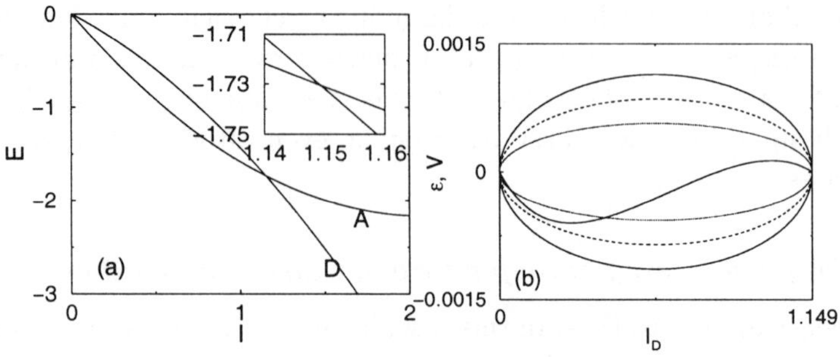

Fig. 11. In (a), energy E_X versus action I_X of the DBs on the uncoupled D and A systems. In this example, each system consists of 10 sites with $E_{D,i}$=0.265,0.346,0.928,0.542,0.165,0.692,0.863,0.097,0.412, 0.700 and $E_{A,i}$=2.655,2.931,2.083,2.748,2.109,2.578,2.308, 2.489,2.851,2.222. The DBs are localized on site $i = 3$ of D and A. The intramolecular coupling is $C = 0.05$. The intersection (zoom) is at $I_T = 1.149297647, E_T = -1.730521602$. In (b), the detuning ϵ and coupling V functions versus I_T are plotted for intermolecular coupling λ=0.001, 0.00075, 0.0005 (solid, dashed, dotted lines, respectively).

action, I_D, I_A, respectively. The DB solutions localized on sites $d = 3$ and $a = 3$, respectively, are calculated by continuation.[27] The on-site energies were randomly chosen in a certain interval. The possibility of ideal TET was detected between the DB on sites d and a, for which $E_{D,d} > E_{D,i}(i \neq d)$,$E_{A,a} < E_{A,i}(i \neq a)$ and $0 < E_{D,i} < 1$,$1 < E_{A,i} < 2$, so that there is a large domain of existence in the gap with no bifurcations for these DBs in continuation of the linear modes $E_{D,3}$ and $E_{A,3}$. The point of intersection (which is better presented in the inset) is where $I_D = I_A = I_T$, i.e., where the TET is produced, if Eq. (17) is satisfied. In the example of Fig. 11, $I_T = 1.149297647, E_T = -1.730521602$.

In Fig. 11(b), we plot the detuning function, Eq. (16), and the coupling function, which in this DNLS model is given by Eq. (36), *i.e.*, $V = H_V$, for three different values of the intermolecular coupling λ=0.001,0.00075,0.0005. Note that λ is two orders of magnitude less than the intramolecular coupling $C = 0.05$. According to (17) complete TET should occur for λ=0.001 and 0.00075.

This is shown in Fig. 12, where the time evolution of the ratio of the energy difference between D and A over the total energy, $r_E = (E_D - E_A)/E_T$, is plotted in (a), and similarly for the action, $r_I = (I_D - I_A)/I_T$, in (b), for the system of Fig. 11. The initial condition is the DB on D at the intersection of the curves of Fig. 11(a). The TET is practically complete for λ=0.001 (solid line) and λ=0.00075 (dashed line), with the ratios oscillating

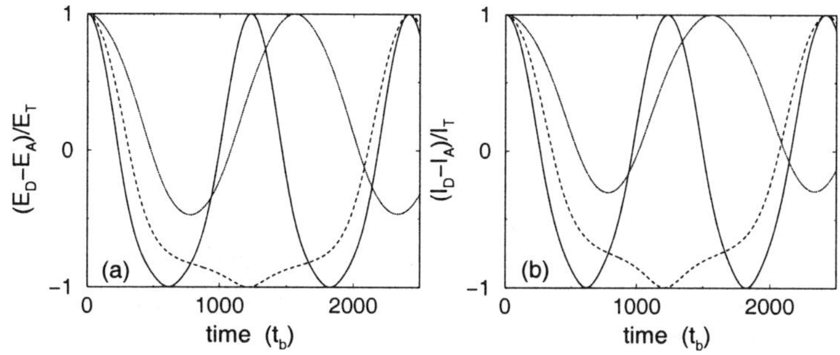

Fig. 12. In (a), $(E_D - E_A)/E_T$, and in (b), $(I_D - I_A)/I_A$, versus time for the system of Fig. 11. The initial condition is the DB on D at the intersection of the curves in Fig. 1(a) with frequency $w_b = 2.0781937765$. The intermolecular coupling is $\lambda = 0.001$, 0.00075, 0.0005 (solid, dashed, dotted lines, respectively). The time unit is the period t_b of the initial DB on D.

between 1 (when $E_D = E_T, I_D = I_T, E_A = 0, I_A = 0$) and -1 (when $E_D = 0, I_D = 0, E_A = E_T, I_A = I_T$). For λ =0.0005 (dotted line) there is a partial transfer at $r_E = -0.472, r_I = -0.300$, and the ratio of the maximum E_A, I_A over E_T, I_T, is $E_{Amax}/E_T = 0.736, I_{Amax}/I_T = 0.650$, respectively. These results are expected from the curves of Fig. 11(b). Although complete TET occurs only when condition eq.(17) is satisfied, we observe that there is a substantial TET even when the detuning function is not completely bounded by the coupling. This is expected since TET proceeds for the values of I_D for which $|V| > |\epsilon|$ and stops when $|V| < |\epsilon|$.

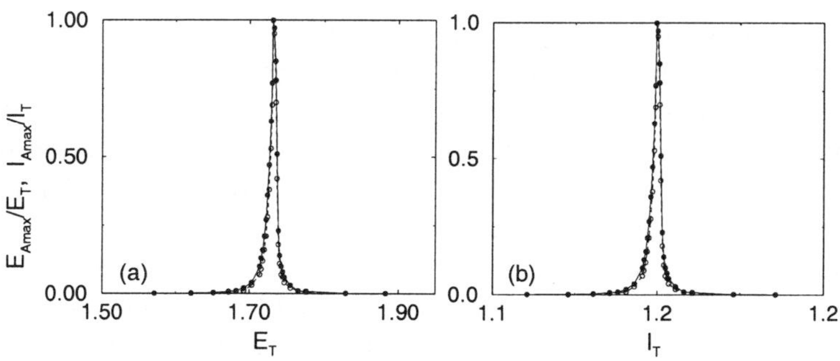

Fig. 13. The ratio of the energy and action transferred from D to A, E_{Amax}/E_T and I_{Amax}/I_T, (solid circles connected by solid line, empty circles connected by dashed line, respectively) as a function of E_T in (a) and of I_T in (b). The DB is initially localized on site d of D.

This TET is very selective, as Fig. 13 demonstrates, where the ratio of energy and action transferred from D to A is plotted, $E_{Amax}/E_T, I_{Amax}/I_T$, as a function of E_T in (a) and I_T in (b). It is clear that TET rapidly vanishes as the energy of the initial DB on D is moving away from the intersection of E_D and E_A of Fig. 11(a). It becomes possible only in a narrow window in E_T or I_T.

The example presented here is quite general since, besides the choice of DB sites d, a and the fact that D is hard and A soft, it does not contain other built-in elements. Still, TET occurs for very weak intermolecular coupling. One can fine tune the model parameters (linear and nonlinear terms) in order to obtain TET for even weaker coupling and even narrower window.

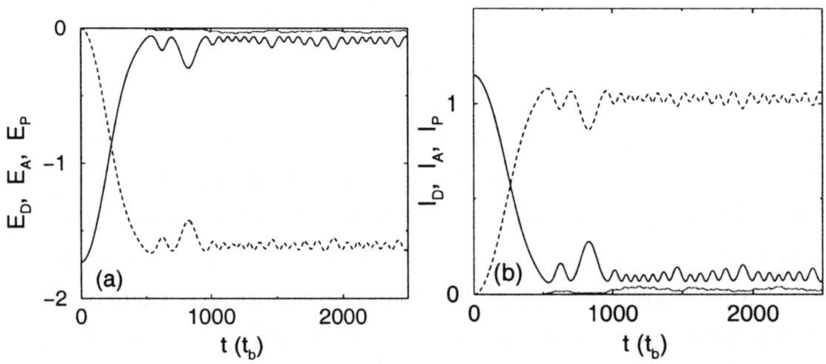

Fig. 14. (a) Energies E_D, E_A, E_P and (b) actions I_D, I_A, I_P versus time (solid, dashed, dotted lines, respectively) for the system of Figs. 11, 12 with $\lambda = 0.001$, but with A coupled to a linear system P with on-site energies $E_{P,i} = 0.933$.

In a real physical or biological system the $D - A$ pair may be interacting with additional degrees of freedom. A direct consequence of this interaction is that, under appropriate conditions, TET becomes irreversible. While the exact resonance condition is kept, most D energy is transferred to A; however, as a result of energy loss and of the high selectivity of TET, the resonance condition can be broken, making a significant energy return to D impossible. As a result, irreversible targeting occurs with almost complete energy transfer. The situation depicted in Fig. 14 represents such a case whereby the system A is coupled to a third linear DNLS system P through the same coupling constant λ=0.001. The initial condition is again the DB on D at the intersection of Fig. 11(a). Indeed, such transfer occurs very efficiently ($E_{Amax} \simeq 0.96$) and subsequently, as a result of the interaction with P, most energy E_A remains localized on A, a small fraction is absorbed by the phonons of P, and a small part returns to D.

One can apply TET in all kinds of models, depending on the physical system. The phenomenon of TET we introduced is very general and relies on nonlinearity and discreteness on one side, that localize energy in the form of hard and soft DBs, and judicious disorder on the other, that exploits specific nonlinear resonances.

4. Ultrafast Electron Transfer

An important physical process where targeted transfer of nonlinear excitations may be applied is the ultrafast electron transfer. While excitations like localized atomic vibrations still have to prove themselves in the real world in the atomic scale, localized electrons are omnipresent at the atomic and molecular level, especially in chemistry and biology. The larger energies involved in electronic transitions indicate that energy transfers via electron transport should be quite substantial and important in bioenergetics and nanotechnology. In fact, Electron Transfer (ET) between weakly interacting molecules is a ubiquitous elementary process in chemical reactions.[48] It has been studied extensively, and with modern techniques like femtosecond spectroscopy, ultrafast ET can be monitored in real time. While ET is mostly considered as a thermally activated process and is well described by Marcus theory[49] in general, we showed that at low temperatures and in the non-adiabatic regime, it may be viewed as a targeted polaron transfer.[50] We will not describe here Marcus theory in detail. We will only present its main features in order to show in what regime and why a different theory is needed which is based on dynamics instead of thermodynamics. We will not derive in detail the model we introduce either. We will only present the main points and results to show the relevance of the TET mechanism for catalytic, ultrafast, low temperature ET.

Our approach and standard ET theories share the following assumptions: We consider the electron within a tight-binding approximation with a few number of states (donor, acceptor, etc). Only electrostatic interactions are taken into account and as a result photon radiation is neglected (a process which is orders of magnitude slower than ultrafast ET and may be calculated with the Fermi's golden rule). Thus, the elecromagnetic field is classical and is included through capacitive potential energy terms in the Hamiltonian. The vibrations of the molecules are assumed to be harmonic, *i.e.*, the phonon bath is considered as a large collection of harmonic oscillators corresponding to the normal modes of the molecules. Since the atomic displacements due to reorganization of the molecules during ET are large, the number of coherent phonons created is very large and the classical approximation for the displacements can be used (except perhaps when protons are involved).

The main differences between our approach and standard theory are: In our ultrafast ET theory, the electron does not follow adiabatically the nuclei (adiabatic limit) and the nuclei do not follow adiabatically the electron (antiadiabatic limit). Nonadiabaticity plays an important role for energy dissipation through phonons and contributes to anharmonicity in the electron dynamics. The tunnelling of the electron is not instantaneous and is not treated in terms of probability of transfer. The true wavefunction dynamics is described, without dropping the phase, coupled with the molecule dynamics. We recover the standard results in the appropriate limits, where coherence of the electronic wavefunction is not important.

Marcus theory is based on transition state theory and considers that the $D - A$ system is initially in the state with the electron in the potential of the donor D, which is weakly coupled to the empty of electron acceptor A (reactants) and that the $D - A$ system's final state is with the electron on A (products). There is a barrier for the electron to overcome in order for the system to reach the product state, *i.e.*, for the electron to be transferred. This is treated like a thermodynamic process and the barrier is overcome by the energy provided to the electron by the fluctuations of the thermalized environment. The characteristic time for the transfer has a standard Arrhenius temperature dependence.

When the coupling between donor and acceptor is small, the electronic eigenstates are localized either on the donor or on the acceptor, except during the time when the electron tunnels from the donor to the acceptor, but this time is short compared to the characteristic time of the ET process. It is then sufficient to consider only two free energy curves of the system as a function of the reaction coordinates, describing the configuration of the nuclei and the environment, when the electron is either on the donor molecule or on the acceptor molecule. One obtains the well-known Marcus scheme represented in Fig. 15 (see Ref. 49 for details). $\Delta G^\star$ appearing in Fig. 1 at the intersection of these two free energy curves, is the energy barrier which the electron has to overcome and it is the characteristic energy of the thermally activated process of ET.

Two regimes appear within this description. In the normal regime, the direct electronic transition from D to A at fixed reaction coordinates (*i.e.*, at fixed nuclei and environment) requires a positive energy $\hbar\omega_{el} = -\Delta_{el}$.

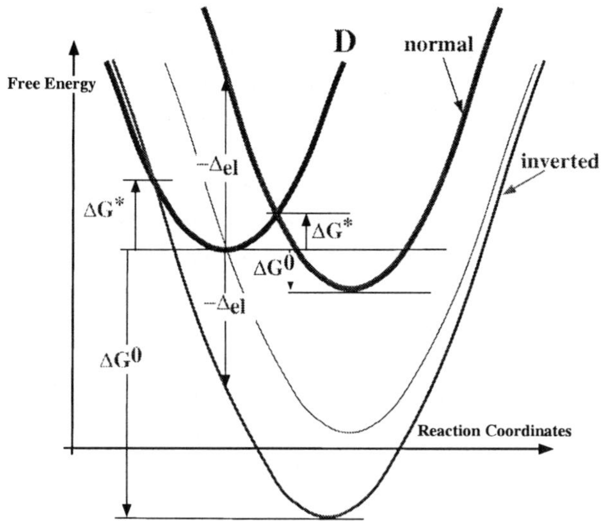

Fig. 15. Free energy versus reaction coordinates of the system donor-acceptor when the electron is on the donor (top left curve D) or on the acceptor for several redox potentials in the normal regime (top right curve), at the inversion point (middle right curve) and in the inverted regime (bottom curve). The chemical reaction energy is the distance between the energy minima ΔG^0. The energy barrier is $\Delta G^\star$. The electronic excitation energy on the donor at fixed reaction coordinates is Δ_{el}.

On the opposite, in the inverted regime, this energy $-\Delta_{el}$ is negative. Although ET in the inverted region could be achieved at low temperature by a photon emission at frequency (energy) $\hbar\omega_{el} = \Delta_{el}$ (photoluminescent chemical reaction,[49]) an activation process above the energy barrier $\Delta G^\star$ becomes more efficient and prevalent at higher temperature. At the top of the energy barrier, the electron has to tunnel from the donor to the acceptor. The time for this process and its probability participate as coefficient for the reaction rate. It could be treated either adiabatically (strong reactants) or diabatic corrections could be made (weak reactants).[48]

$\Delta G^\star$ is zero at the inversion point when $\Delta_{el} = 0$ (see Fig. 1). This is the regime where ET is expected to become effective at low temperature. Because of the absence of energy barrier, no thermal fluctuations are required and at low temperature ET is essentially reduced to a direct quantum tunneling (followed by a relaxation of the environment once the electron is on the acceptor). Since the coupling between D and A is supposed to be small, the characteristic time of this quantum tunneling is nevertheless relatively long and could be comparable to the phonon characteristic times. In this case, the deformability of the self-consistent local potential followed by the electron during the intermediate stage of the tunneling should be taken into account. In other words, in the vicinity of this inversion point, the validity of the adiabatic hypothesis breaks down since the characteristic energy $|\Delta_{el}|$ of the direct electronic excitation (at fixed environment) becomes small and necessarily comparable to the phonon characteristic energies (note that Δ_{el} is calculated at zero coupling but is not significantly changed if the coupling is small). In order to be valid, the adiabatic assumption requires that the largest phonon energy $\hbar\omega_{ph}$ involved be much smaller than the smallest excitation energy of the electron. In our case, this electronic excitation energy is essentially the distance Δ_{el} between the two electronic levels at fixed environment. This condition is well fulfilled in many real ETs which are far enough from the inversion point but fails close to it.

Although it is clear that the characteristic time for ET should be minimum in the vicinity of the Marcus inversion point, and that the conditions for ultrafast ET require to be in this vicinity, ultrafast ET is not well described within the original Marcus theory because of its adiabatic assumption. We propose a non-adiabatic theory for the dimer model which recovers the Marcus theory far from the inversion point (when adiabaticity is recovered) and which improves it close to it. Moreover, our approach can be extended when more electronic sites are involved and opens new perspectives for understanding enzymatic catalysis. In particular, we show that a weakly interacting but well tuned third site (catalyst) could trigger selectively ultrafast ET a low temperature which otherwise would require a huge activation energy and would not occur. The same theory may open new perspectives to quantum transitions in general.

4.1. *Nonlinear dynamical model for ET*

Our model consists of weakly interacting molecules, with electronic levels that depend on the occupation, coupled with vibrational modes that provide the thermal bath and the energy dissipation mechanism. Norm preserving dissipation is important for irreversible ET. We consider two large donor and acceptor molecules, D and A with many vibrational degrees of freedom (our approach extends to any number of molecules). We assume for simplicity that one electronic state from each molecule is involved with wavefunction $|\mathbf{\Psi}_\alpha >= \mathbf{\Psi}_\alpha(\mathbf{r}; \{u_{\alpha i}\})$, where , $\alpha = D$ or A, $\mathbf{r}$ is the space coordinate and $u_{\alpha i}$ the phonon coordinates, and that the Born-Oppenheimer (adiabatic) approximation is valid for the isolated molecules (this is true when state $|\Psi_\alpha >$ is well separated from the other electronic states of the molecule, at the scale of phonon energies). In other words, the wavefunction is a function of the whole environment. In a tight-binding representation, the total wavefunction of the $D-A$ system takes the form $\sum_\alpha \psi_\alpha |\Psi_\alpha >$. We use the complex amplitudes ψ_α of the electronic wavefunction as Kramers reaction coordinates[51] with the normalization condition:

$$|\psi_D|^2 + |\psi_A^2| = 1\,. \tag{38}$$

This choice of reaction coordinates for the whole system, although very relevant in the sense that electronic density is crucial in the problem and the rest of (atomic) variables may be treated as thermal bath, neglects electronic variables internal to each molecule, but these are assumed to be very fast.

The Hamiltonian consists of two parts: One is the dimer of Sec. 3.2 described by Eq. (18), which describes an infinitely slow ET in the limit of zero coupling. All fast variables follow adiabatically the slow collective electronic variables ψ_α. We call this limit of antiadiabatic ET, as opposed to the adiabatic ET where the electron has no intrinsic dynamics and is driven by the atomic fluctuations (Marcus theory). The energy of each isolated molecule turns out to be a phase independent nonlinear function of the electronic density $I_\alpha = |\psi_\alpha|^2$, *i.e.*, to lowest order

$$H_\alpha(|\psi_\alpha|^2) = \mu_\alpha|\psi_\alpha|^2 + \frac{1}{2}\chi_\alpha|\psi_\alpha|^4 \tag{39}$$

and thus, the electronic level is not fixed but depends on its electronic occupation density

$$E_\alpha = \frac{\partial H_\alpha}{\partial I_\alpha} = \mu_\alpha + \chi_\alpha|\psi_\alpha|^2\,. \tag{40}$$

The coefficient μ_α of the linear term is the electronic level at zero occupation and the nonlinear coefficient is the sum of two contributions $\chi_\alpha = \chi_\alpha^C + \chi_\alpha^R$. The positive coefficient χ_α^C takes into account the local electronic dielectric constant of the environment, *i.e.*, the fast dielectric relaxation of the other electrons of the molecule and environment screening effects. The energy required to put an electron in state α without any relaxation of the molecular environment is then $\mu_\alpha + \chi_\alpha^C/2$. The negative coefficient χ_α^R characterizes the reorganization energy of the molecule. This is the energy gain from the relaxation of the surrounding atoms of the molecule. In our model it can be calculated as the energy dissipated through the phonon bath during complete relaxation of the molecule. Both the positive electrostatic energy and the negative reorganization energy term may be of the order of several eV. Depending on the molecule and the environment, the sum χ_α may be positive, when there is an overall energy cost to accommodate an electron, or negative, when energy decreases after complete relaxation. We expect, for instance, the positive term χ_α^C to be prevalent when the electronic state α belongs to the inner d shell of a transition metal ion, which could be embedded in a biomolecule. The negative term χ_α^R should prevail when the state belongs to bonds in organic molecules. The former case is an example of a "hard" electronic level while the latter is a "soft" level. Besides this type of empirical arguments, the determination of coefficient χ_α with experiments and *ab initio* calculations would be highly desirable. Appropriate use of concepts that emerge in studies of chemical reactivity with density functional theory, such as hardness which is directly related to our coefficients χ_α, may prove useful.[52] The accurate prediction of χ_α is crucial for our model of catalytic transfer, as becomes obvious in what follows. It was explained in Sec. 3 that the combination of weakly interacting hard and soft nonlinear systems is the key to selective resonant transfers.

The coupling term in (18) represents the small but essential overlap of the D and A electronic orbitals. The coupling coefficient $\lambda = \lambda_{DA}$ is the standard overlap integral of the tight-binding representation, often referred to as transfer integral. Although weak, of the order of the phonon energies, this term is responsible for the ET.

In the (unrealistic) antiadiabatic limit, the transfer integral λ_{DA} is much smaller that the phonon energies and ET is much slower than phonon dynamics. As we noted earlier, slow ET may be described by the Hamiltonian term (18) while the dynamics of the fast phonon variables are described by the second part of the Hamiltonian, which consists of the phonon and electron-phonon interaction terms:

$$H_\alpha^{ph}(\{|\psi_\alpha|^2, u_{\alpha i}, p_{\alpha i}\}) = \sum_i \frac{1}{2} m_{\alpha i}\omega_{\alpha i}^2 \left(u_{\alpha i} - k_{\alpha i}|\psi_\alpha|^2\right)^2 + \frac{1}{2m_{\alpha i}} p_{\alpha i}^2 . \quad (41)$$

The electronic density $|\psi_\alpha|^2$ on each molecule α couples to the coordinates of the same molecule assumed to be harmonic and thus consisting of a collection of independent harmonic oscillators i with position-momentum coordinates $u_{\alpha i}, p_{\alpha i}$, mass $m_{\alpha i}$, frequencies $\omega_{\alpha i}$, and electron-oscillator coupling $k_{\alpha i}$. It is assumed that all possible dynamical interactions of the electron with the atomic coordinates are involved in Eq. (41), including chemical and electrostatic energies.

Thus, the complete nonadiabatic Hamiltonian of the ultrafast ET consists of the terms (18) and (41). Ultrafast ET at very low temperature cannot be obtained neither at the antiadiabatic limit of zero λ_{DA} nor at the adiabatic limit. It is well described by our nonadiabatic Hamiltonian which provides the appropriate dissipation mechanism so that the final electronic energy be lower than the initial (something that cannot happen in the antiadiabatic case). Dynamical coupling of the electron with the electromagnetic field is not considered since it is very weak (lifetimes of electronic states obtained with Fermi's golden rule are long compared to the times of ET.[50]) Quantum emission of photons can be neglected as a dissipation mechanism in this case.

Since the reorganization in the molecule due the electron presence involves large atomic displacements and a lot of degrees of freedom, it is a good approximation to consider that the phonon variables $\{u_{\alpha i}\}$ are classical while the electronic amplitudes are quantum variables.[50] The dynamical equations for the coupled electron-phonon system are:

$$i\dot{\psi}_\alpha = (\mu_\alpha + \chi_\alpha|\psi_\alpha|^2)\psi_\alpha - \lambda_{\alpha\beta}\psi_\alpha - \sum_i m_{\alpha i}k_{\alpha i}\omega_{\alpha i}^2(u_{\alpha i} - k_{\alpha i}|\psi_\alpha|^2)\psi_\alpha\,, \tag{42}$$

$$\ddot{u}_{\alpha i} + \omega_{\alpha i}^2(u_{\alpha i} - k_{\alpha i}|\psi_\alpha|^2) = 0\,, \tag{43}$$

(where $\hbar = 1$ in rescaled units). The harmonic motions $u_{\alpha i}(t)$ can be explicitly obtained[50] from the linear differential equations (43) as the sum of functions of the time dependent driving force $|\psi_\alpha(t)|^2$ and a solution of the equation without driving force. Then, substituting $u_{\alpha i}(t)$ in Eq. (42) yields the fundamental equation for non-adiabatic electron dynamics (which conserves the norm $\sum_\alpha |\psi_\alpha|^2$):

$$i\dot{\psi}_\alpha = (\mu_\alpha + \chi_\alpha|\psi_\alpha|^2)\psi_\alpha - \lambda_{\alpha\beta}\psi_\alpha + \left(\int_{-\infty}^t \Gamma_\alpha(t-\tau)\frac{d|\psi_\alpha|^2}{d\tau}d\tau + \zeta_\alpha(t)\right)\psi_\alpha\,, \tag{44}$$

where $\Gamma_\alpha(t) = \sum_i m_{\alpha i}\omega_{\alpha i}^2 k_{\alpha i}^2 \cos(\omega_{\alpha i}t)$. If there are many phonon modes with a rather uniform distribution, $\Gamma_\alpha(t)$ can be assumed to be a smooth decaying function of time. It generates energy dissipation as a kernel in Eq. (44) (The absorption rate in energy of a charge fluctuation at site α at frequency ω is nothing but the product of the square of its amplitude with the Fourier transform of $\Gamma_\alpha(t)$.) We also have $\Gamma_\alpha(0) = -\chi_\alpha^R = \sum_i m_{\alpha i}\omega_{\alpha i}^2 k_{\alpha i}^2$. The time dependent potential $\zeta_\alpha(t)$ is produced by the thermal fluctuations of the lattice. It is a colored random Langevin force with correlation function which fulfils $< \zeta_\alpha(t+\tau)\zeta_\alpha(\tau) >_\tau = \Gamma_\alpha(t)\ k_B T$ at temperature T.[50]

Thus, the effect of non-adiabaticity is to transform the standard linear Schrödinger equation describing the dynamics of the electron into a nonlinear Schrödinger equation (44) with norm preserving energy dissipation terms and with random colored time dependent potentials generated by atomic thermal fluctuations.

4.2. *ET in the Dimer*

The basic results of the standard ET theory are recovered far from the inversion point. Using our model, the free energy surfaces shown in Fig. 15 are obtained.[50]

The system of equations (44) has a very rich behavior, even in the dimer case, where some analytic results are obtained. It is shown that for most initial conditions (which are not eigenstates of Hamiltonian (18)) dynamical equations (44) converge to a lower energy eigenstate.[50]

We discuss here only the zero temperature case, when there are no thermal fluctuations, i.e., $\zeta_\alpha(t) = 0$ in eq.(44). Within our approach, the inversion point is obtained when the D and A electronic levels at the beginning of ET are resonant, *i.e.*, from Eq. (40)

$$\mu_D + \chi_D = \mu_A \,. \tag{45}$$

The variation of the energy on the dimer $D-A$ as a function of the electronic density $I_A = |\psi_A|^2$ on the acceptor (taking into account normalization condition (38) and the local Hamiltonians (39)) is $E_T(I_A) = H_D(1-I_A) + H_A(I_A) - H_D(1)$, *i.e.*,

$$E_T(I_A) = (\mu_A - \mu_D - \chi_D)I_A + \frac{1}{2}(\chi_D + \chi_A)I_A^2 \tag{46}$$

When at the beginning of ET $\frac{dE_T}{dI_A}(0) > 0$, or equivalently $\mu_A < \mu_D + \chi_D$, there is always an energy barrier. When $\frac{dE_T}{dI_A}(0) = 0$ we obtain the inversion point, Eq. (45), but still, there is no energy barrier only when $\chi_D + \chi_A \leq 0$. In this case, complete ET becomes possible as long as $\frac{dE_T}{dI_A}$ remains negative with a modulus that does not exceed the phonon cutoff frequency (in order to have efficient energy dissipation). In Fig. 16 we present two examples of ET at the inversion point where we assume that the phonon frequency cutoff is large compared to the characteristic ET electronic frequency, so that in dynamical equations (44) it is $\Gamma(t) = 2\gamma_\alpha\delta(t)$. Note that in the examples of Fig. 16 ET is faster in the case of large dissipation but the transfer time does not increase monotonically with γ_α.[50]

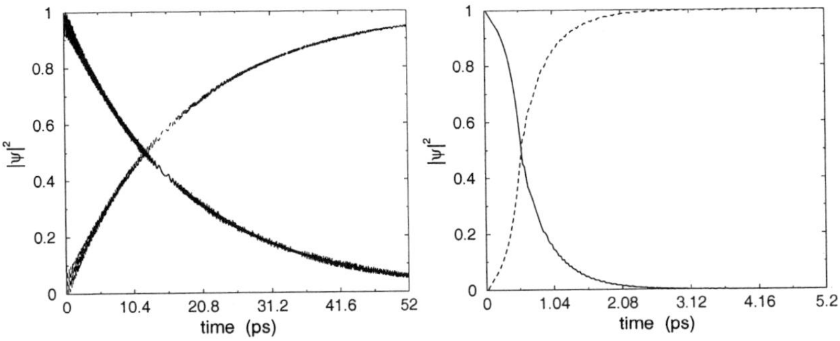

Fig. 16. Electron density on the donor (solid) and the acceptor (dashed) versus time for the dimer model at the inversion point and zero degree K where $\mu_D = 2$, $\chi_D = -1$, $\mu_A = 1$, $\chi_A = -0.75$, $\lambda_{DA}=10^{-2}$, $\gamma_D = \gamma_A = 1$ (left) or 40 (right) (the time unit is 1 ps$=10^{-12}$ s for energy in units of eV).

4.3. *Catalytic ET in a trimer*

As explained in the previous Subsec. 4.2, ET may be obtained at the inversion point when the chemical reaction energy is small compared to the phonon energy cutoff. We propose that when the $D - A$ dimer is far from the inversion point, a third site may be used to induce oscillations that bring D and A in resonance. Instead of thermal fluctuations, an appropriately tuned molecule, which we call "catalyst" C, is used. Following the TET theory described in Sec. 3, the molecule C is chosen under TET conditions with the D. The large electronic fluctuations that are generated, while not strongly binding D and C, trigger an ultrafast ET to A, which has an electronic level in the range of variation of these fluctuations.

We illustrate the basic principle in a trimer model which is basically the same as the dimer of the previous Subsec. 4.2 but with one additional "catalytic" site, with local Hamiltonian, Eq. (39), $\mu_C|\psi_C|^2 + \frac{1}{2}\chi_C|\psi_C|^4$, weakly coupled to A and D, with λ_{AC} and λ_{DC}, respectively. Assume that D and A are soft, *i.e.*, $\chi_D < 0$,$\chi_A < 0$; and that the reaction is exothermic, *i.e.*, $\mu_A+\chi_A/2 < \mu_D+\chi_D/2$. In the absence of the catalyst ($\lambda_{DA} = \lambda_{AC} = 0$), there is a large energy barrier, *i.e.*, $\mu_D + \chi_D < \mu_A$ and no transfer takes place. The propeties of C are determined from the TET conditions Eqs. (31),(33), which together with the normalization $I_T = 1$, Eq. (38), give $\mu_C = \mu_D + \chi_D$ and $\chi_C = -\chi_D$. Since D and A are chosen soft, C has to be hard. The initial electronic level μ_A should belong to the variation interval of the electronic level of the $D-C$ system, *i.e.*, $\mu_D+\chi_D < \mu_A < \mu_D$ at weak damping or $\mu_D+\chi_D/2 < \mu_A < \mu_D$ at strong damping. In Fig. 17 we present

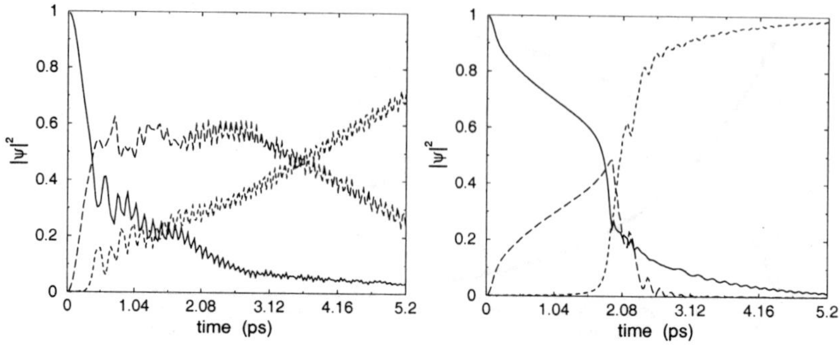

Fig. 17. Electron density on donor (solid), acceptor (dashed), and catalyst (long dashed) versus time in the trimer $\mu_D = 2$, $\chi_D = -1$, $\mu_C = 1$, $\chi_C = 1$, $\mu_A = 1.5$, $\chi_A = -0.75$, $\lambda_{DA} = \lambda_{AC} = \lambda_{DC} = 10^{-2}$ and damping $\gamma_D = \gamma_C = \gamma_A = 2$ (left) or 10 (right).

one such example for two different dissipation values. For $\gamma_\alpha = 0$ (not shown here), large fluctuations are induced but the chaotic time evolution cannot lead to the electron transferred in the ground state of the system, *i.e.*, to the acceptor. When dissipation is introduced, the electron falls on A while C only transitively takes a fraction of electronic charge (Fig. 17).

4.4. *The example of bacterial photosynthetic reaction center*

The ET is highly sensitive to small perturbations of the donor-catalyst system which easily break the TET conditions. We have shown for example that relatively small electric fields are sufficient for blocking ET at 0°K.[50] These principles may be extended to many-site networks of electronic levels where the electron can choose a specific path very selectively. This path can be blocked and switched under small perturbations. Logical functions with one or few electrons could be built at the molecular level suggesting potential nanodevice applications and complex biological functions to be studied in living cells.

One example where we think that our proposed mechanism applies directly is the primary charge separation in the bacterial photosynthetic reaction center,[53,54] which has been studied extensively with laser pump-probe experiments and protein mutations.[55,56] The observed features of this system are overall explained by our model, at least qualitatively at this point. Photosynthesis[54] occurs in the chloroplasts of the photosynthetic cells. These chloroplasts contain stacking of many tykaloids which are small vesicles. The membrane of these tykaloids contains many com-

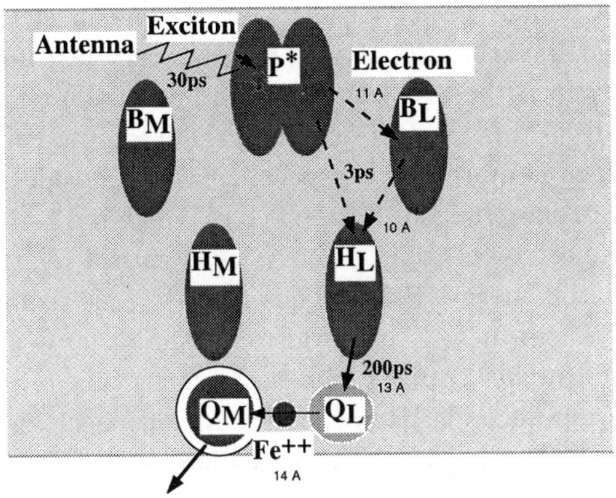

Fig. 18. Sketch of the reaction center embedded in the tykaloid membrane. Its structure consists of a scaffold of polypeptides, a special pair of bacteriochlorophyll a forming the dimer (P) and an almost symmetric structure consisting of the two ancillary bacteriochlorophyll a (B_L) and (B_M), the two bacteriopheophytins (H_L) and (H_M) and the menaquinone (Q_L) and the ubiquinone(Q_M) bridged by Fe^{++}. Electron transfer between ($P^\star$) and H_L is ultrafast and occurs in 3 picosecond (room temperature) or one picosecond (low temperature).

plex structures of biomolecules. Among them pigment molecules (bacteriochlorophylls a and b, carotenes, *etc.*) are organized into complexes LH1, LH2, etc, forming in a series of rings as an onion. Their global structure operates as a kind of antenna capturing photons as electronic excitations (excitons) which systematically funnel to the center of the antenna at the photosynthetic reaction center.

The excitons collected in the pigments are transferred very fast within 100 fs through the pigment molecules to the inner ring of the antenna. Then, the exciton is transferred to the the photosynthetic reaction center within about 30 ps. This reaction center consists of many assembled polypeptides forming the scaffold of the system maintaining the cofactors (bacteriochlorophylls, bacteriopheophytin) and the metallic ions in appropriate positions and interactions (see Fig. 18). The role of the reaction center is to use the exciton energy for pumping electrons from the outside of the tykaloid membrane (the periplasm) to the inside of the tykaloid (the cytoplasm).

The exciton is captured by the strongly bonded dimer (P) of bacteriochlorophyll a which is very close to the periplasm (P) $\longrightarrow$ (P*). This excited state (P*) has the property to transfer very fast an electron to the bacteriopheophytin BPhea ($H_L A$) in the middle of the tykaloid membrane $P^* H_L A \to P^+ H_L^-$. This transfer occurs at room temperature within a half-time of 3 ps which is quite short, especially when considering the distance between the electronic sites on (P*) and (H_L) which is 17 Å. The free energy variation of the electron is about 0.25 eV which is still relatively small compared to the exciton energy (P*) which is 1.38 eV. Surprisingly, this transfer time becomes shorter at low temperature of the order of $\approx$ 1 ps at 10 K. In the absence of further transfer, the electron is stable on BPhea molecule (H_L) over few nanoseconds. However, it has also been demonstrated experimentally that the presence of the ancillary bacteriochlorophyll BChla (B_L) is necessary for the electron transfer to BPhea (H_L) to occur, but its exact role is controversial because no substantial accumulation of BChla$^-$ has ever been observed in femtosecond experiments.

There is a subsequent series of electron transfers which become slower and slower till the electron completely crosses the tykaloid membrane. Although we do not focus our interest on them because they are not ultrafast, we shortly describe them. Still in the reaction center, the electron arrives at menaquinone (Q_L) within a half-time of 200 ps. The charge separation is now stable for about 100 ms. Next again, the electron transfers from menaquinone (Q_L) to ubiquinone (Q_M) through Fe^{++} within a half-time of 30-100 μs, and now is stable for about 1 s. The final result is that an electron has been transferred from one side of the reaction center to the other side across the tykaloid membrane. Next, electron transfers continue elsewhere in the solvent with diffusive molecules involving cytochrome. P^+ recovers its electron from the water of the periplasm, which is decomposed releasing electrons, protons and neutral oxygen molecules. The proton gradient between the two sides (cytoplasm and periplasm) of the tykaloid membrane fuels ATPase which is another complex structure in the tykaloid membrane. The flow of proton through this structure generates ATP from ADP. At this stage, a substantial part of the light energy which was harvested, has been stored in a stable form. Further chemical reactions then use this ATP energy for fixing carbon dioxide as hydrocarbons through the Calvin-Benson cycle.

We are here especially interested in understanding the ultrafast electron transfer occurring between (P*) and (H_L) with the help of the ancillary molecule (B_L). There are other puzzling phenomena associated with this transfer. It has been observed that the excited state (P*) is associated with low frequency coherent molecule vibrations (with period 500 fs and possibly 2 ps) which strongly depend on point mutations of the protein cofactor matrix.[55,56] The electron transfer is unexpectedly sensitive to minor mutations of the molecules involved and, in particular, to macroscopic electric fields,[57] which however involve electron energy changes of a few meV at most, *i.e.*, rather small compared to electron energy differences counted in eV. Another puzzling feature is that despite the almost symmetric structure of the RC, the pathway used by the electron systematically breaks this symmetry. These observed features could be interpreted within our approach considering that the pair of bacteriochlorophyll (P) is the donor. The two bacteriochlorophyll BChla (B) are slightly nonequivalent and only one (B_L) is well-tuned as a catalyst on the donor for triggering ET to the nearby bacteriopheophytin BPhea (H_L). The biomolecule (B_L) plays the role of the catalyst in our trimer model for triggering ET from the donor ($P^\star$) to the acceptor (H_L). Actually, as it can be seen above in our numerical results, there is only a partial and transitory ET between ($P^\star$) and (B_L) which could explain why (B_L^-) is not observed. Moreover, ET occurs with oscillations associated with important molecular reorganization, which could be observed.

Despite the near symmetry, small differences in the environment of (B_M) compared to (B_L) are sufficient to detune (B_M) from ($P^\star$) so that ET practically does not occur to (H_M). This effect can be observed numerically when considering ET in a more realistic model with 5 sites which is almost symmetric, thus modeling the real system shown in Fig. 18 (not presented here due to lack of space, see Ref. 50).

In real living systems, long evolution and natural selection have optimized the tuning of the whole system so that ET is faster at low temperature. In this case, thermal fluctuations slow down ET, although it is still very efficient at room temperature. Again, this can be observed numerically (but will not be shown here).

Since ET is very well optimized, minor mutations at crucial points which affect the quality of the tuning, could sharply reduce the efficiency of ET. In particular, the relatively small electric fields (in terms of energy at the molecular scale) strongly affect ET by detuning (B_L) from the donor ($P^\star$). Moreover, we observed numerically that the effect of the electric field is asymmetric (as observed experimentally) because an energy barrier for the (partial) electron transfer between ($P^\star$) and (B_L) is created when the electric field has a certain sign (ET moves towards the normal Marcus regime) while when its sign is opposite, there is no barrier but only a slowing down of ET (ET moves towards the Marcus inverted regime).[50]

5. Conclusions and perspectives

In this discussion we presented DBs as an important paradigm and as the starting point for investigating nonlinear localized excitations in discrete systems. They provide a general framework for exploring a variety of phenomena and applications in physics, chemistry, materials science, biochemistry and biophysics. Once nonlinear localization mechanisms are understood in these systems, efficient and selective energy transfer mechanisms may be conceived. In the short term, many of our results may be applied to diverse and interesting physical systems, such as Bose-Einstein condensates, Josephson junctions, nonlinear photonic crystals, optical waveguides, magnetic compounds, biomolecules, etc. We only discussed atomic-scale vibrational and electronic excitations.

In section 2, after briefly reviewing the general theory with illustrations in toy models, we showed that DBs are highly relevant in atomistic materials simulations, even when realistic models are used. Besides the technical self-consistent numerical calculations that we briefly outlined, DBs may be found by molecular dynamics simulations. They are formed spontaneously, e.g., through modulational instabilities, and especially in conditions out of equilibrium. "Thermal shocks", i.e., very rapid quenching, result in DBs creating "hot spots" where energy is trapped and the the system does not thermalize. We presented in some detail DB creation in a different out of equilibrium situation, i.e., when a relatively large amount of energy is deposited locally. This may happen by photon absorption, after irradiation, atomic collisions, chemical reactions, etc. The ramifications of DBs formation in energy relaxation and redistribution in the nanoscale have only recently started to being explored. Experimental data that point indirectly to DBs existence at this scale need to be studied or revisited with the insight obtained from DB theory and simulations. For the specific examples we presented, carbon-hydrogen systems and semiconductors, potential applications range from molecular electronics to understanding biomolecules and interstellar medium. For this latter case, we have argued that anomalies observed in infrared radiation from interstellar dust may be related to vibrational energy being trapped in DBs and emitted by photons. The discussion on disordered systems was based on toy models. Investigations of IDBs with realistic models are underway. Phenomena typically explained with distributions of two-level systems, such as low temperature thermal conductivity and spectral hole burning, can be analyzed in terms of DBs. The long lifetime of IDBs may explain many anomalous energy relaxations in disordered materials (glasses, polymers, biopolymers, etc). When there are no linear propagating modes, thermalization is impossible at low temperatures. Energy propagation is possible through extended IDBs. Glassy behavior is obtained with weak nonlinearity. The methods we mentionned for the studies of random nonlinear discrete systems are not only useful for understanding properties of amorphous materials and glasses. The bifuraction diagrams for instance, are interesting for small systems as well, in the context of vibrational motion of highly excited molecules. Accurate quantum chemical calculations, the structure of classical phase space for the corresponding potential energy surfaces, and spectroscopic data for small molecules confirm that stable periodic orbits correspond to the quantum wavefunctions and that the appearence of new quantum states is related to bifurcations in the classical phase space. Although not feasible for a large number of atoms, these studies answer to a lot of questions regarding quantum mechanics and DBs. Besides helping us understand random systems by providing us with DBs solutions *in situ*, the NLRM method can be applied to situations where nonlinear oscillator arrays are subjected to external fields, for example, to understand self-induced transparency.[58] Some materials become transparent to ultrasound waves above a certain critical value of the field. MultiDBs are related to the appearence of propagating modes. Nonlinear wave propagation phenomena in periodic and disordered systems are understood by studying multiDBs and their stability.

In section 3 we presented a general mechanism for targeted transfer of nonlinear localized excitations. TET induces highly selective and efficient energy transport and focusing in nonlinear systems. Based on the concept of nonlinear resonance, it was illustrated on a dimer and on two weakly coupled random chains. Complete exchange of specific amount of energy occurs under very well defined conditions. Moreover, irreversibility was illustrated by coupling with a third system. Since it can be applied to models describing atomic vibrations, polarons, Davydov solitons, excitons, spin excitations, etc, we expect TET to be relevant in a variety of atomic-scale phenomena in bioenergetics, biological motors, chemical reactions, biological and synthetic catalysts, quantum dots, etc. Other areas for applications include the forementioned mesoscopic systems where nonlinear localized modes are observed. Further studies should include several donors for the same acceptor (funnels), cascades of targeted transfer, design of artificial devices, etc. The selectivity and robustness of TET may prove useful in engineering functional materials and nanodevices. For instance, it is straightforward to implement logic gates and circuits using simple models and the TET principles but the main effort has to be towards specific, realistic applications.

In section 4 we applied TET to the problem of low temperature ultrafast electron transfer. We presented a non-adiabatic theory for ET, explained the regime where dynamics takes over from thermodynamics as well as where our model recovers the standard theory of thermally activated ET. We then showed that our approach opens new perspectives for understanding ultrafast ET and enzymatic catalysis, especially at relatively low temperatures when it is not thermally activated. The appropriate "catalyst" for a certain donor-acceptor dimer is found by applying the TET theory. In donors and acceptors that do not react at low temperature because of a large activation barrier, the addition of the "catalyst", tuned to the donor, triggers an ultrafast ET. Dissipation, originating from the coupling of the electron with the phonon bath, is important in this case, since it brings the electron to the acceptor (ground state). We discussed in some detail ultrafast ET in the photosynthetic reaction center, which is a very well studied system. Our approach qualitatively reproduces experimental observations concerning times and paths of transfer, influence of mutations and external fields, etc. However, more elaborate quantitative investigations are needed. In order to show the general validity of our theory, realistic calculations in several enzymatic reactions, and on the biomolecules involved, are necessary.

Acknowledgments

During the course of the work desrcibed here, G. K. has been supported by Greek GSRT under grant ΕΠΕΤ II, 97ΕΛ-52. This work was supported by EC through network LOCNET, TMR contract HPRN-CT-1999-00163.

References

1. L. Landau, Phys. Z. Sowjetunion **3** 664 (1933).
2. T. Holstein, Ann. Phys. **8** 325 (1959).
3. A. S. Davydov, J. Theor. Biol. **38**, 559 (1973); ibid. **66** 379 (1977); Biology and Quantum Mechanics, (Pergamon, Oxford 1982); A. C. Scott, Phys. Rep. **217**, 1 (1992).
4. Alwyn Scott, Nonlinear Science, Emergence and Dynamics of Coherent Structures, Oxford University Press (1999) and references therein.
5. J. Edler, P. Hamm, and A. C. Scott, Phys. Rev. Lett. **88**, 067403 (2002).
6. R.T. Birge, H. Sponer, Phys. Rev. **28** 259 (1926); J.W. Ellis, Phys. Rev. **33** 27 (1929).

7. A.A. Ovchinnikov, N.S. Erikhman, Usp. Fiz. Nauk **138** 289 (1982) and Sov. Phys. Usp. **25** 738-755.
8. M. Joyeux, S.C. Farantos, R. Schinke, J. Phys. Chem. A **106** 5407 (2002) and references therein.
9. S. Aubry, Physica D **103** 201 (1997).
10. S. Flach and C. R. Willis, Phys. Rep. **295** 182 (1998).
11. H.S. Eisenberg, Y. Silberberg, R. Morandotti, A.R. Boyd, J.S. Aitchison, Phys. Rev. Lett. **81** 3383 (1998).
12. E. Trias, J.J. Mazo, T.P. Orlando, Phys. Rev. Lett. **84** 741 (2000); P. Binder, D. Abraimov, A.V. Ustinov, S.Flach, Y. Zolotaryuk, Phys. Rev. Lett. **84** 745 (2000).
13. M. Sato, B.E. Hubbard, A.J. Sievers, B. Ilic, D.A. Czaplewski, H.G. Craighead, Phys. Rev. Lett. **90** 044102 (2003).
14. B.I. Swanson, J.A. Brozik, S.P. Love, G.F. Strouse, A.P. Shreeve, A.R. Bishop, W-Z. Wang, M.I. Salkola, Phys. Rev. Lett. **82** 3288 (1999).
15. U.T. Schwartz, L.Q. English, A.J. Sievers, Phys. Rev. Lett. **83** 223-226 (1999).
16. A. Xie, L. van der Meer, W. Hoff, R.H. Austin, Phys. Rev. Lett. **84** 5435 (2000).
17. G. Kopidakis and S. Aubry, Physica **B** 237 (2001).
18. A. J. Sievers and S. Takeno, Phys. Rev. Lett. **61**, 970 (1988).
19. R. S. MacKay and S. Aubry, Nonlinearity **7** 1623 (1994).
20. R.S. MacKay, J-A. Sepulchre, Physica **D** **82** 243 (1995).
21. J-A. Sepulchre, R.S. MacKay, Nonlinearity **10** 679 (1997).
22. R. Livi, M. Spicci, R.S. MacKay, Nonlinearity **10** 1421 (1997).
23. S. Aubry, Ann. Inst. H. Poincaré, Phys. Théor. **68** 381 (1998).
24. S. Aubry, G. Kopidakis, and V. Kadelburg, Discrete Cont. Dyn-B **1**, 271 (2001).
25. G. James, J. Nonlinear Sci. **13** 27 (2003).
26. J.L. Mariñ, S. Aubry, Nonlinearity **9** 1501 (1996).
27. G. Kopidakis, S. Aubry, Physica D **130** 155 (1999).
28. G. Kopidakis, S. Aubry, Physica D **139** 247 (2000).
29. G. P. Tsironis and S. Aubry, Phys. Rev. Lett. **77**, 5225 (1996); A. Bikaki, N.K. Voulgarakis, S. Aubry, G.P. Tsironis, Phys. Rev. **E** **59** 1234 (1999).
30. J. Tersoff, Phys. Rev. B **37** 6991 (1988).
31. J. Tersoff, Phys. Rev. Lett. **61** 2879 (1988).
32. C.H. Xu, C.Z. Wang, C.T. Chan and K.M. Ho, J.Phys.:Condens. Matter **4** 6047 (1992).
33. G. Kopidakis, C.Z. Wang, C.M. Soukoulis, and K.M. Ho, in 'Microcrystalline and Nanocrystalline Semiconductors", Proceedings of the 1994 MRS Fall Meeting, Volume **358** 73 (1995).
34. C.Z. Wang, K.M. Ho, C.T. Chan, Comp. Mat. Sc. **2** 93 (1994).
35. G. Kopidakis, C.Z. Wang, C.M. Soukoulis, and K.M. Ho, Phys. Rev. B **58** 14106 (1998).
36. G. Kopidakis, C.Z. Wang, C.M. Soukoulis, and K.M. Ho, J. of Phys: Cond. Matt. **9** 7071 (1997).

37. A.M. Morgante, M. Johansson, G. Kopidakis, S. Aubry, Phys. Rev. Lett. **85** 550 (2000); *ibid*, Physica D **162** 53 (2002); *ibid*, Eur. Phys. J. B **29** 279 (2002).
38. A.M. Morgante, M. Johansson, S. Aubry, G. Kopidakis, J. Phys. A: Math. Gen **35** 4999 (2002).
39. G. Kopidakis, S. Aubry, Phys. Rev. Lett. **84** 3236 (2000).
40. P.W. Anderson, Phys. Rev. **109** 1492 (1958).
41. C. Albanese and J. Fröhlich, Commun. Math. Phys. **138** 193 (1991).
42. S. Aubry, G. Kopidakis, A.M. Morgante, G. Tsironis, Physica B **296** 222 (2001).
43. G. Kopidakis, S. Aubry, G. Tsironis, Phys. Rev. Lett. **87** 165501 (2001).
44. X. Hu, A. Damjanovic, T. Ritz and K. Schulten, Proc. Nat. Acad. Sci. (USA) **95** 5935 (1998).
45. F. Julicher, A. Ajdari, J. Prost, Rev. Mod. Phys. **69** 1269 (1997).
46. P. S. Lomdahl and W. C. Kerr, Phys. Rev. Lett. **55**, 1235 (1985); X. D. Wang, D. W. Brown and K. Lindenberg, Phys. Rev. Lett. **62** 1796 (1989).
47. P. Maniadis, G. Kopidakis, S. Aubry, to appear in Physica D; *ibid*, in the proceedings of *Localization and Energy Transfer in Nonlinear Systems*, San Lorenzo de El Escorial, Spain, Jun. 2002 (World Scientific, 2003), p. 268.
48. A.M. Kuznetzov and J. Ulstrup, *Electron Transfer in Chemistry and Biology: An introduction to the theory*, Wiley series in Theoretical Chemistry (1999).
49. R.A. Marcus, Rev. Mod. Phys. **65** 599 (1993).
50. S. Aubry and G. Kopidakis, *Preprint* cond-mat/0210215; S. Aubry and G. Kopidakis, in the proceedings of *Localization and Energy Transfer in Nonlinear Systems*, San Lorenzo de El Escorial, Spain, Jun. 2002 (World Scientific, 2003), p.1; S. Aubry and G. Kopidakis, to be submitted.
51. H.A. Kramers, Physica **7** 284 (1940).
52. *e.g.* R.G. Parr and W. Yang, *Density functional theory of atoms and molecules*, Oxford University Press, New York.; Eds. J.D. Dunitz *et al*, Top. Curr. Chem. **183** (1996); P. Senet, J.Chem.Phys. **107** (1997) 2516.
53. *e.g.* http://photoscience.la.asu.edu/photosyn/education/antenna.html and R.E. Blankenship, *Molecular Mechanisms of Photosynthesis*, Blackwell Science (2002).
54. K. Schulten, *From Simplicity to Complexity and Back: Function, Architecture and Mechanism of Light Harvesting Systems in Photosynthetic Bacteria* in *Simplicity and Complexity in Proteins and Nucleic Acids* Eds. H. Frauenfelder, J. Deisenhofer and P.G. Wolynes, Dahlem University Press (1999).
55. M.H. Vos, J.L. Lambry, S.J. Robles, D.C. Youvan, J. Breton and J-L. Martin, Proc. Nat. Acad. Sci. (USA) **88** 8885 (1991).
56. C.Rischel, D. Spiedel, J.P. Ridge, M.R. Jones, J. Breton, J-C. Lambry, J-L. Martin and M.H. Vos, Proc. Nat. Acad. Sci. (USA) **95** 12306 (1998).
57. S. Tanaka and R.A. Marcus, J. Phys. Chem **B101** 5031 (1997).
58. P. Maniadis, G. Kopidakis, S. Aubry, in preparation.

INDEX